Biochemistry and Biotechnology: A Laboratory Manual

Biochemistry and Biotechnology: A Laboratory Manual

Shiv Naz

Biochemistry and Biotechnology: A Laboratory Manual

ISBN 978-93-5111-205-1

Published in 2014 in India by

RANDOM PUBLICATIONS

4376-A/4B, Gali Murari Lal, Ansari Road
New Delhi-110 002
Phone : +91-11-43580356, +91-11-23289044
e-mail: randomexports@gmail.com, sales@randompublications.com,
info@randompublications.com

Reprinted 2026

Type Setting by : Keystoneprintads, Delhi-110051

Preface

Biochemistry is the science in which chemistry is applied to the study of living organisms and the atoms and molecules which comprise living organisms. Take a closer look at what biochemistry is and why the science is important. Biochemistry is the study of the chemistry of living things. This includes organic molecules and their chemical reactions. Most people consider biochemistry to be synonymous with molecular biology.

Biochemistry is the study of the chemistry of life. Biochemists study the physical and chemical properties of molecules that make up living organisms using the same techniques and theoretical framework that other types of chemists use to study molecules in other settings. What sets biochemistry apart from other areas of chemistry is that it focuses on the chemical reactions that occur in living organisms. What sets it apart from other areas of biology is that it zooms in on the molecular and atomic details of living organisms. Cellular and molecular biologists are also interested in molecules, but they generally do not study things like how chemical bonds are made and broken in an enzymatic reaction.

At its simplest, biotechnology is technology based on biology - biotechnology harnesses cellular and biomolecular processes to develop technologies and products that help improve our lives and the health of our planet. We have used the biological processes of microorganisms for more than 6,000 years to make useful food products, such as bread and cheese, and to preserve dairy products.

Biotechnology is an applied science that uses the tools of biochemistry, molecular biology and other fields of biology to solve problems that affect humans. One major area of biotechnology is the development of genetically engineered plants which carry genes that make them more useful to us. For example it is possible to use techniques and background knowledge from genetics, molecular biology, microbiology and biochemistry to transfer a gene from one organism into another. If one plant has a gene that allows it to survive harsh conditions, scientists can isolate that gene and transfer it into another type of plant, so that the second plant gains a new ability to survive in areas where it would have died without the new gene. This laboratory manual in biochemistry and biotechnology has been written to provide a book of technical information on a number of experiments routinely performed by the biology, biotechnology and agriculture graduate and post graduate students in practical classes.

I thank all members of my team who have helped in the preparation of the book. My special thanks go to "Random Publications" who have published the book.

– *Shiv Naz*

Contents

1

Biochemistry and Human Biology

Our understanding of biochemistry has had and will continue to have extensive effects on many aspects of human endeavor. *First, biochemistry is an intrinsically beautiful and fascinating body of knowledge.*

We now know the essence and many of the details of the most fundamental processes in biochemistry, such as how a single molecule of DNA replicates to generate two identical copies of itself and how the sequence of bases in a DNA molecule determines the sequence of amino acids in an encoded protein.

Our ability to describe these processes in detailed, mechanistic terms places a firm chemical foundation under other biological sciences. Moreover, the realization that we can understand essential life processes, such as the transmission of hereditary information, as chemical structures and their reactions has significant philosophical implications. What does it mean, biochemically, to be human? What are the biochemical differences between a human being, a chimpanzee, a mouse, and a fruit fly? Are we more similar than we are different?

Second, biochemistry is greatly influencing medicine and other fields. The molecular lesions causing sickle-cell anemia, cystic fibrosis, hemophilia, and many other genetic diseases have been elucidated at the biochemical level.

Some of the molecular events that contribute to cancer development have been identified. An understanding of the underlying defects opens the door to the discovery of effective therapies. Biochemistry makes possible the rational design of new drugs, including specific inhibitors of enzymes required for the replication of viruses such as human immunodeficiency virus (HIV).

Genetically engineered bacteria or other organisms can be used as "factories" to produce valuable proteins such as insulin and stimulators of blood-cell development. Biochemistry is also contributing richly to clinical diagnostics. For example, elevated levels of telltale enzymes in the blood reveal whether a patient has recently had a myocardial infarction (heart attack).

DNA probes are coming into play in the precise diagnosis of inherited disorders, infectious diseases, and cancers. Agriculture, too, is benefiting from advances in biochemistry with the development of more effective, environmentally safer herbicides and pesticides and the creation of genetically engineered plants that are, for example, more resistant to insects. All of these endeavors are being accelerated by the advances in genomic sequencing.

Third, advances in biochemistry are enabling researchers to tackle some of the most exciting questions in biology and medicine. How does a fertilized egg give rise to cells as different as those in muscle, brain, and liver? How do the senses work? What are the molecular bases for mental disorders such as Alzheimer disease and schizophrenia? How does the immune system distinguish between self and nonself? What are the molecular mechanisms of short-term and long-term memory? The answers to such questions, which once seemed remote, have been partly uncovered and are likely to be more thoroughly revealed in the near future.

Because all living organisms on Earth are linked by a common origin, evolution provides a powerful organizing theme for biochemistry. This book is organized to emphasize the unifying principles revealed by evolutionary considerations. We begin in the next chapter with a brief tour along a plausible evolutionary path from the formation of some of the chemicals that we now associate with living organisms through the evolution of the processes essential for the development of complex, multicellular organisms.

The remainder of Part I of the book more fully introduces the most important classes of biochemicals as well as catalysis and regulation. Part II, Transducing and Storing Energy, describes how energy from chemicals or from sunlight is converted into usable forms and how this conversion is regulated.

As we will see, a small set of molecules such as adenosine triphosphate (ATP) act as energy currencies that allow energy, however captured, to be utilized in a variety of biochemical processes. This part of the text examines the important pathways for the conversion of environmental energy into molecules such as ATP and uncovers many unifying principles.

Part III, Synthesizing the Molecules of Life, illustrates the use of the molecules discussed in Part II to synthesize key molecular building blocks, such as the bases of DNA and amino acids, and then shows how these precursors are assembled into DNA, RNA, and proteins. In Parts II and III, we will highlight the relation between the reactions within each pathway and between those in different pathways so as to suggest how these individual reactions may have combined early in evolutionary history to produce the necessary molecules.

From the student's perspective, the existence of features common to several pathways enables material mastered in one context to be readily applied to new contexts. Part IV, Responding to Environmental Changes, explores some of the mechanisms that cells and multicellular organisms have evolved to detect and respond to changes in the environment. The topics range from general mechanisms, common to all organisms, for regulating the expression of genes to the sensory systems used by human beings and other complex organisms. In many cases, we can now see how these elaborate systems evolved from pathways that existed earlier in evolutionary history. Many of the s in Part IV link biochemistry with other fields such as cell biology, immunology, and neuroscience. We are now ready to begin our journey into biochemistry with events that took place more than 3 billion years ago.

WHAT IS BIOCHEMISTRY?

Biochemistry is the study of the chemistry of living things. This includes organic molecules and their chemical reactions. Most people consider biochemistry to be synonymous with molecular biology.

TYPES OF MOLECULES

The principal types of biological molecules, or biomolecules are:

- carbohydrates
- lipids
- proteins
- nucleic acids.

Many of these molecules are complex molecules called polymers, which are made up of monomer subunits. Biochemical molecules are based oncarbon.

HISTORY OF BIOCHEMISTRY

It once was generally believed that life and its materials had some essential property or substance distinct from any found in non-living matter, and it was thought that only living beings could produce the molecules of life. Then, in 1828, Friedrich Wöhler published a paper on the synthesis of urea, proving that organic compounds can be created artificially.

The dawn of biochemistry may have been the discovery of the first enzyme, diastase (today called amylase), in 1833 by Anselme Payen.Eduard Buchner contributed the first demonstration of a complex biochemical process outside of a cell in 1896: alcoholic fermentation in cell extracts of yeast. Although the term "biochemistry" seems to have been first used in 1882, it is generally accepted that the formal coinage of biochemistry occurred in 1903 by Carl Neuberg, a German chemist. Previous to this time, this area would have been referred to asphysiological chemistry. Since then, biochemistry has advanced, especially since the mid-20th century, with the development of new techniques such as chromatography, X-ray diffraction, dual polarisation interferometry, NMR spectroscopy, radioisotopic labeling, electron microscopy, and molecular dynamics simulations. These techniques allowed for the discovery and detailed analysis of many molecules andmetabolic pathways of the cell, such as glycolysis and the Krebs cycle (citric acid cycle).

Another significant historic event in biochemistry is the discovery of the gene and its role in the transfer of information in the cell. This part of biochemistry is often called molecular biology. In the 1950s, James D. Watson, Francis Crick, Rosalind Franklin, and Maurice Wilkins were instrumental in solving DNA structure and suggesting its relationship with genetic transfer of information. In 1958, George Beadle and Edward Tatum received the Nobel Prize for work in fungi showing that one gene produces one enzyme. In 1988, Colin Pitchfork was the first person convicted of murder with DNA evidence, which led to growth of forensic science. More recently, Andrew Z. Fire and Craig C. Mello received the 2006 Nobel Prize for discovering the role of RNA interference (RNAi), in the silencing of gene expression.

STARTING MATERIALS: THE CHEMICAL ELEMENTS OF LIFE

Around two dozen of the 92 naturally occurring chemical elements are essential to various kinds of biological life. Most rare elements on Earth are not needed by life (exceptions being selenium and iodine), while a few common ones (aluminum and titanium) are not used. Most organisms share element needs, but there are a few differences between plants and animals. For example ocean algae use bromine but land plants and animals seem to need none. All animals require sodium, but some plants do not. Plants need boron and silicon, but animals may not (or may need ultra-small amounts).

Just six elements—carbon, hydrogen, nitrogen, oxygen, calcium, and phosphorus—make up almost 99 per cent of the mass of a human body (see composition of the human body for a complete list). In addition to the six major elements that compose most of the human body, humans require smaller amounts of possibly 18 more.

Biomolecules

The four main classes of molecules in biochemistry are carbohydrates, lipids, proteins, and nucleic acids. Many biological molecules are polymers: in this terminology,*monomers* are relatively small micromolecules that are linked together to create large macromolecules, which are known as *polymers*. When monomers are linked together to synthesize a biological polymer, they undergo a process called dehydration synthesis. Different macromolecules can assemble in larger complexes, often needed for biological activity.

Carbohydrates

Carbohydrates are made from monomers called *monosaccharides*. Some of these monosaccharides include glucose ($C_6H_{12}O_6$), fructose($C_6H_{12}O_6$), and deoxyribose ($C_5H_{10}O_4$). When two monosaccharides undergo dehydration synthesis, water is produced, as twohydrogen atoms and one oxygen atom are lost from the two monosaccharides' hydroxyl group.

Lipids

Lipids are usually made from one molecule of glycerol combined with other molecules. In triglycerides, the main group of bulk lipids, there is one molecule of glycerol and three fatty acids. Fatty acids are considered the monomer in that case, and may be saturated (no double bonds in the carbon chain) or unsaturated (one or more double bonds in the carbon chain).

Lipids, especially phospholipids, are also used in various pharmaceutical products, either as co-solubilisers (*e.g.*, in parenteral infusions) or else as drug carrier components (*e.g.*, in a liposome or transfersome).

Proteins

Proteins are very large molecules – macro-biopolymers – made from monomers called *amino acids*. There are 20 standard amino acids, each containing a carboxyl group, an amino group, and a side-chain (known as an "R" group). The "R" group is what makes each amino acid different, and the properties of the side-chains greatly influence the overall three-dimensional conformation of a protein. When amino acids combine, they form a special bond called a peptide bond through dehydration synthesis, and become a *polypeptide*, or protein.

In order to determine whether two proteins are related, or in other words to decide whether they are homologous or not, scientists use sequence-comparison methods. Methods like Sequence Alignments and Structural Alignments are powerful tools that help scientists identify homologies between related molecules.

The relevance of finding homologies among proteins goes beyond forming an evolutionary pattern of protein families. By finding how similar two protein sequences are, we acquire knowledge about their structure and therefore their function.

Nucleic Acids

Nucleic acids are the molecules that make up DNA, an extremely important substance that all cellular organisms use to store their genetic information. The most common nucleic acids are deoxyribonucleic acid and ribonucleic acid. Their monomers are callednucleotides. The most common nucleotides are adenine, cytosine, guanine, thymine, and uracil. Adenine binds with thymine and uracil; Thymine binds only with adenine; and cytosine and guanine can bind only with each other.

Monosaccharides

The simplest type of carbohydrate is a monosaccharide, which among other properties contains carbon, hydrogen, and oxygen, mostly in a ratio of 1:2:1 (generalized formula $C_nH_{2n}O_n$, where n is at least 3). Glucose, one of the most important carbohydrates, is an example of a monosaccharide. So is fructose, the sugar commonly associated with the sweet taste of fruits. Some carbohydrates (especially after condensation to oligo- and polysaccharides) contain less carbon relative to H and O, which still are present in 2:1 (H:O) ratio. Monosaccharides can be grouped into aldoses (having an aldehydegroup at the end of the chain, *e.g.* glucose) and ketoses (having a keto group in their chain; *e.g.* fructose). Both aldoses and ketoses occur in an equilibrium (starting with chain lengths of C4) cyclic forms. These are generated by bond formation between one of the hydroxyl groups of the sugar chain with the carbon of the aldehyde or keto group to form a hemiacetal bond. This leads to saturated five-membered (in furanoses) or six-membered (in pyranoses) heterocyclic rings containing one O as heteroatom.

Disaccharides

Two monosaccharides can be joined together using dehydration synthesis, in which a hydrogen atom is removed from the end of one molecule and a hydroxyl group (—OH) is removed from the other; the remaining residues are then attached at the sites from which the atoms were removed. The H—OH or H_2O is then released as a molecule of water, hence the term *dehydration*. The new molecule, consisting of two monosaccharides, is called a *disaccharide* and is conjoined together by a glycosidic or ether bond. The reverse reaction can also occur, using a molecule of water to split up a disaccharide and break the glycosidic bond; this is termed*hydrolysis*. The most well-known disaccharide is sucrose, ordinary sugar (in scientific contexts, called *table sugar* or *cane sugar* to differentiate it from other sugars). Sucrose consists of a glucose molecule and a fructose molecule joined together. Another important disaccharide is lactose, consisting of a glucose molecule and a galactose molecule. As most humans age, the production of lactase, the enzyme that hydrolyzes lactose back into glucose and galactose, typically decreases. This results in lactase deficiency, also called *lactose intolerance*.

Sugar polymers are characterised by having reducing or non-reducing ends. A reducing end of a carbohydrate is a carbon atom that can be in equilibrium with the open-chainaldehyde or keto form. If the joining of monomers takes place at such a carbon atom, the free hydroxy group of the pyranose or furanose form is exchanged with an OH-side-chain of another sugar, yielding a full acetal. This prevents opening of the chain to the aldehyde or keto form and renders the modified residue non-reducing. Lactose contains a reducing end at its glucose moiety, whereas the galactose moiety form a full acetal with the C4-OH group of glucose. Saccharose does not have a reducing end because of full acetal formation between the aldehyde carbon of glucose (C1) and the keto carbon of fructose (C2).

DNA IS CONSTRUCTED FROM FOUR BUILDING BLOCKS

DNA is a *linear polymer* made up of four different monomers. It has a fixed backbone from which protrude variable substituents. The backbone is built of repeating sugar-phosphate units. The sugars are molecules of *deoxyribose* from which DNA receives its name. Joined to each deoxyribose is one of four possible bases: adenine (A), cytosine (C), guanine (G), and thymine (T).

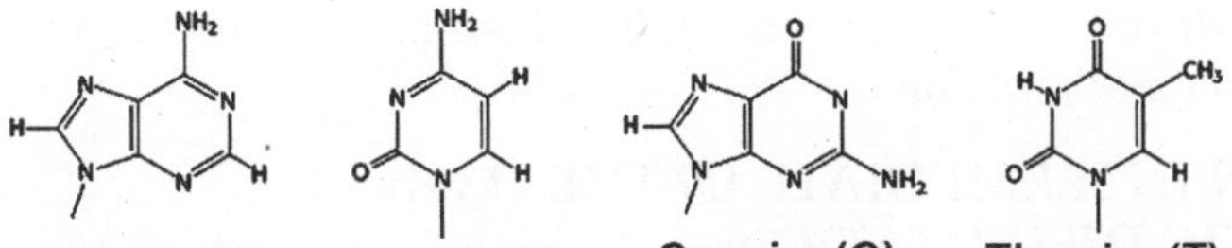

Adenine(A) Cytosine(C) Guanine(G) Thymine(T)

All four bases are planar but differ significantly in other respects. Thus, the monomers of DNA consist of a sugar-phosphate unit, with one of four bases attached to the sugar. *These bases can be arranged in any order along a strand of DNA*. The order of these bases is what is displayed in the sequence that begins this chapter. For example, the first base in the sequence shown is G (guanine), the second is A (adenine), and so on. *The sequence of bases along a DNA strand constitutes the genetic information*—the instructions for assembling proteins, which themselves orchestrate the synthesis of a host of other biomolecules that form cells and ultimately organisms.

TWO SINGLE STRANDS OF DNA COMBINE TO FORM A DOUBLE HELIX

Most DNA molecules consist of not one but two strands. How are these strands positioned with respect to one another? In 1953, James Watson and Francis Crick deduced the arrangement of these strands and proposed a three-dimensional structure for DNA molecules.

This structure is a *double helix* composed of two intertwined strands arranged such that the sugar-phosphate backbone lies on the outside and the bases on the inside. The key to this structure is that the bases form *specific base pairs* (bp) held together by *hydrogen bonds*: adenine pairs with thymine (A-T) and guanine pairs with cytosine (G-C), as shown. Hydrogen bonds are much weaker than covalent bonds such as the carbon-carbon or carbon-nitrogen bonds that define the structures of the bases themselves. Such weak bonds are crucial to biochemical systems; they are weak enough to be reversibly broken in biochemical processes, yet they are strong enough, when many form simultaneously, to help stabilize specific structures such as the double helix.

The structure proposed by Watson and Crick has two properties of central importance to the role of DNA as the hereditary material.

First, the structure is compatible with *any sequence of bases*. The base pairs have essentially the same shape and thus fit equally well into the centre of the double-helical structure. Second, because of base-pairing, *the sequence of bases along one strand completely determines the sequence along the other strand*.As Watson and Crick so coyly wrote: "It has not escaped our notice that the specific pairing we have postulated immediately suggests a possible copying mechanism for the genetic material." Thus, if the DNA double helix is separated into two single strands, each strand can act as a template for the generation of its partner strand through specific base-pair formation. *The three-dimensional structure of DNA beautifully illustrates the close connection between molecular form and function.*

RNA IS AN INTERMEDIATE IN THE FLOW OF GENETIC INFORMATION

An important nucleic acid in addition to DNA is *ribonucleicacid (RNA).* Some viruses use RNA as the genetic material, and even those organisms that employ DNA must first convert the genetic information into RNA for the information to be accessible or functional. Structurally, RNA is quite similar to DNA. It is a linear polymer made up of a limited number of repeating monomers, each composed of a sugar, a phosphate, and a base. The sugar is ribose instead of deoxyribose (hence, *R*NA) and one of the bases is uracil (U) instead of thymine (T).

Unlike DNA, an RNA molecule usually exists as a single strand, although significant segments within an RNA molecule may be double stranded, with G pairing primarily with C and A pairing with U. This intrastrand base-pairing generates RNA molecules with complex structures and activities, including catalysis.

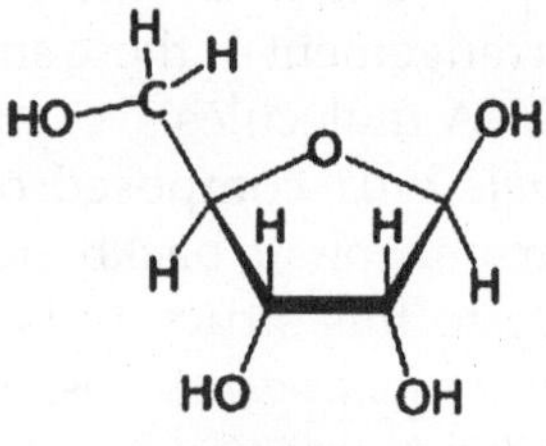

Ribose

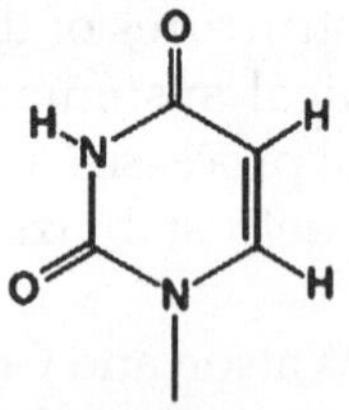

Uracil (U)

RNA has three basic roles in the cell. First, it serves as the intermediate in the flow of information from DNA to protein, the primary functional molecules of the cell. The DNA is copied, or *transcribed,* into messenger RNA (mRNA), and the mRNA is *translated* into protein. Second, RNA molecules serve as adaptors that translate the information in the nucleic acid sequence of mRNA into information designating the sequence of constituents that make up a protein. Finally, RNA molecules are important functional components of the molecular machinery, called ribosomes, that carries out the translation process. As will be discussed the unique position of RNA between the storage of genetic information in DNA and the functional expression of this information as protein as well as its potential to combine genetic and catalytic capabilities are indications that RNA played an important role in the evolution of life.

PROTEINS, ENCODED BY NUCLEIC ACIDS, PERFORM MOST CELL FUNCTIONS

A major role for many sequences of DNA is to encode the sequences of *proteins,* the workhorses within cells, participating in essentially all processes. Some proteins are key structural components, whereas others are specific catalysts (termed *enzymes*) that promote chemical reactions. Like DNA and RNA, proteins are linear polymers. However, proteins are more complicated in that they are formed from a selection of 20 building blocks, called *amino acids,* rather than 4.

The functional properties of proteins, like those of other biomolecules, are determined by their three-dimensional structures. Proteins possess an extremely important property: a protein spontaneously folds into a welldefined and elaborate three-dimensional structure that is dictated entirely by the sequence of amino acids along its chain. *The self-folding nature of proteins constitutes the transition from the one-dimensional world of sequence information to the three-dimensional world of biological function.* This marvelous ability of proteins to self assemble into complex structures is responsible for their dominant role in biochemistry.

How is the sequence of bases along DNA translated into a sequence of amino acids along a protein chain? We will consider the details of this process in later chapters, but the important finding is that *three bases along a DNA chain encode a single amino acid.* The specific correspondence between a set of three bases and 1 of the 20 amino acids is called the *genetic code.* Like the use of DNA as the genetic material, the genetic code is essentially universal; the same sequences of three bases encode the same amino acids in all life forms from simple microorganisms to complex, multicellular organisms such as human beings.

Knowledge of the functional and structural properties of proteins is absolutely essential to understanding the significance of the human genome sequence.

For example, the sequence at the beginning of this chapter corresponds to a region of the genome that differs in people who have the genetic disorder *cystic fibrosis.* The most common mutation causing cystic fibrosis, the loss of three consecutive Ts from the gene sequence, leads to the loss of a single amino acid within a protein chain of 1480 amino acids.

This seemingly slight difference—a loss of 1 amino acid of nearly 1500—creates a life-threatening condition. What is the normal function of the protein encoded by this gene? What properties of the encoded protein are compromised by this subtle defect?

Can this knowledge be used to develop new treatments? These questions fall in the realm of biochemistry. Knowledge of the human genome sequence will greatly accelerate the pace at which connections are made between DNA sequences and disease as well as other human characteristics. However, these connections will be nearly meaningless without the knowledge of biochemistry necessary to interpret and exploit them.

2

Proteins

Proteins are any of a group of complex organic compounds containing carbon, hydrogen, nitrogen and sulfur. Proteins, the principle constituents of the protoplasm of all cells, are of high molecular weight and consist of alpha-amino acids joined by peptide linkages. Different amino acids are commonly found in proteins, each protein having a unique, genetically defined amino-acid sequence, which determines its specific shape and function. They serve as enzymes, structural elements, hormones, immunoglobulins, etc. and are involved in oxygen transport, muscle contraction, electron transport, and other activities.

The importance of adequate protein intake to proper immune function has been extensively studied. The most severe effects of"Protein-Calorie Malnutrition" (PCM) are on cell-mediated immunity, although all facets of immune function are ultimately affected. PCM is not, however, a single nutrient deficiency. It is normally associated with multiple nutrient deficiencies, and some immune dysfunctions attributed to PCM are most likely due to these other factors. Partial deficiencies of dietary vitamins produce a comparatively greater depression on immune functions than do partial protein deficiencies. Non-the less, adequate protein is essential for optimal immune function.

High protein diets are not recommended for individuals with kidney or liver disease. There are no other known side effects.

A macromolecule made from amino acids; some consist of one, long polypeptide chain while others are made from several intertwined polypeptides (proteios = first place) Proteins are about 50 per cent of the dry weight of most cells, and are the most structurally complex macromolecules known. Each type of protein has its own unique structure and function.

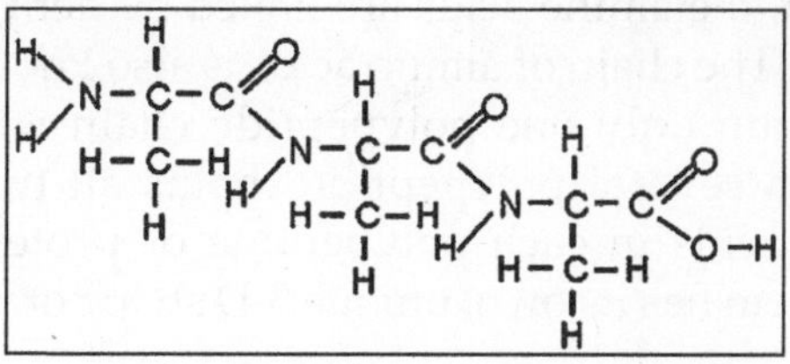

LARGE MOLECULES OF POLYMER

A large molecule made of many, smaller, repeating, identical or similar subunits (poly = many; mer = a part) Polymers are any kind of large molecules made of repeating identical or similar subunits called monomers. The starch and cellulose we previously discussed are polymers of glucose, which in that case, is the monomer. Proteins are polymers of about 20 amino acids (the monomer).

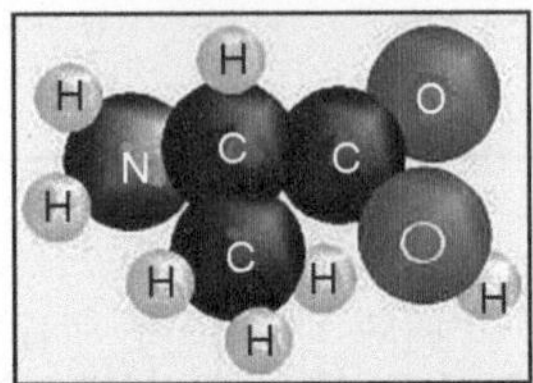

Amino acids are built from a central carbon bonded to four different groups. (run your mouse over the amino acid to see the names of the parts).

- Hydrogen (–H),
- Amino group (–NH_2),
- Carboxyl group (–COOH),

and some side chain symbolized by"R".

In the approximately-20 amino acids found in our bodies, what varies is the side chain. Some side chains are hydrophilic while others are hydrophobic.

Since these side chains stick out from the backbone of the molecule, they help determine the properties of the protein made from them.

Most naturally-occurring amino acids are the l- form, whereas synthetically-produced amino acids give a 50:50 mixture. Notice that these molecules are mirror images of each other, thus there is no way you can rotate one molecule to make it look like the other.

To form protein, the amino acids are linked by dehydration synthesis to form peptide bonds. The chain of amino acids is also known as a polypeptide. Some proteins contain only one polypeptide chain while others, such as hemoglobin, contain several polypeptide chains all twisted together. The sequence of amino acids in each polypeptide or protein is unique to that protein, so each protein has its own, unique 3-D shape or native conformation.

If even one amino acid in the sequence is changed, that can potentially change the protein's ability to function. For example, sickle cell anemia is caused by a change in only one nucleotide in the DNA sequence that causes just one amino acid in one of the hemoglobin polypeptide molecules to be different. Because of this, the whole red blood cell ends up being deformed and unable to carry oxygen properly.

As a polypeptide chain forms, it naturally twists and bends into its native conformation. One of the things that helps determine the native conformation of a protein is the side chains of all the amino acids involved. Remember some amino acid side chains are hydrophobic while others are hydrophilic. In this case, the"likes" attract: all the hydrophobic side chains (here represented by yellow beads) try to"get together" in the Centre of the molecule, away from the watery environment, while the hydrophilic side chains are attracted to the outside of the molecule, near the watery environment. Additionally, some of the hydrophilic side chains have groups of atoms attached that make them acidic, while others have groups attached that make them basic. Side chains with acidic ends are attracted to side chains with basic ends, and can form ionic bonds.

Thus, the side chains interacting with each other help to hold the protein in its native conformation.

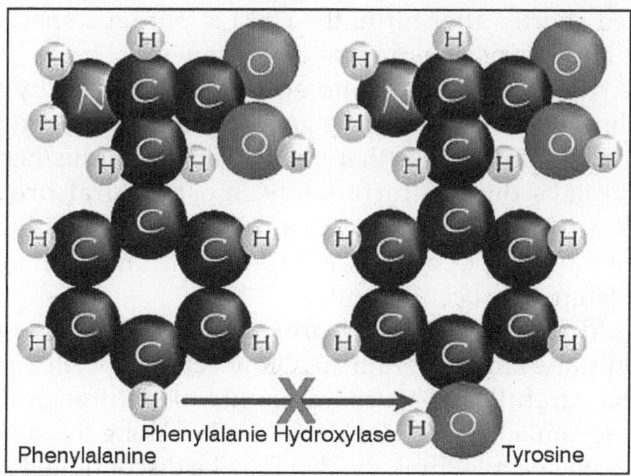

Denaturation is when a protein loses its native conformation. For example, egg white (also called albumen) contains a protein called albumin which is water-soluble. However, if heated, albumin becomes denatured and loses its ability to be water-soluble. There are several possible things that can denature proteins including changing the temperature (adding heat), changing the pH or salt concentration of the solution, or putting the protein into a hydrophobic solvent. In a hydrophobic solvent, the amino acids with hydrophobic side chains (the yellow beads) would all try to go to the outside of the molecule, and all those with hydrophilic side chains would cluster in the Centre of the molecule. If a protein remains water-soluble when denatured (unlike albumin), it can return to its native conformation if/when placed back into a"normal" environment.

PKU or phenylketonuria is a genetic disorder for which, by law, all newborns must be tested. People with this disorder are lacking an enzyme called phenylalanine hydroxylase needed to turn the amino acid, phenylalanine into tyrosine so it can be eliminated via the urine. Since phenylalanine cannot be eliminated, it builds up and in children whose nervous systems are still developing, causes mental retardation.

According to the Merck Manual, the body is slowly able to turn it into several other compounds which can be eliminated via the urine, but these chemical reactions are too inefficient to get rid of the excess phenylalanine. If PKU is detected shortly after birth, the child is put on a special diet low in phenylalanine and has normal mental development.

Since phenylalanine is an amino acid that is needed by the body for normal growth and development, the goal is to limit intake to just what the child needs without any excess that could build up in his/her system. The Merck Manual says that, unfortunately, most natural protein sources, including milk, are too high in phenylalanine for children with PKU, so these babies/children must be put on a special formula from which most, if not all, of the phenylalanine has been removed.

As the child grows, some low-protein, natural foods such as fruits, vegetables, and some kinds of grain are OK to eat. The needed phenylalanine is supplied via carefully-measured amounts of natural protein plus the residual in the formula. One area of"debate" is how long the special diet must be continued. Some people think the diet may be discontinued after the brain finishes developing at about age 5. Others have noted some Behaviour problems and learning difficulties that appear to be related to going off the diet, and recommend that people with PKU stay on the diet for life.

Staying on the special diet is especially important if a woman who has PKU is even thinking about trying to become pregnant. Because any excess phenylalanine in her system could cross the placenta and could cause mental retardation in the baby, especially in the first few weeks while the nervous system is forming, she has to make sure that she adheres to the special diet from before conception.

Note that this is not a major concern for a normal woman carrying a baby with PKU (she wouldn't know that until after the baby was born and tested) because the woman's body can digest phenylalanine, including any excess from the baby's body that is sent back, via the placenta, to her body.

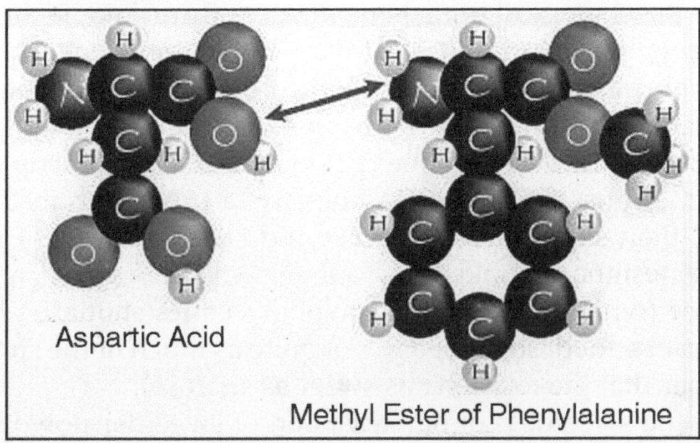

Nutrasweet has the chemical name aspartame and is made of two amino acids: l-aspartic acid and the methyl ester of phenylalanine, thus people with PKU can't have it. Look at a can of diet soft drink that is sweetened with aspartame. Somewhere on the can, there's always a cautionary label warning people with PKU not to drink it.

But what about the rest of us? That methyl group on the phenylalanine can break off pretty easily, and at temperatures as"low" as 86° F (= 30° C - lower than body temperature or a summer day), aspartame starts to disintegrate, releasing methanol (which in larger quantity causes blindness, etc. - it's toxic).

As the liver tries to deal with the methanol using normal metabolic pathways, it ends up turning the methanol into formaldehyde, which is a known carcinogen. Interestingly, because of this and several other concerns, originally there was controversy over FDA approval of aspartame. Even still, its safety is controversial. More recently, some people have suggested a possible link between aspartame consumption and an increase in brain tumors in this country.

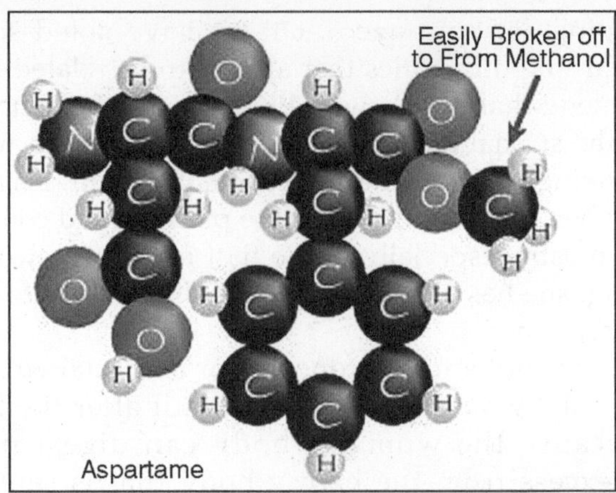

Equal® is not the same as Nutrasweet, but does contain some of it. There's an interesting psychological thing here: in our culture, people think they need sweetener by the teaspoon, but Nutrasweet is so much sweeter than sugar that only a tiny pinch would be needed to gain the equivalent sweetness. Artificial sweetner manufacturers know that if they don't make a product such that one teaspoon is as sweet as one teaspoon of the sugar people are used to using, people will use one teaspoon anyway (a tiny pinch just doesn't look like enough), then say it was too sweet and stop buying it. Thus, artificial sweetners are designed to take up about a teaspoon of space, and Equal® is mostly glucose (which diabetics and hypoglycemics shouldn't have) with a little Nutrasweet added, so in theory, not quite as much of the sugar is needed (also remember that glucose isn't as sweet as sucrose).

To make the sweetness of one teaspoon of sweetner powder equivalent to that of one teaspoon of sucrose, various fillers (things like cellulose) are added. These products also contain a number of other chemicals, including silicon dioxide (SiO2)"to prevent caking." This is the chemical name for sand or glass (although, from what I've heard, a bit of silicon is needed for healthy fingernails)! Pre-mixed, pre-packaged items, such as artificially-sweetened soft drinks, can contain all aspartame as their sweetener because the manufacturers of these products do not have to be concerned with the"one teaspoon" psychological factor.

Remember, because aspartame degrades into methanol at warm temperatures, it should not be used in anything hot or anything that will be cooked (coffee sweetener, cake mixes, cocoa, etc.), and bottles of artificially-sweetened soft drinks that have been sitting out in the hot sun at the local gas station are suspect.

STRUCTURE OF MEMBRANE PROTEINS

The lipid bilayer constitutes the fundamental structural unit of all biological membranes. Proteins, in contrast, carry out essentially all of the active functions of membranes, including transport activities, receptor functions, and other related processes. As suggested by Singer and Nicolson, most membrane proteins can be classified as peripheral or integral. The peripheral proteins are globular proteins that interact with the membrane mainly through electrostatic and hydrogen-bonding interactions with integral proteins. Although peripheral proteins are not discussed further here, many proteins of this class are described in the context of other discussions throughout this textbook. Integral proteins are those that are strongly associated with the lipid bilayer, with a portion of the protein embedded in, or extending all the way across, the lipid bilayer. Another class of proteins not anticipated by Singer and Nicolson, the lipid-anchored proteins, are important in a variety of functions in different cells and tissues. These proteins associate with membranes by means of a variety of covalently linked lipid anchors.

INTEGRAL MEMBRANE PROTEINS

Despite the diversity of integral membrane proteins, most fall into two general classes. One of these includes proteins attached or anchored to the membrane by only a small hydrophobic segment, such that most of the protein extends out into the water solvent on one or both sides of the membrane. The other class includes those proteins that are more or less globular in shape and more totally embedded in the membrane, exposing only a small surface to the water solvent outside the membrane. In general, those structures of integral membrane protein within the nonpolar core of the lipid bilayer are dominated by a-helices or ß-sheets because these secondary structures neutralize the highly polar N —H and C = O functions of the peptide backbone through H-bond formation.

A Protein with a Single Transmembrane Segment

In the case of the proteins that are anchored by a small hydrophobic polypeptide segment, that segment often takes the form of a single a-helix. One of the best examples of a membrane protein with such an a-helical structure is glycophorin. Most of glycophorin's mass is oriented on the outside surface of the cell, exposed to the aqueous milieu (Figure).

The C-terminus of the peptide, whose sequence is shown here, faces the cytosol of the erythrocyte; the N-terminal domain is extracellular. Points of attachment of carbohydrate groups are indicated.

A variety of hydrophilic oligosaccharide units are attached to this extracellular domain. These oligosaccharide groups constitute the ABO and MN blood group antigenic specificities of the red cell. This extracellular portion of the protein also serves as the receptor for the influenza virus. Glycophorin has a total molecular weight of about 31,000 and is approximately 40 per cent protein and 60 per cent carbohydrate. The glycophorin primary structure consists of a segment of 19 hydrophobic amino acid residues with a short hydrophilic sequence on one end and a longer hydrophilic sequence on the other end. The 19-residue sequence is just the right length to span the cell membrane if it is coiled in the shape of an a-helix. The large hydrophilic sequence includes the amino terminal residue of the polypeptide chain.

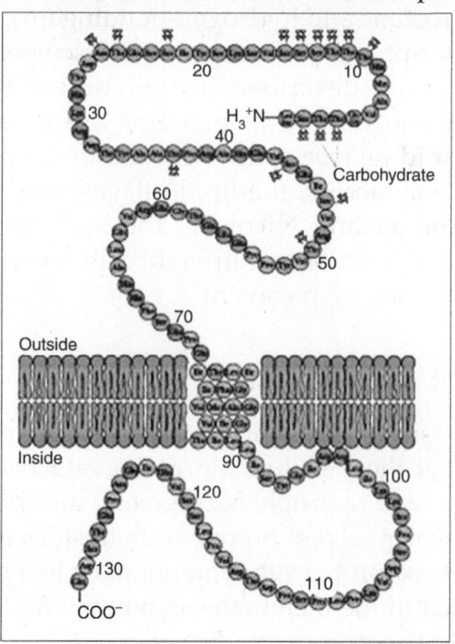

Fig. Glycophorin A Spans the Membrane of the Human Eerythrocyte via a Single a-helical Transmembrane Segment.

Numerous other membrane proteins are also attached to the membrane by means of a single hydrophobic a-helix, with hydrophilic segments extending into either the cytoplasm or the extracellular space. These proteins often function as receptors for extracellular molecules or as recognition sites that allow the immune system to recognize and distinguish the cells of the host organism from invading foreign cells or viruses. The proteins that represent the major transplantation antigens H2 in mice and human leukocyte associated (HLA) proteins in humans are members of this class. Other such proteins include thesurface immunoglobulin receptors on B lymphocytes and the spike proteins of many membrane viruses. The function of many of these proteins depends primarily on their extracellular domain, and thus the segment facing the intracellular surface is often a shorter one.

Bacteriorhodopsin: A 7-Transmembrane Segment Protein

Membrane proteins that take on a more globular shape, instead of the rodlike structure previously described, are often involved with transport activities and other functions requiring a substantial portion of the peptide to be embedded in the membrane. These proteins may consist of numerous hydrophobic a-helical segments joined by hinge regions so that the protein winds in a zig-zag pattern back and forth across the membrane. A well-characterized example of such a protein is bacteriorhodopsin, which clusters in purple patches in the membrane of the bacterium Halobacterium halobium.

The name Halobacterium refers to the fact that this bacterium thrives in solutions having high concentrations of sodium chloride, such as the salt beds of San Francisco Bay. Halobacterium carries out a light-driven proton transport by means of bacterio-rhodopsin, named in reference to its spectral similarities to rhodopsin in the rod outer segments of the mammalian retina. When this organism is deprived of oxygen for oxidative metabolism, it switches to the capture of energy from sunlight, using this energy to pump protons out of the cell. The proton gradient generated by such light-driven proton pumping represents potential energy, which is exploited elsewhere in the membrane to synthesize ATP.

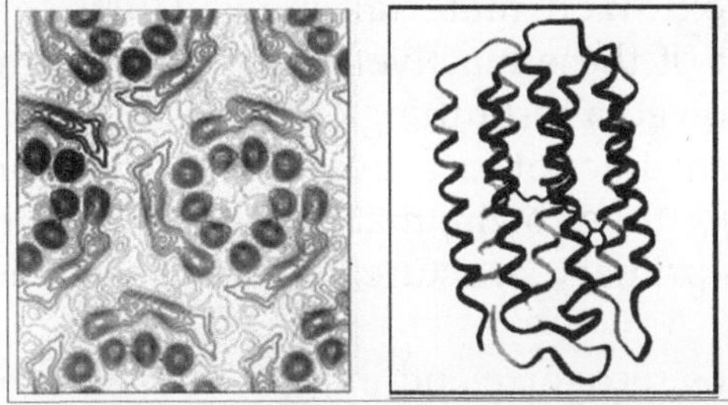

Fig. An Electron Density Profile Illustrating the Three Centers of Threefold Symmetry in Arrays of Bacterio-Rhodopsin in the Purple Membrane of Halobacterium Halobium, Together with a Computer-generated Model Showing the Seven a-helical Transmembrane Segments in Bacteriorhodopsin.

Bacteriorhodopsin clusters in hexagonal arrays in the purple membrane patches of Halobacterium, and it was this orderly, repeating arrangement of proteins in the membrane that enabled Nigel Unwin and Richard Henderson in 1975 to determine the bacteriorhodopsin structure. The polypeptide chain crosses the membrane seven times, in seven a-helical segments, with very little of the protein exposed to the aqueous milieu. The bacteriorhodopsin structure has become a model of globular membrane protein structure. Many other integral membrane proteins contain numerous hydro-phobic sequences that, like those of bacteriorhodopsin, could form a-helical transmembrane segments. For example, the amino acid sequence of the sodium - potassium transport ATPase contains ten hydrophobic segments of length sufficient to span the plasma membrane. By analogy with bacterio-rhodopsin, one would expect that these segments form a globular hydrophobic core that anchors the ATPase in the membrane. The helical segments may also account for the transport properties of the enzyme itself.

AMINO ACID SEQUENCE IN PROTEIN STRUCTURE

It can be inferred from the first section of this chapter that many different forces work together in a delicate balance to determine the overall three–dimensional structure of a protein. These forces operate both within the protein structure itself and between the protein and the water solvent. How, then, does nature dictate the manner of protein folding to generate the three–dimensional structure that optimizes and balances these many forces? *All of the information necessary for folding the peptide chain into its "native" structure is contained in the amino acid sequence of the peptide.* This principle was first appreciated by C. B. Anfinsen and F. White, whose work in the early 1960s dealt with the chemical denaturation and subsequent renaturation of bovine pancreatic ribonuclease. Ribonuclease was first denatured with urea and mercaptoethanol, a treatment that cleaved the four covalent disulfide (S–S) cross–bridges in the protein. Subsequent air oxidation permitted random formation of disulfide cross–bridges, most of which were incorrect.

Thus, the air–oxidized material showed little enzymatic activity. However, treatment of these inactive preparations with small amounts of mercaptoethanol allowed a reshuffling of the disulfide bonds and permitted formation of significant amounts of active native enzyme. In such experiments, the only road map for the protein, that is, the only "instructions" it has, are those directed by its primary structure, the linear sequence of its amino acid residues.

Just how proteins recognize and interpret the information that is stored in the polypeptide sequence is not well understood yet. It may be assumed that certain loci along the peptide chain act as nucleation points, which initiate folding processes that eventually lead to the correct structures. Regardless of how this process operates, it must take the protein correctly to the final native structure, without getting trapped in a local energy–minimum state which, although stable, may be different from the native state itself. A long–range goal of many researchers in the protein structure field is the prediction of three–dimensional conformation from the amino acid sequence.

As the details of secondary and tertiary structure are described in this chapter, the complexity and immensity of such a prediction will be more fully appreciated. This area is perhaps the greatest uncharted frontier remaining in molecular biology.

SECONDARY STRUCTURE IN PROTEINS

Any discussion of protein folding and structure must begin with the *peptide bond,* the fundamental structural unit in all proteins. As we saw, the resonance structures experienced by a peptide bond constrain the oxygen, carbon, nitrogen, and hydrogen atoms of the peptide group, as well as the adjacent α–carbons, to all lie in a plane. The resonance stabilization energy of this planar structure is approximately 88 kJ/mol, and substantial energy is required to twist the structure about the C–N bond. A twist of θ degrees involves a twist energy of 88 $\sin^2 \theta$ kJ/mol.

CONSEQUENCES OF THE AMIDE PLANE

The planarity of the peptide bond means that there are only two degrees of freedom per residue for the peptide chain. Rotation is allowed about the bond linking the α–carbon and the carbon of the peptide bond and also about the bond linking the nitrogen of the peptide bond and the adjacent α–carbon.

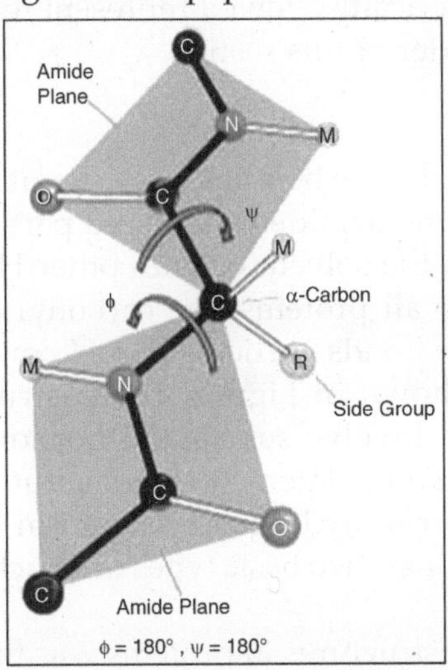

Fig. The Amide or Peptide Bond Planes are Joined by the Tetrahedral Bonds of the α –carbon

The rotation parameters are ϕ and ψ. The conformation shown corresponds to ϕ = 180° and ψ = 180°. Note that positive values of *f* and *c* correspond to clockwise rotation as viewed from C_{α}. As shown in Figure, each α–carbon is the joining point for two planes defined by peptide bonds.

The angle about the C_{α} —N bond is denoted by the Greek letter ϕ (phi) and that about the C_{α}—— C_{o} is denoted by ψ (psi). For either of these bond angles, a value of 0° corresponds to an orientation with the amide plane bisecting the H__C_{α} __ R (sidechain) plane and a cis configuration of the main chain around the rotating bond in question (Figure).

Several noteworthy examples are shown here. In any case, the entire path of the peptide backbone in a protein is known if the ϕ and ψ rotation angles are all specified. Some values of ϕ and ψ are not allowed due to steric interference between nonbonded atoms. As shown in Figure, values of ϕ = 180° and ψ = 0° are not allowed because of the forbidden overlap of the N–H hydrogens. Similarly, ϕ = 0° and ψ = 180° are forbidden because of unfavourable overlap between the carbonyl oxygens.

The shaded regions indicate particularly favourable values of these angles. Dots in purple indicate actual angles measured for 1000 residues (excluding glycine, for which a wider range of angles is permitted) in eight proteins. G. N. Ramachandran and his coworkers in Madras, India, first showed that it was convenient to plot ϕ values against y values to show the distribution of allowed values in a protein or in a family of proteins. A typical Ramachandran plot is shown in Figure. Note the clustering of ϕ and ψ values in a few regions of the plot. Most combinations of ϕ and ψ are sterically forbidden, and the corresponding regions of the Ramachandran plot are sparsely populated. The combinations that are sterically allowed represent the subclasses of structure described in the remainder of this section.

THE ALPHA–HELIX

The discussion of hydrogen bonding pointed out that the carbonyl oxygen and amide hydrogen of the peptide bond could participate in H bonds either with water molecules in the solvent or with other H–bonding groups in the peptide chain. In nearly all proteins, the carbonyl oxygens and the amide protons of many peptide bonds participate in H bonds that link one peptide group to another, as shown in Figure. These structures tend to form in cooperative fashion and involve substantial portions of the peptide chain. Structures resulting from these interactions constitute secondary structure for proteins. When a number of hydrogen bonds form between portions of the peptide chain in this manner, two basic types of structures can result: α–*helices* and β–*pleated sheets*.

Evidence for helical structures in proteins was first obtained in the 1930s in studies of fibrous proteins. However, there was little agreement at that time about the exact structure of these helices, primarily because there was also lack of agreement about interatomic distances and bond angles in peptides. In 1951, Linus Pauling, Robert Corey, and their colleagues at the California Institute of Technology summarized a large volume of crystallographic data in a set of dimensions for polypeptide chains.

With these data in hand, Pauling, Corey, and their colleagues proposed a new model for a helical structure in proteins, which they called the α–helix. The report from Caltech was of particular interest to Max Perutz in Cambridge, England, a crystallographer who was also interested in protein structure. By taking into account a critical but previously ignored feature of the X–ray data, Perutz realised that the α–helix existed in keratin, a protein from hair, and also in several other proteins. Since then, the α–helix has proved to be a fundamentally important peptide structure. Several representations of the α–helix. One turn of the helix represents 3.6 amino acid residues. (A single turn of the α–helix involves 13 atoms from the O to the H of the H bond. For this reason, the α–helix is sometimes referred to as the 3.6 13 helix.)

The formation of H–bonds with other nearby donor and acceptor groups is referred to as helix capping. Capping may also involve appropriate hydrophobic interactions that accomodate non–polar side chains at the ends of helical segments.

Careful studies of the polyamino acids, polymers in which all the amino acids are identical, have shown that certain amino acids tend to occur in α–helices, whereas others are less likely to be found in them. Polyleucine and polyalanine, for example, readily form α–helical structures. In contrast, polyaspartic acid and polyglutamic acid, which are highly negatively charged at pH 7.0, form only random structures because of strong charge repulsion between the R groups along the peptide chain. At pH 1.5 to 2.5, however, where the side chains are protonated and thus uncharged, these latter species spontaneously form a–helical structures. In similar fashion, polylysine is a random coil at pH values below about 11, where repulsion of positive charges prevents helix formation. At pH 12, where polylysine is a neutral peptide chain, it readily forms an α–helix.

The tendencies of the amino acids to stabilize or destabilize a–helices are different in typical proteins than in polyamino acids. The occurrence of the common amino acids in helices is summarized in Table. Notably, proline (and hydroxyproline) act as helix breakers due to their

Table. Helix–forming and Helix–breaking Behaviour of the Amino Acids

Amino Acid		Helix Behavior*		Amino Acid		Helix Behavior*	
A	Ala	H	(I)	M	Met	H	
C	Cys	Variable		N	Asn	C	(I)
D	Asp	Variable		P	Pro	B	
E	Glu	H		Q	Gln	H	(I)
F	Phe	H		R	Arg	H	(I)
G	Gly	I	(B)	S	Ser	C	(B)
H	His	H	(I)	T	Thr	Variable	
I	Ile	H	(C)	V	Val	Variable	
K	Lys	Variable		W	Trp	H	(C)
L	Leu	H		Y	Tyr	H	(C)

- *H = helix former; I = indifferent; B = helix breaker; C = random coil; ( ) = secondary tendency.

Unique structure, which fixes the value of the C_α–N–C bond angle. Helices can be formed from either $_D$– or $_L$–amino acids, but a given helix must be composed entirely of amino acids of one configuration. α–Helices cannot be formed from a mixed copolymer of $_D$– and $_L$–amino acids. An α–helix composed of $_D$–amino acids is left–handed.

OTHER HELICAL STRUCTURES

There are several other far less common types of helices found in proteins. The most common of these is the 3_{10} helix, which contains 3.0 residues per turn (with 10 atoms in the ring formed by making the hydrogen bond three residues up the chain). It normally extends over shorter stretches of sequence than the α–helix. Other helical structures include the 2_7 ribbon and the π–helix, which has 4.4 residues and 16 atoms per turn and is thus called the 4.416 helix.

THE BETA–PLEATED SHEET

Another type of structure commonly observed in proteins also forms because of local, cooperative formation of hydrogen bonds. That is the pleated sheet, or β–structure, often called the β–pleated sheet. This structure was also first postulated by Pauling and Corey in 1951 and has now been observed in many natural proteins. A β–pleated sheet can be visualized by laying thin, pleated strips of paper side by side to make a "pleated sheet" of paper (Figure). Each

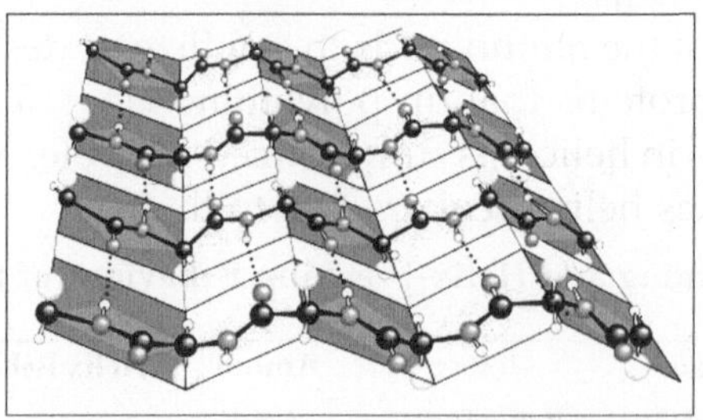

Fig. A "Pleated Sheet" of Paper with an Antiparallel β –sheet Drawn on it

Strip of paper can then be pictured as a single peptide strand in which the peptide backbone makes a zigzag pattern along the strip, with the α–carbons lying at the folds of the pleats. The pleated sheet can exist in both parallel and antiparallel forms. In the parallel β–pleated sheet, adjacent chains run in the same direction (N→ C or C → N). In the antiparallel β–pleated sheet,adjacent strands run in opposite directions.

Each single strand of the β–sheet structure can be pictured as a twofold helix, that is, a helix with two residues per turn. The arrangement of successive amide planes has a pleated appearance due to the tetrahedral nature of the C_{α} atom. It is important to note that the hydrogen bonds in this structure are essentially *inter*strand rather than *intra*strand. The peptide backbone in the β–sheet is in its most extended conformation (sometimes called the ε–conformation). The optimum formation of H bonds in the parallel pleated sheet results in a slightly less extended conformation than in the antiparallel sheet. The H bonds thus formed in the parallel β–sheet are bent significantly. The distance between residues is 0.347 nm for the antiparallel pleated sheet, but only 0.325 nm for the parallel pleated sheet.

Figure shows examples of both parallel and antiparallel β–pleated sheets. Note that the side chains in the pleated sheet are oriented perpendicular or normal to the plane of the sheet, extending out from the plane on alternating sides. Parallel β–sheets tend to be more regular than antiparallel β–sheets. The range of ϕ and ψ angles for the peptide bonds in parallel sheets is much smaller than that for antiparallel sheets. Parallel sheets are typically large structures; those composed of less than five strands are rare. Antiparallel sheets, however, may consist of as few as two strands. Parallel sheets characteristically distribute hydrophobic side chains on both sides of the sheet, while antiparallel sheets are usually arranged with all their hydrophobic residues on one side of the sheet.

This requires an alternation of hydrophilic and hydrophobic residues in the primary structure of peptides involved in antiparallel β–sheets because alternate side chains project to the same side of the sheet.

Antiparallel pleated sheets are the fundamental structure found in silk, with the polypeptide chains forming the sheets running parallel to the silk fibres. The silk fibres thus formed have properties consistent with those of the β–sheets that form them.

They are quite flexible but cannot be stretched or extended to any appreciable degree. Antiparallel structures are also observed in many other proteins, including immunoglobulin G, superoxide dismutase from bovine erythrocytes, and concanavalin A. Many proteins, including carbonic anhydrase, egg lysozyme, and glyceraldehyde phosphate dehydrogenase, possess both a–helices and β–pleated sheet structures within a single polypeptide chain.

THE BETA–TURN

Most proteins are globular structures. The polypeptide chain must therefore possess the capacity to bend, turn, and reorient itself to produce the required compact, globular structures. A simple structure observed in many proteins is the β–turn (also known as the *tight turn*or β*–bend*), in which the peptide chain forms a tight loop with the carbonyl oxygen of one residue hydrogen–bonded with the amide proton of the residue three positions down the chain. This H bond makes the β–turn a relatively stable structure.

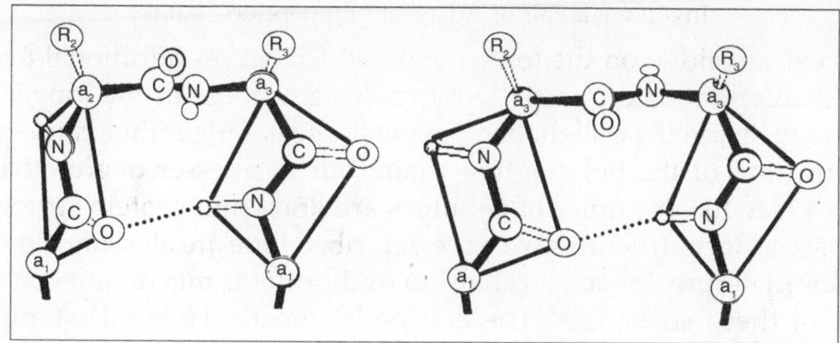

Fig. The Structures of Two Kinds of β –turns (also Called Tight Turns or β –bends)

As shown in Figure, the β–turn allows the protein to reverse the direction of its peptide chain. This figure shows the two major types of β–turns, but a number of less common types are also found in protein structures. Certain amino acids, such as proline and glycine, occur frequently in β–turn sequences, and the particular conformation of the β–turn sequence depends to some extent on the amino acids composing it. Due to the absence of a side chain, glycine is sterically the most adaptable of the amino acids, and it accommodates conveniently to other steric constraints in the β–turn. Proline, however, has a cyclic structure and a fixed ϕ angle, so, to some extent, it forces the formation of a β–turn, and in many cases this facilitates the turning of a polypeptide chain upon itself. Such bends promote formation of antiparallel β–pleated sheets.

THE BETA–BULGE

One final secondary structure, the β–bulge, is a small piece of nonrepetitive structure that can occur by itself, but most often occurs as an irregularity in antiparallel β–structures. A β–bulge occurs between two normal β–structure hydrogen bonds and comprises two residues on one strand and one residue on the opposite strand. Figure illustrates typical β–bulges.

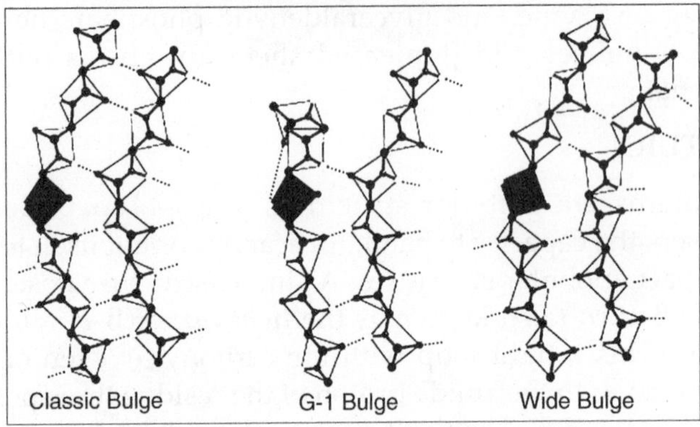

Fig. Three Different Kinds of β–bulge Structures Involving a Pair of Adjacent Polypeptide Chains

The extra residue on the longer side, which causes additional backbone length, is accommodated partially by creating a bulge in the longer strand and partially by forcing a slight bend in the β–sheet. Bulges thus cause changes in the direction of the polypeptide chain, but to a lesser degree than tight turns do. Over 100 examples of β–bulges are known in protein structures.

The secondary structures we have described here are all found commonly in proteins in nature. In fact, it is hard to find proteins that do not contain one or more of these structures. The energetic (mostly H–bond) stabilization afforded by α–helices, β–pleated sheets, and β–turns is important to proteins, and they seize the opportunity to form such structures wherever possible.

PROTEIN FOLDING AND TERTIARY STRUCTURE

The folding of a single polypeptide chain in three–dimensional space is referred to as its tertiary structure. As discussed, all of the information needed to fold the protein into its native tertiary structure is contained within the primary structure of the peptide chain itself. With this in mind, it was disappointing to the biochemists of the 1950s when the early protein structures did not reveal the governing principles in any particular detail. It soon became apparent that the proteins knew how they were supposed to fold into tertiary shapes, even if the biochemists did not. Vigourous work in many laboratories has slowly brought important principles to light.

First, secondary structures—helices and sheets—form whenever possible as a consequence of the formation of large numbers of hydrogen bonds. Second, α–helices and β–sheets often associate and pack close together in the protein. No protein is stable as a single–layer structure, for reasons that become apparent later. There are a few common methods for such packing to occur. Third, because the peptide segments between secondary structures in the protein tend to be short and direct, the peptide does not execute complicated twists and knots as it moves from one region of a secondary structure to another. A consequence of these three principles is that protein chains are usually folded so that the secondary structures are arranged in one of a few common patterns. For this reason, there are families of proteins that have similar tertiary structure, with little apparent evolutionary or functional relationship among them. Finally, proteins generally fold so as to form the most stable structures possible.

The stability of most proteins arises from:

- The formation of large numbers of intramolecular hydrogen bonds and
- The reduction in the surface area accessible to solvent that occurs upon folding.

FIBROUS PROTEINS

Fibrous proteins contain polypeptide chains organized approximately parallel along a single axis, producing long fibres or large sheets. Such proteins tend to be mechanically strong and resistant to solubilization in water and dilute salt solutions. Fibrous proteins often play a structural role in nature.

α–Keratin

As their name suggests, the structure of the α–keratins is dominated by α–helical segments of polypeptide. The amino acid sequence of α–keratin subunits is composed of central α–helix—rich rod domains about 311 to 314 residues in length, flanked by nonhelical N– and C–terminal domains of varying size and composition. The structure of the central rod domain of a typical α–keratin is shown in Figure. It consists of four helical strands arranged as twisted pairs of two–stranded coiled coils. X–ray diffraction patterns show that these structures resemble a–helices, but with a pitch of 0.51 nm rather than the expected 0.54 nm.

The primary structure of the central rod segments of α–keratin consists of quasi–repeating seven–residue segments of the form $(a\text{–}b\text{–}c\text{–}d\text{–}e\text{–}f\text{–}g)_n$. These units are not true repeats, but residues *a* and *d* are usually nonpolar amino acids. In α–helices, with 3.6 residues per turn, these nonpolar residues are arranged in an inclined row or stripe that twists around the helix axis.

These nonpolar residues would make the helix highly unstable if they were exposed to solvent, but the association of hydrophobic strips on two coiled coils to form the two–stranded rope effectively buries the hydrophobic residues and forms a highly stable structure. The helices clearly sacrifice some stability in assuming this twisted conformation, but they gain stabilization energy from the packing of side chains between the helices. In other forms of keratin, covalent disulfide bonds form between cysteine residues of adjacent molecules, making the overall structure rigid, inextensible, and insoluble—important properties for structures such as claws, fingernails, hair, and horns in animals. How and where these disulfides form determines the amount of curling in hair and wool fibres. When a hairstylist creates a permanent wave (simply called a "permanent") in a hair salon, disulfides in the hair are first reduced and cleaved, then reorganized and reoxidized to change the degree of curl or wave. In contrast, a "set" that is created by wetting the hair, setting it with curlers, and then drying it represents merely a rearrangement of the hydrogen bonds between helices and between fibres.

Fibroin and β–Keratin: β–Sheet Proteins

The fibroin proteins found in silk fibres represent another type of fibrous protein. These are composed of stacked antiparallel β–sheets, as shown in Figure. The primary structure of fibroin molecules consists of long stretches of alternating glycine and alanine or serine residues. When the sheets stack, the more bulky alanine and serine residues on one side of a sheet interdigitate with similar residues on an adjoining sheet. Glycine hydrogens on the alternating faces interdigitate in a similar manner, but with a smaller intersheet spacing.

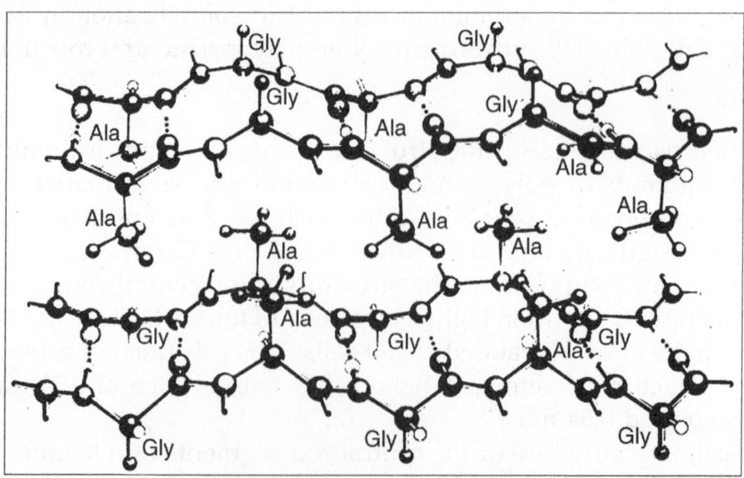

Fig. Silk Fibroin Consists of a Unique Stacked Array of β –sheets

In the polypeptide sequence of silk proteins, there are large stretches in which every other residue is a glycine. As previously mentioned, the residues of a β–sheet extend alternately above and below the plane of the sheet. As a result, the glycines all end up on one side of the sheet and the other residues (mainly alanines and serines) compose the opposite surface of the sheet. Pairs of β–sheets can then pack snugly together (glycine surface to glycine surface or alanine—serine surface to alanine—serine surface). The β–keratins found in bird feathers are also made up of stacked β–sheets.

Collagen: A Triple Helix

Collagen is a rigid, inextensible fibrous protein that is a principal constituent of connective tissue in animals, including tendons, cartilage, bones, teeth, skin, and blood vessels. The high tensile strength of collagen fibres in these structures makes possible the various animal activities such as running and jumping that put severe stresses on joints and skeleton. Broken bones and tendon and cartilage injuries to knees, elbows, and other joints involve tears or hyperextensions of the collagen matrix in these tissues.

The basic structural unit of collagen is tropocollagen, which has a molecular weight of 285,000 and consists of three intertwined polypeptide chains, each about 1000 amino acids in length. Tropocollagen molecules are about 300 nm long and only about 1.4 nm in diameter. Several kinds of collagen have been identified. *Type I collagen,* which is the most common, consists of two identical peptide chains designated α 1(I) and one different chain designated α 2(I). Type I collagen predominates in bones, tendons, and skin. *Type II collagen,* found in cartilage, *and type III collagen,* found in blood vessels, consist of three identical polypeptide chains. Collagen has an amino acid composition that is unique and is crucial to its three–dimensional structure and its characteristic physical properties. Nearly one residue out of three is a glycine, and the proline content is also unusually high. Three unusual modified amino acids are also found in collagen: 4–hydroxy–proline (Hyp), 3–hydroxyproline, and 5–hydroxylysine (Hyl). Proline and Hyp together compose up to 30 per cent of the residues of collagen.

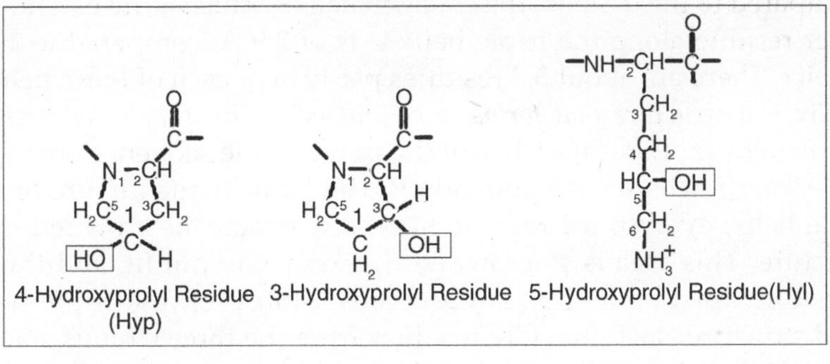

4-Hydroxyprolyl Residue (Hyp) 3-Hydroxyprolyl Residue 5-Hydroxyprolyl Residue(Hyl)

Fig. The Hydroxylated Residues Typically Found in Collagen

Interestingly, these three amino acids are formed from normal proline and lysine *after* the collagen polypeptides are synthesized. The modifications are effected by two enzymes: *prolyl hydroxylase* and *lysyl hydroxylase*. The prolyl hydroxylase reaction (Figure) requires molecular oxygen,

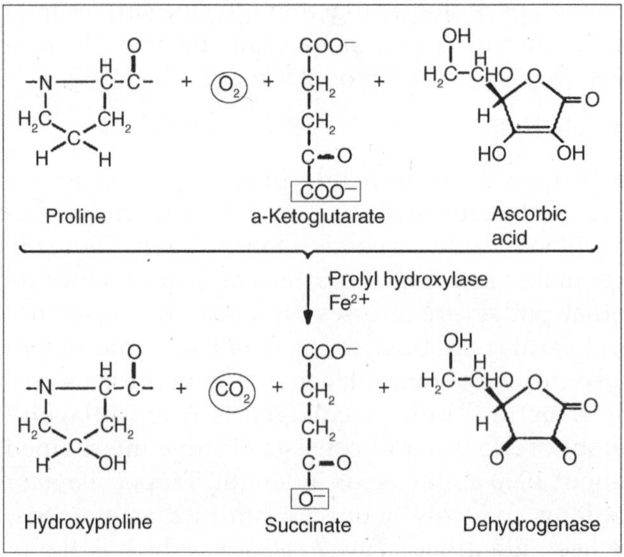

Fig. Hydroxylation of Proline Residues is Catalyzed by Prolyl Hydroxylase

The reaction requires α –ketoglutarate and ascorbic acid (vitamin C).

α–ketoglutarate, and ascorbic acid (vitamin C) and is activated by Fe^{2+}. The hydroxylation of lysine is similar. These processes are referred to as posttranslational modifications because they occur after genetic information from DNA has been translated into newly formed protein.

Because of their high content of glycine, proline, and hydroxyproline, collagen fibres are incapable of forming traditional structures such as α–helices and β–sheets. Instead, collagen polypeptides intertwine to form a unique triple helix, with each of the three strands arranged in a helical fashion.

Compared to the α–helix, the collagen helix is much more extended, with a rise per residue along the triple helix axis of 2.9 Å, compared to 1.5 Å for the α–helix. There are about 3.3 residues per turn of each of these helices. *The triple helix is a structure that forms to accommodate the unique composition and sequence of collagen.* Long stretches of the polypeptide sequence are repeats of a Gly–x–y motif, where x is frequently Pro and y is frequently Pro or Hyp. In the triple helix, every third residue faces or contacts the crowded center of the structure. This area is so crowded that only Gly can fit, and thus every third residue must be a Gly (as observed). Moreover, the triple helix is a *staggered* structure, such that Gly residues from the three strands stack along the center of the triple helix and the Gly from one strand lies adjacent to an x residue from the second strand and to a y from the third. This allows the N–H of each Gly residue to hydrogen bond with the C=O of the adjacent x residue. The triple helix structure is further stabilized and strengthened by the formation of interchain H bonds involving hydroxyproline.

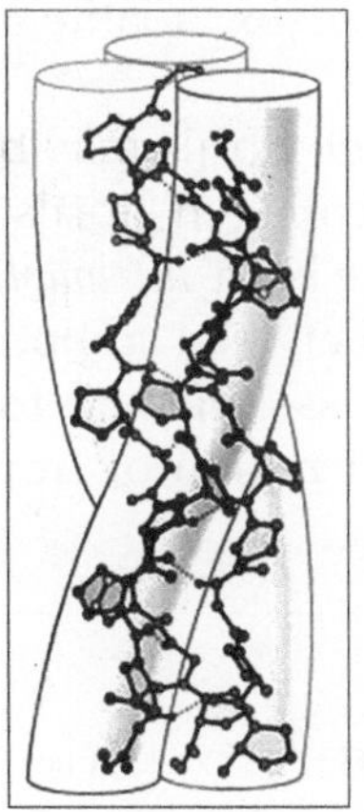

Fig. Poly(Gly–pro–pro), a Collagen–like Right–handed Triple Helix Composed of Three Left–handed Helical Chains

The periodic arrangement of triple helices in a head–to–tail fashion results in banded patterns in electron micrographs. The banding pattern typically has a periodicity (repeat distance) of 68 nm. Because collagen triple helices are 300 nm long, 40–nm gaps occur between adjacent collagen molecules in a row along the long axis of the fibrils and the pattern repeats every five rows (5 × 68 nm = 340 nm).

The 40–nm gaps are referred to as *hole regions,* and they are important in at least two ways. First, sugars are found covalently attached to 5–hydroxylysine residues in the hole regions of collagen.

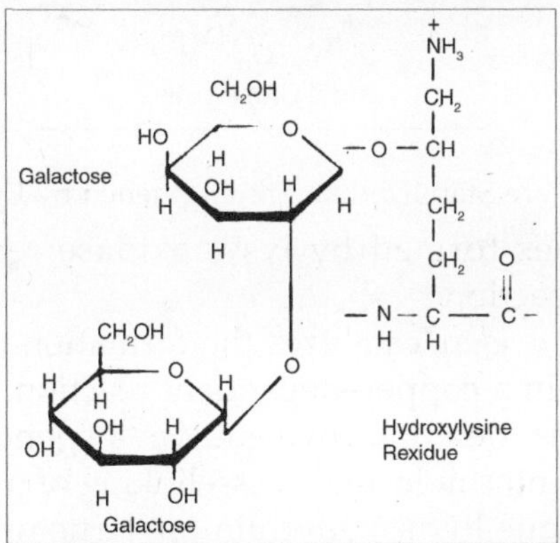

Fig. A Disaccharide of Galactose and Glucose is Covalently Linked to the 5–hydroxyl Group of Hydroxylysines in Collagen by the Combined Action of the Enzymes Galactosyl Transferase and Glucosyl Transferase

The occurrence of carbohydrate in the hole region has led to the proposal that it plays a role in organizing fibril assembly. Second, the hole regions may play a role in bone formation. Bone consists of microcrystals of hydroxyapatite, $Ca_5(PO_4)_3OH$, embedded in a matrix of collagen fibrils. When new bone tissue forms, the formation of new hydroxyapatite crystals occurs at intervals of 68 nm.

The hole regions of collagen fibrils may be the sites of nucleation for the mineralization of bone. The collagen fibrils are further strengthened and stabilized by the formation of both *intramolecular* (within a tropocollagen molecule) and *intermolecular* (between tropocollagen molecules in the fibril) cross–links. Intramolecular cross–links are formed between lysine residues in the (nonhelical) N–terminal region of tropocollagen in a unique pair of reactions shown in Figure.

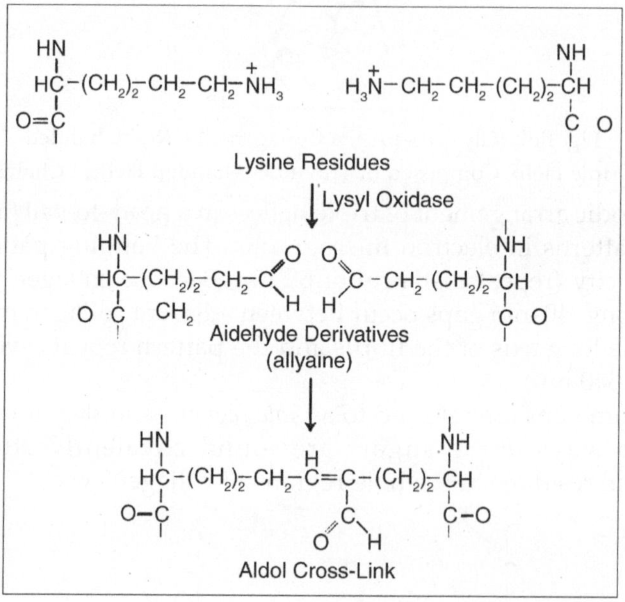

Lysine Residues

Lysyl Oxidase

Aidehyde Derivatives
(allyaine)

Aldol Cross-Link

Fig. Collagen Fibres are Stabilized and Strengthened by Lys – Lys Cross–links

Aldehyde moieties formed by lysyl oxidase react in a spontaneous nonenzymatic aldol reaction.

The enzyme *lysyl oxidase* catalyzes the formation of aldehyde groups at the lysine side chains in a copper–dependent reaction. The aldehyde groups of two such side chains then link covalently in a spontaneous nonenzymatic *aldol condensation*. The intermolecular cross–linking of tropocollagens involves the formation of a unique hydroxypyridinium structure from one lysine and two hydroxylysine residues (Figure). These cross–links form between the N–terminal region of one tropocollagen and the C–terminal region of an adjacent tropocollagen in the fibril.

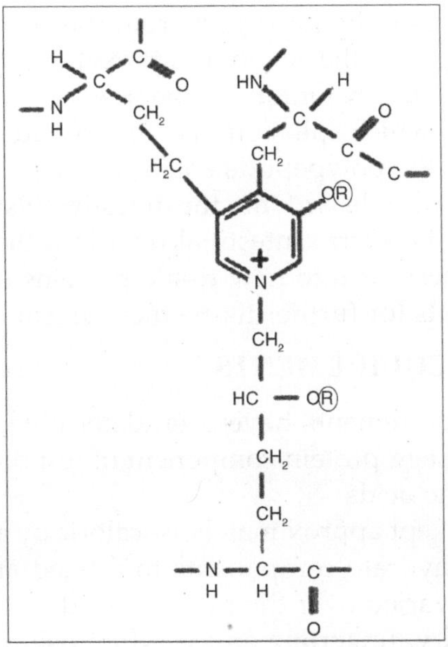

Fig. The Hydroxypyridinium Structure Formed by the Cross–linking of a Lys and Two Hydroxy Lys Residues

GLOBULAR PROTEINS

Fibrous proteins, although interesting for their structural properties, represent only a small percentage of the proteins found in nature. Globular proteins, so named for their approximately spherical shape, are far more numerous.

Packing Considerations

The secondary and tertiary structures of myoglobin and ribonuclease A illustrate the importance of packing in tertiary structures. Secondary structures pack closely to one another and also intercalate with (insert between) extended polypeptide chains. If the sum of the van der Waals volumes of a protein's constituent amino acids is divided by the volume occupied by the protein, packing densities of 0.72 to 0.77 are typically obtained. This means that, even with close packing, approximately 25 per cent of the total volume of a protein is not occupied by protein atoms. Nearly all of this space is in the form of very small cavities. Cavities the size of water molecules or larger do occasionally occur, but they make up only a small fraction of the total protein volume. It is likely that such cavities provide flexibility for proteins and facilitate conformation changes and a wide range of protein dynamics.

PROTEIN DIGESTION AND METABOLISM

Ingested proteins are first split into smaller fragments by pepsin in the stomach or by trypsin or chymotrypsin from the pancreas. These peptides are then further reduced by the action of carboxypeptidase which hydrolyzes off one amino acid at a time beginning at the free carboxyl end of the molecule or by aminopeptidase which splits off one amino acid at a time beginning at the free amino end of the polypeptide chain.

The free amino acids released into the digestive system are then absorbed through the walls of the gastro intestinal tract into the blood stream where they are then resynthesized into new tissue proteins or are catabolyzed for energy or for fragments for further tissue metabolism.

GROSS PROTEIN REQUIREMENTS

Gross protein requirements have been determined for a few species of fish. Simulated whole egg protein component of test diets contains an excess of indispensable amino acids.

These diets were kept approximately isocaloric by adjusting total protein plus digestible carbohydrate components to a fixed amount as the protein diet treatments were varied over the ranges tested.

Tests in feeding fry, fingerling and yearling fish have shown that gross protein requirements are highest in initial feeding fry and that they decrease as fish size increases.

To grow at the maximum rate, fry must have a diet in which nearly half of the digestible ingredients consist of balanced protein; at 6-8 weeks this requirement is decreased to about 40 per cent of the diet for salmon and trout and to about 35 per cent of the diet for yearling salmonids raised at standard environmental temperature (SET).

Initially feeding fry require that about 50 per cent of the digestible components of the ration be protein and the requirement decreases with size. Some feeding trials with salmon have indicated direct relationships between changes in the protein requirements of young fish and changes in water temperature.

Chinook salmon in 7 C water require about 40 per cent whole egg protein for maximum growth; the same fish in 15 C water require about 50 per cent protein. Salmon, trout and catfish can use more protein than required for maximum growth because of efficiency in eliminating nitrogenous wastes in the form of soluble ammonia compounds through the gill tissue directly into the water environment.

This system for eliminating nitrogen is more efficient than that available to fowl and mammals. Fowl and mammals consume energy to synthesize urea, uric acid, or other nitrogen compounds which are excreted through the kidney tissue and expelled in urine.

Digestible carbohydrate and fat will spare excess protein in the diet as long as the protein requirement for maximum growth is met (Figures 1 and 2).

Table. Estimated Dietary Protein Requirement of Certain Fish

Species	Crude protein level in diet for optimal growth (g/kg)
Rainbow trout (Salmo gairdneri)	400-460
Carp (Cyprinus carpio)	380
Chinook salmon (Oncorhynchus tshawytscha)	400
Eel (Anguilla japonica)	445
Plaice (Pleuronectes platessa)	500
Gilthead bream (Chrysophrys aurata)	400
Grass carp (Ctenopharyngodon idella)	410-430
Brycon sp.	356
Red sea bream (Chrysophrys major)	550
Yellowtail (Seriola quinqueradiata)	550

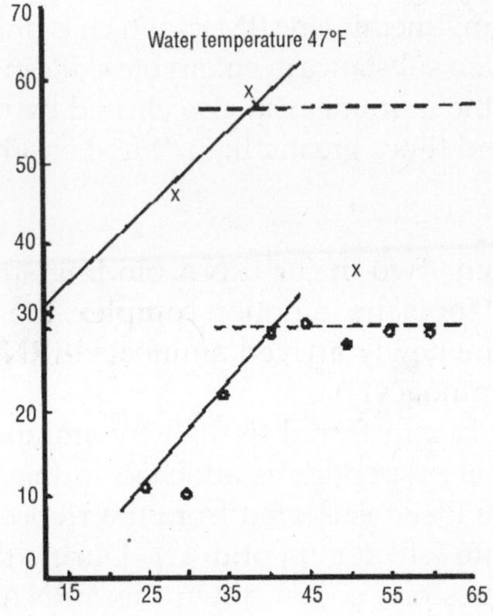

Fig. Protein Requirement of Chinook Salmon at 47°F. Top Curve: Initial Individual Average Weight of Fish, 1.5g. Bottom Curve: Initial Individual Average Weight of Fish, 5.6g.

Basically the fish must be given a diet containing graded levels of high quality protein and energy and adequate balances of essential fatty acids, vitamins and minerals over a prolonged period. From the resulting dose/

response curve the protein requirement is usually obtained by an Almquist plot. These differences in apparent protein requirement are thought to be due to differences in culture techniques and diet composition.

The relatively high dietary protein levels required for maximal growth of certain fish such as grass carp, Ctenopharyngodon idella and Brycon spp. are surprising as these fish are omnivorous. Brycon spp. are grown on unwanted fruit and other plant material of low protein content and under these conditions there is presumably a substantial contribution to their protein intake from a natural food chain. Protein requirement of eurythaline fish such as the rainbow trout, Salmo gairdneri and the coho salmon, Oncorhynchus kisutch, reared in water of salinity 20 ppt is about the same as the requirement in freshwater. No data are available for the protein requirement of these species in full strength sea water.(35 ppt).

PROCEDURE OF PROTEIN SYNTHESIS

Initiation

Peptide synthesis always starts from methionine (Met). Therefore, the initial aminoacyl-tRNA is Met-tRNAiMet, where the subscript "iΔ specifies "initiation". In bacteria, the methionine of the initial aminoacyl-tRNA has been modified by the addition of a formyl group (HCO) to its amino group. The modified methionine is called formylmethionine (fMet), which is unique for bacteria. Thus, fMet is an obvious foreign substance in eukaryotes. It can elicit a strong immune response. In humans, the immune response elicited by the peptide "fMet-Leu-Phe" is about a thousand times greater than "Met-Leu-Phe".

Elongation

A ribosome contains two major tRNA-binding sites: A site and P site. After the large subunit joins the initiation complex, the initial Met-tRNAiMet enters the P site and the newly arrived aminoacyl-tRNA is always placed at the A site ("A" for "aminoacyl").

Then, methionine is transferred to the new aminoacyl-tRNA, forming a "peptidyl-tRNA" where a peptide is attached to the tRNA. Subsequently, the empty tRNA at the P site is ejected from the ribosome and the peptidyl-tRNA jumps to the P site ("P" for "peptidyl"). During this translocation step, the ribosome also moves one codon down the mRNA chain. Similar steps are repeated in the next cycles of elongation.

Termination

Protein synthesis will terminate when the ribosome arrives at one of three stop codons. The termination process is assisted by special proteins called termination factors which recognize the stop codons. Their association stimulates the release of the peptidyl-tRNA from the ribosome. Subsequently, the released peptidyl-tRNA divides into tRNA and a newly synthesized peptide chain. The ribosome also divides into the large and small subunits, ready for synthesizing another peptide.

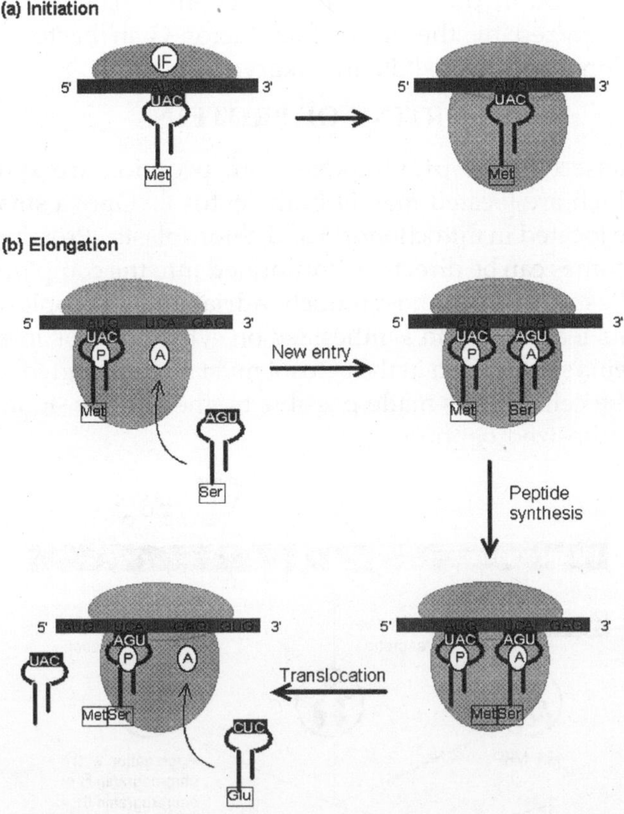

Fig. The Steps Involved in Protein Synthesis

- In the absence of mRNA, the large and small subunits of a ribosome are separated. At the beginning of peptide synthesis, initiation factors (IF) first assist the assembly of the small subunit, mRNA and the initial aminoacyl-tRNA. Then, the large subunit is recruited to join the complex.
- In the elongation process, one cycle involves the following steps:
  - *New entry.* A new aminoacyl-tRNA with a correct anticodon is brought to the A site. This step is catalyzed by elongation factors Tu and Ts in prokaryotes and by elongation factors EF1 and EF1b in eukaryotes..
  - *Peptide synthesis.* The peptide attached to the peptidyl-tRNA at the P site is transferred to the new aminoacyl-tRNA at the A site, generating a peptidyl-tRNA with a longer peptide. This step is catalyzed by peptidyl transferase.
  - *Translocation.* The empty tRNA at the P site is ejected from the

ribosome and the peptidyl-tRNA generated at the A site takes over the vacant P site. In the mean time, the ribosome moves one codon down the mRNA chain. The A to P switch is catalyzed by the elongation factor G in bacteria and by the elongation factor EF2 in eukaryotes.

SORTING OF PROTEINS

As discussed in the previous section, proteins are synthesized on ribosomes which are located mainly in the cytosol. Only a small number of ribosomes are located in mitochondria and chloroplasts. Proteins synthesized on these ribosomes can be directly incorporated into the compartments within these organelles. However, most mitochondrial and chloroplast proteins are encoded by nuclear DNA and synthesized on cytosolic ribosomes. These and all other proteins synthesized in the cytosol must be transported to appropriate locations in the cell. This is made possible by the specific signal sequence in the newly synthesized peptide.

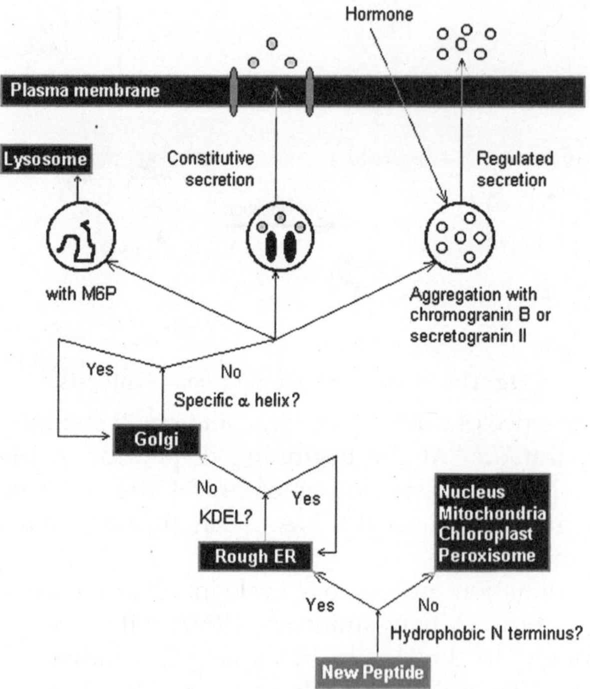

Fig. Protein Sorting

- If the N terminus of the new peptide contains a stretch of hydrophobic residues, it is sent to the rough endoplasmic reticulum (ER) for further sorting. Otherwise, it goes to non-ER pathways.
- The new peptide is retained in the rough ER if its C-terminus

contains the sequence "Lys-Asp-Glu-Leu" (KDEL in one-letter code). Otherwise, it will move to the Golgi apparatus.

- Proteins containing a specific transmembrane a helix will be localized in the Golgi apparatus.
- After glycosylation at the Golgi apparatus, the modified protein containing a mannose 6-phosphate (M6P) will be delivered to the lysosome.
- Proteins that can aggregate with chromogranin B (secretogranin I) or secretogranin II will be sorted into regulated secretory vesicles where proteins are released upon specific stimulation. Otherwise, they are sorted into another type of vesicles which continuously move to the plasma membrane or outside the cell.

GENE AND PROTEIN STRUCTURES

Characteristics of the human and mouse members of the ACAT/DGAT gene family are shown in Table. The human ACAT1 gene encodes four mRNAs of 7.0, 4.3, 3.6, and 2.8 kilobases (kb) (12, 18-20); all contain the same translational reading frame but differ in the length of the untranslated regions.

The two shorter mRNAs are products of a proximal ACAT promoter. The 4.3-kb mRNA is derived from an unusual RNA recombination mechanism involving *trans*-splicing of two discontinuous precursor RNAs produced from chromosomes 1 and 7 (21). The origin of the 7.0-kb transcript is unknown. The human ACAT2 gene encodes a single mRNA of 2.2 kb and the human DGAT gene encodes two mRNAs of 2.0 and 2.4 kb. Regions of shared sequence identity are present throughout the ACAT and DGAT proteins, but the greatest similarity is found in their C termini. Human and mouse ACAT2 are ~40 per cent identical to human ACAT1 and DGAT is ~20 per cent identical to ACAT1.

Little is known about the relationship between protein structure and function in ACAT/DGAT family members. One motif, FYXDWWN (amino acids 403-409 of human ACAT1), is highly conserved in all family members and may be involved in fatty acyl-CoA binding.

Another motif, MKXXSF (amino acids 265-270 of human ACAT1), is conserved in ACAT family members. The serine of this motif is required for ACAT activity. All three family members have a potential tyrosine phosphorylation motif and at least one *N*-linked glycosylation site, both of unknown significance. ACAT1 and ACAT2 contain an N-terminal leucine zipper motif. Although ACAT1 functions as a homotetramer it is unknown whether this leucine zipper motif participates in tetramer formation or whether ACAT2 and DGAT act as multimers. ACAT1 and ACAT2 do not form hetero-oligomeric complexes.

All ACAT/DGAT proteins possess multiple hydrophobic regions predicted to serve as transmembrane domains, although there is not yet a consensus structural model for the proteins. Investigations of human ACAT1 topology have yielded different results. One study, performed by inserting epitopes at different positions in the expressed protein found seven transmembrane domains. Another study, performed by expressing truncated forms of the protein found five transmembrane domains.

In both studies, the N terminus of the protein localized to the cytoplasm and the C terminus to the ER lumen. Topologic studies of human ACAT2 performed with the truncation mutant method also found five transmembrane domains with the N terminus in the cytoplasm and the C terminus in the ER lumen. The latter study found that a serine residue essential for ACAT activity was located in the cytoplasm for ACAT1 and in the ER lumen for ACAT2, suggesting that the active sites of these enzymes reside on opposite sides of the ER membrane. Several putative transmembrane sequences are conserved in ACAT1 and ACAT2 but not in DGAT suggesting that these regions are involved in cholesterol binding. The topology of DGAT has not been reported, but a putative diacylglycerol binding motif has been identified

TISSUE AND CELLULAR DISTRIBUTIONS

The tissue distributions of ACAT1 and ACAT2 are largely complementary. ACAT1 mRNA is present in many tissues, with the highest levels in macrophages and in adrenal and sebaceous glands all of which store cholesterol esters in cytoplasmic droplets. ACAT1 is also highly expressed in human atherosclerotic lesions, particularly in macrophage foam cells. ACAT2 is expressed predominantly in the liver and small intestine. In humans, nonhuman primates and mice, ACAT2 appears to be the major ACAT in the small intestine and the predominant isozyme in hepatocytes of nonhuman primates and mice. Whether ACAT1 or ACAT2 is the predominant enzyme in human hepatocytes is controversial. Both are expressed in human hepatocytes and HepG2 cells. However, immunodepletion experiments found that ACAT1 accounts for ~90 per cent of ACAT activity in human liver, whereas ACAT2 accounts for most of the ACAT activity in adult small intestine and fetal liver but only 10-20 per cent of activity in adult liver.

It is unclear whether the hepatic ACAT1 activity resides in hepatocytes or macrophage-derived Kupffer cells. ACAT1 expression was detected immunohistochemically in both cell types, with stronger staining in Kupffer cells. In nonhuman primates, ACAT1 was expressed only in Kupffer cells in the liver. A significant proportion of the ACAT1 activity in human liver thus may be attributable to Kupffer cells. Another study found little evidence of ACAT2 expression in human hepatocytes.

DGAT mRNA and activity are ubiquitous in mouse and human tissues, with the highest levels in liver, small intestine and adipose tissue and somewhat lower levels in testis and adrenal gland. This wide range of tissue expression is consistent with the involvement of DGAT in the glycerol phosphate pathway of triglyceride synthesis, which is common to most cells.

DGAT protein expression in tissues has not been studied because of the lack of suitable antibodies. The intracellular localization of enzymes in this class has been examined only for ACAT1. In human melanoma cells and fibroblasts and in mouse macrophages, ACAT1 is located primarily in the ER. In mouse macrophages, a small portion of ACAT1 immunoreactivity localizes to a region near the *trans*-Golgi network. In macrophages, ACAT1 localization may change with different conditions

BIOCHEMICAL ACTIVITY AND REGULATION

ACAT and DGAT substrates have been identified by expressing the proteins in insect cells or yeast devoid of sterol esterification activity. In insect cells, mouse and human ACAT1 and ACAT2 utilize a variety of oxysterols in addition to cholesterol as substrates. Ergosterol and plant sterols are poor substrates. Both enzymes also utilize a wide variety of long chain fatty acyl-CoAs as substrates although arachidonyl-CoA appears to be a better substrate for ACAT1 than ACAT2. In insect cells, DGAT utilizes only diacylglycerol as the acyl acceptor. DGAT has a broad fatty acyl-CoA specificity.

ACAT1 is regulated primarily by post-translational mechanisms. In cholesterol-loaded cells, esterification increases without changes in the expression level of ACAT1 mRNA or protein. Purified ACAT1 is activated allosterically by cholesterol or oxysterols.

ACAT1 activity may also be regulated by its intracellular localization. For example, a portion of ACAT1 immunoreactivity in macrophages is found near the *trans*-Golgi network and the endocytic recycling compartment.

These cholesterol-rich organelles are involved in membrane traffic with the plasma membrane and the proximity of ACAT1 to these organelles may be important for cholesterol ester synthesis during the formation of macrophage foam cells. The observation that 30-40 per cent of ACAT1 redistributed from the ER to small vesicles during foam cell formation is consistent with this possibility. Regulation by phosphorylation has not been demonstrated.

ACAT1 is also transcriptionally regulated; however, only the expression levels of the shorter ACAT1 mRNAs (2.8 and 3.6 kb) change. In mice and rabbits a fat and cholesterol-rich diet increases ACAT1 mRNA levels 2-3-fold in the liver. In human HepG2 cells, free fatty acids, but not 25-hydroxycholesterol, increase ACAT1 mRNA levels 1.5-2-fold. ACAT1 mRNA and protein expression levels also increase during the differentiation of monocytes into macrophages. Two factors that induce gene expression during monocyte-macrophage differentiation1,25-dihydroxyvitamin D_3 and 9-*cis*-retinoic acid, also increased ACAT1 mRNA, protein [a]nd cholesterol esterification activity in THP-1 cells.

Little is known about the regulation of ACAT2 and DGAT. Levels of DGAT mRNA increase markedly in parallel with enzyme activity in NIH 3T3-

L1 cells during their differentiation into adipocytes. This finding suggests that transcription factors involved in adipogenesis activate DGAT expression. Like other family members, DGAT possesses a putative tyrosine phosphorylation site. The use of this site in DGAT regulation has not been demonstrated, although two reports have indicated that the enzyme may be posttranslationally regulated by a tyrosine kinase.

FUNCTIONAL ANALYSIS OF ACAT1 AND ACAT2

Gene knockout experiments in mice have provided insights into the biological roles of ACAT1 and ACAT2. In ACAT1-deficient (ACAT1$^{/}$) mice, plasma cholesterol levels, hepatic cholesterol esterification activity and intestinal cholesterol absorption are normal, consistent with the presence of a second ACAT enzyme in the small intestine and liver of mice. ACAT1 deficiency has its greatest effect in the adrenal glands and macrophages.

Cholesterol esters and cholesterol esterification activity are virtually undetectable in the adrenal cortex of ACAT1$^{/}$ mice, although the depletion of cholesterol esters does not affect corticosterone synthesis. In addition, peritoneal macrophages from ACAT1$^{/}$ mice do not accumulate cholesterol esters when incubated with cholesterol-containing lipoproteins. This finding provides a unique tool for testing the role of ACAT1 in macrophage foam cell formation and atherosclerosis.

Another ACAT1 gene disruption study in mice confirmed general aspects of the ACAT1$^{/}$ phenotype and showed that ACAT1 deficiency reduces cholesterol esterification activity by >95 per cent in testis and ovary and results in atrophy of the meibomian glands of the eyelids. A naturally occurring mutation in the mouse ACAT1 gene is present in the AKR strain of inbred mice which are homozygous for the *adrenocortical lipid depletion* (*ald*) allele.

The analysis of ACAT2-deficient mice (ACAT2$^{/}$) (34) complemented findings from ACAT1$^{/}$ mice. ACAT2$^{/}$ mice have nearly undetectable ACAT activity and lack cholesterol esters in the liver and small intestine. Because they cannot synthesize cholesterol esters in the small intestine, ACAT2$^{/}$ mice have a reduced capacity to obtain cholesterol from the diet.

As a result, they do not develop hypercholesterolemia or form cholesterol gallstones when fed a diet high in fat, cholesterol ^{a}nd cholic acid. ACAT2 deficiency markedly decreases the cholesterol ester content of apolipoprotein (apo) B-containing lipoproteins in mice fed either a chow diet or the above diet. Thus, ACAT2 is the cholesterol esterification enzyme in mouse liver and small intestine and plays a major role in mediating the response to dietary cholesterol.

Gene disruption studies suggest that both ACAT enzymes synthesize cholesterol esters: ACAT1 for storage in cytosolic droplets and ACAT2 for secretion in apoB-containing lipoproteins. However, these results may reflect their different tissue distributions rather than inherent biochemical differences. Both ACAT1 and ACAT2, when expressed in a cell line deficient in cholesterol esterification activity (AC29 cells), synthesize cholesterol esters that are stored in cytoplasmic droplets (25). Also, overexpression of ACAT1 in mouse liver causes both cholesterol ester storage and secretion.

3

Structure of Carbohydrates

STRUCTURE

Formerly the name "carbohydrate" was used in chemistry for any compound with the formula $C_m(H_2O)_n$. Following this definition, some chemists considered formaldehyde CH_2O to be the simplest carbohydrate, while others claimed that title for glycolaldehyde Today the term is generally understood in the biochemistry sense, which excludes compounds with only one or two carbons.

Natural saccharides are generally built of simple carbohydrates called monosaccharides with general formula $(CH_2O)_n$ where n is three or more.

A typical monosaccharide has the structure H-$(CHOH)_x$(C=O) – $(CHOH)_y$–H, that is, an aldehyde or ketone with many hydroxyl groups added, usually one on each carbon atom that is not part of the aldehyde or ketone functional group.

Examples of monosaccharides are glucose, fructose, and glyceraldehyde. However, some biological substances commonly called "monosaccharides" do not conform to this formula (*e.g.*, uronic acids and deoxy-sugars such as fucose), and there are many chemicals that do conform to this formula but are not considered to be monosaccharides (*e.g.*, formaldehyde CH_2O and inositol $(CH_2O)_6$).

The open-chain form of a monosaccharide often coexists with a closed ring form where the oxygen of the carbonyl group C=O is replaced by an internal -O- bridge. Monosaccharides can be linked together into what are called polysaccharides (or oligosaccharides) in a large variety of ways. Many carbohydrates contain one or more modified monosaccharide units that have had one or more groups replaced or removed. For example, deoxyribose, a component of DNA, is a modified version of ribose; chitin is composed of repeating units of N-acetylglucosamine, a nitrogen-containing form of glucose.

MONOSACCHARIDES

Monosaccharides are the simplest carbohydrates in that they cannot be hydrolyzed to smaller carbohydrates. They are aldehydes or ketones with two or more hydroxyl groups. The general chemical formula of an unmodified mono saccharide is $(CH_2O)_n$, literally a "carbon hydrate."

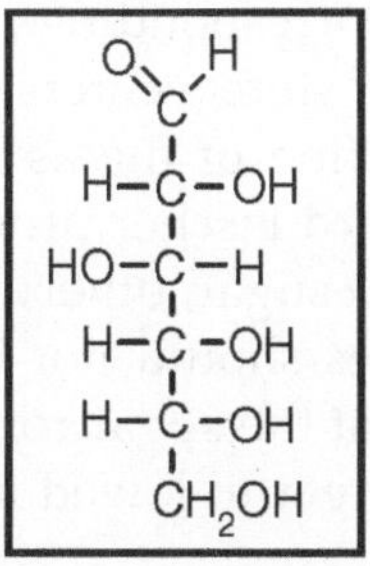

Fig. D-glucose is an Aldohexose with the Formula $(C{\cdot}H_2O)_6$.

Monosaccharides are important fuel molecules as well as building blocks for nucleic acids. The smallest mono-saccharides, for which n = 3, are dihydroxyacetone and D- and L-glyceraldehyde.

Classification of Monosaccharides

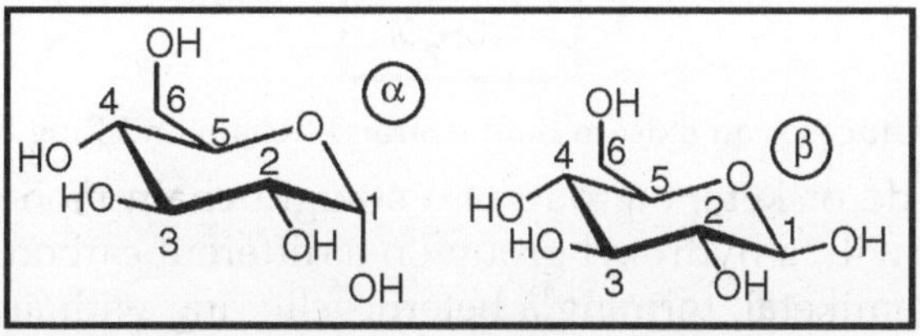

Fig. The α and β Anomers of Glucose.

Note the position of the anomeric carbon (red or green) relative to the CH_2OH group bound to carbon 5: they are either on the opposite sides (α), or the same side (β).

Monosaccharides are classified according to three different characteristics: the placement of its carbonyl group, the number of carbon atoms it contains, and its chiral handedness. If the carbonyl group is an aldehyde, the monosaccharide is an aldose; if the carbonyl group is a ketone, the monosaccharide is a ketose. Monosaccharides with three carbon atoms are called trioses, those with four are called tetroses, five are called pentoses, six are hexoses, and so on. These two systems of classification are often combined. For example, glucose is an aldohexose (a six-carbon aldehyde), ribose is an aldopentose (a five-carbon aldehyde), and fructose is a ketohexose (a six-carbon ketone).

Each carbon atom bearing a hydroxyl group (-OH), with the exception of the first and last carbons, are asymmetric, making them stereocenters with two possible configurations each (R or S). Because of this asymmetry, a number of isomers may exist for any given monosaccharide formula. The aldohexose D-glucose, for example, has the formula $(C{\cdot}H_2O)_6$, of which all but two of its six carbons atoms are stereogenic, making D-glucose one of $2^4 = 16$ possible stereoisomers. In the case of glyceraldehyde, an aldotriose, there is one pair of possible stereoisomers, which are enantiomers and epimers. 1,3-dihydroxyacetone, the ketose corresponding to the aldose glyceraldehyde, is a symmetric molecule with no stereocenters). The assignment of D or L is made according to the orientation of the asymmetric carbon furthest from the carbonyl group: in a standard Fischer projection if the hydroxyl group is on the right the molecule is a D sugar, otherwise it is an L sugar.

The "D-" and "L-" prefixes should not be confused with "d-" or "l-", which indicate the direction that the sugar rotates plane polarized light. This usage of "d-" and "l-" is no longer followed in carbohydrate chemistry.

Ring-straight Chain Isomerism

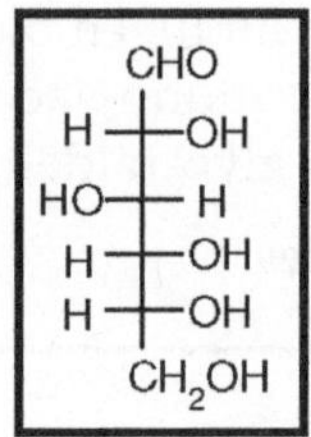

Fig. Glucose can Exist in Both a Straight-chain and Ring form.

The aldehyde or ketone group of a straight-chain monosaccharide will react reversibly with a hydroxyl group on a different carbon atom to form a hemiacetal or hemiketal, forming a heterocyclic ring with an oxygen bridge between two carbon atoms. Rings with five and six atoms are called furanose and pyranose forms, respectively, and exist in equilibrium with the straight-chain form. During the conversion from straight-chain form to cyclic form, the carbon atom containing the carbonyl oxygen, called the anomeric carbon, becomes a stereogenic centre with two possible configurations: The oxygen atom may take a position either above or below the plane of the ring.

The resulting possible pair of stereoisomers are called anomers. In the *a anomer*, the -OH substituent on the anomeric carbon rests on the opposite side (trans) of the ring from the CH_2OH side branch. The alternative form, in which the CH_2OH substituent and the anomeric hydroxyl are on the same side (cis) of the plane of the ring, is called the *b anomer*. You can remember that the b anomer is cis by the mnemonic, "It's always better to be up". Because the ring and straight-chain forms readily interconvert, both anomers exist in equilibrium.

Use in Living Organisms

Monosaccharides are the major source of fuel for metabolism, being used both as an energy source (glucose being the most important in nature) and in biosynthesis. When monosaccharides are not immediately needed by many cells they are often converted to more space efficient forms, often polysaccharides. In many animals, including humans, this storage form is glycogen, especially in liver and muscle cells. In plants, starch is used for the same purpose.

DISACCHARIDES

It is composed of two monosaccharides: D-glucose (left) and D-fructose (right). Two joined monosaccharides are called a disaccharide and these are the simplest polysaccharides. Examples include sucrose and lactose. They are composed of two monosaccharide units bound together by a covalent bond known as a glycosidic linkage formed via a dehydration reaction, resulting in the loss of a hydrogen atom from one monosaccharide and a hydroxyl group from the other.

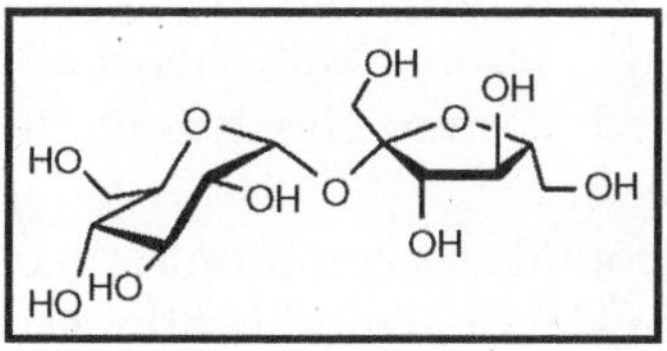

Fig. Sucrose, also Known as Table Sugar, is a Common Disaccharide.

The formula of unmodified disaccharides is $C_{12}H_{22}O_{11}$. Although there are numerous kinds of disaccharides, a handful of disaccharides are particularly notable. Sucrose, pictured to the right, is the most abundant disaccharide, and the main form in which carbohydrates are transported in plants. It is composed of one D-glucose molecule and one D-fructose molecule. The systematic name for sucrose, *O*-α-D-glucopyranosyl-(1→2)-D-fructofuranoside, indicates four things:

Its monosaccharides: Glucose and fructose.

Their ring types: glucose is a pyranose, and fructose is a furanose. How they are linked together: the oxygen on carbon number 1 (C1) of α-D-glucose is linked to the C2 of D-fructose. The *-oside* suffix indicates that the anomeric carbon of both monosaccharides participates in the glycosidic bond. Lactose, a disaccharide composed of one D-galactose molecule and one D-glucose molecule, occurs naturally in mammalian milk. The systematic name for lactose is *O*-β-D-galactopyranosyl-(1→4)-D-glucopyranose. Other notable disaccharides include maltose (two D-glucoses linked α-1,4) and cellulobiose (two D-glucoses linked β-1,4).

Oligosaccharides and Polysaccharides

It can be made of several thousands of glucose units. It is one of the two components of starch, the other being amylopectin. Oligosaccharides and polysaccharides are composed of longer chains of monosaccharide units bound together by glycosidic bonds. The distinction between the two is based upon the number of monosaccharide units present in the chain.

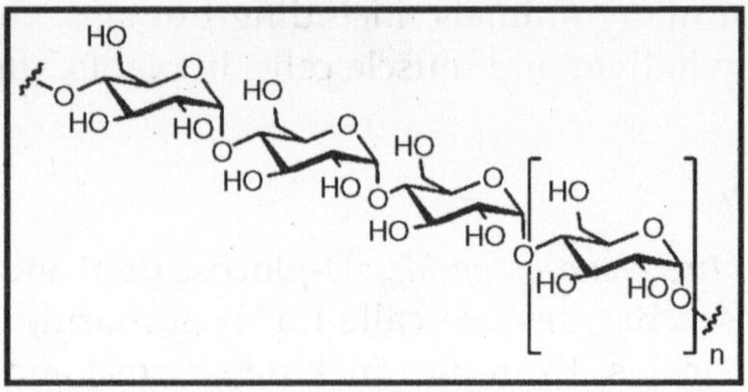

Fig. Amylose is a Linear Polymer of Glucose Mainly Linked with a(1$\rightarrow$4) Bonds.

Oligosaccharides typically contain between two and nine monosaccharide units, and polysaccharides contain greater than ten monosaccharide units. Definitions of how large a carbohydrate must be to fall into each category vary according to personal opinion. Examples of oligosaccharides include the disaccharides mentioned above, the trisaccharide raffinose and the tetrasaccharide stachyose.

Oligosaccharides are found as a common form of protein posttranslational modification. Such posttranslational modifications include the Lewis and ABO oligosaccharides responsible for blood group classifications and so of tissue incompatibilities, the alpha-Gal epitope responsible for hyperacute rejection in xenotransplanation, and O-GlcNAc modifications.

Polysaccharides represent an important class of biological polymers. Their function in living organisms is usually either structure- or storage-related.

Starch (a polymer of glucose) is used as a storage polysaccharide in plants, being found in the form of both amylose and the branched amylopectin. In animals, the structurally-similar glucose polymer is the more densely-branched glycogen, sometimes called 'animal starch'. Glycogen's properties allow it to be metabolized more quickly, which suits the active lives of moving animals. Cellulose and chitin are examples of structural polysaccharides. Cellulose is used in the cell walls of plants and other organisms, and is claimed to be the most abundant organic molecule on earth.

It has many uses such as a significant role in the paper and textile industries, and is used as a feedstock for the production of rayon (via the viscose process), cellulose acetate, celluloid, and nitrocellulose.

Chitin has a similar structure, but has nitrogen-containing side branches, increasing its strength. It is found in arthropod exoskeletons and in the cell walls of some fungi. It also has multiple uses, including surgical threads.

Other polysaccharides include callose or laminarin, chrysolaminarin, xylan, mannan, fucoidan, and galacto-mannan.

NUTRITION

Foods high in carbohydrates include breads, pastas, beans, potatoes, bran, rice, and cereals. Most such foods are high in starch. Carbohydrates are the most common source of energy in living things. Proteins and fat are necessary building components for body tissue and cells, and are also a source of energy for most organisms. Carbohydrates are not essential nutrients in humans: the body can obtain all its energy from protein and fats. However, the brain and neurons generally cannot burn fat and need glucose for energy; the body can make some glucose from a few of the amino acids in protein and also from the glycerol backbone in triglycerides.

Carbohydrate contains 15.8 kilojoules (3.75 kilocalories) and proteins 16.8 kilojoules (4 kilocalories) per gram, while fats contain 37.8 kilojoules (9 kilocalories) per gram. In the case of protein, this is somewhat misleading as only some amino acids are usable for fuel. Likewise, in humans, only some carbohydrates are usable for fuel, as in many monosaccharides and some disaccharides. Other carbohydrate types can be used, but only with the assistance of gut bacteria. Ruminants and termites can even process cellulose, which is indigestible to other organisms.

Based on the effects on risk of heart disease and obesity, the Institute of Medicine recommends that American and Canadian adults get between 45-65 per cent of dietary energy from carbohydrates. The Food and Agriculture Organization and World Health Organization jointly recommend that national dietary guidelines set a goal of 55-75 per cent of total energy from carbohydrates, but only 10 per cent directly from sugars (their term for simple carbohydrates).

Classification

For dietary purposes, carbohydrates can be classified as simple (monosaccharides and disaccharides) or complex (oligosaccharides and polysaccharides). The term complex carbohydrate was first used in the U.S. Senate Select Committee on Nutrition and Human Needs publication Dietary Goals for the United States (1977), where it denoted "fruit, vegetables and whole-grains".

Dietary guidelines generally recommend that complex carbohydrates, and such nutrient-rich simple carbohydrate sources such as fruit (glucose or fructose) and dairy products (lactose) make up the bulk of carbohydrate consumption. This excludes such sources of simple sugars as candy and sugary drinks. The USDA's Dietary Guidelines for Americans 2005 dispensed with the simple/complex distinction, instead recommending fibre-rich foods and whole grains. The glycemic index and glycemic load concepts have been developed to characterize food behaviour during human digestion.

They rank carbohydrate-rich foods based on the rapidity of their effect on blood glucose levels. The insulin index is a similar, more recent classification method that ranks foods based on their effects on blood insulin levels, which are caused by glucose (or starch) and some amino acids in food. Glycemic index is a measure of how quickly food glucose is absorbed, while glycemic load is a measure of the total absorbable glucose in foods.

MONOSACCHARIDES STRUCTURES

Sugars can be defined as polyhydroxy aldehydes or ketones. Hence the simplest sugars contain at least three carbons. The most common are the aldo- and keto-trioses, tetroses, pentoses and hexoses. The simplest 3C sugars are glyceraldehye and dihydroxyacetone. Glucose, an aldo-hexose, is a central sugar in metabolism. It and other 5 and 6C sugars can cyclize through intramolecular nucleophilic attack of one of the OH's on the carbonyl C of the aldehyde or ketone. Such intramolecular reactions occur if stable 5 or 6 member rings can form. The resulting rings are labeled furanose (5 member) or pyranose (6 member) based on their similarity to furan and pyran. On nucleophilic attack to form the ring, the carbonyl O becomes an OH which points either below the ring (a anomer) or above the ring (β anomer). Monosaccharides in solution exist as equilbrium mixtures of the straight and cyclic forms. In solution, glucose (Glc) is mostly in the pyranose form, fructose is 67 per cent pyranose and 33 per cent furanose and ribose is 75 per cent furanose and 25 per cent pyranose. However, in polysaccharides, Glc is exclusively pyranose and fructose and ribose are furanoses. Sugars can be drawn in the straight chain form as either Fisher projections or perspective structural formulas.

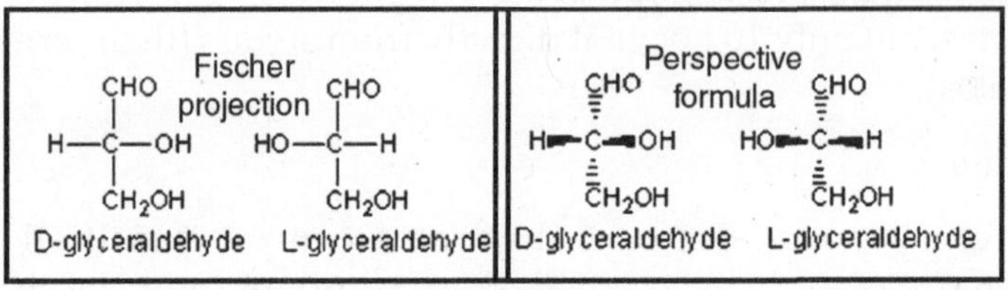

Fischer projection: D-glyceraldehyde, L-glyceraldehyde | Perspective formula: D-glyceraldehyde, L-glyceraldehyde

Cyclic forms can be drawn either as the Haworth projections, which shows the molecule as cyclic and planar with substituents above or below the ring) or the more plausible bent forms (showing Glc in the chair or boat conformations, for example). b-D-glucopyranose is the only aldohexose which can be drawn with all its bulky substituents (OH and CH_2OH) in equatorial positions, which probably accounts for its widespread prevalence in nature.

Haworth projections are more realistic than the Fisher projections, but you should be able to draw both structures. In general, if a substituents points to the right in the Fisher structure, it points down in the Haworth. if it points left, it points up. In general, the OH on the a-anomer points down (ants down) while on the b-anomer it points up (butterflys up).

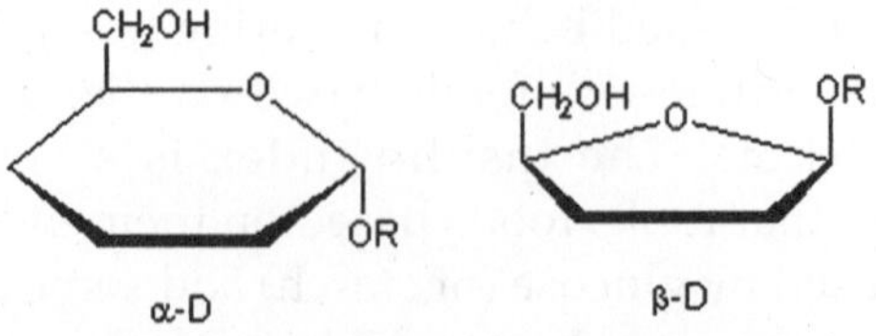

α-D β-D

Fig. A more Rigourous view of the Relationship Between the Anomeric OH and the OH on the Last Chiral C of a Sugar

The most common monosaccharides (other than glyceraldehyde and dihydroxyacetone) which you need to know are shown below.

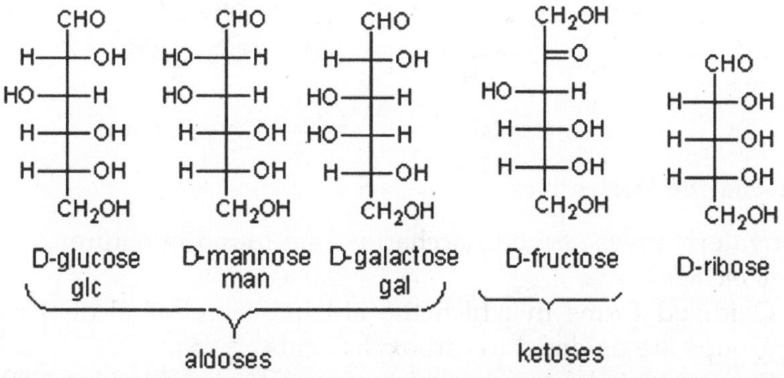

D-glucose (glc)	D-mannose (man)	D-galactose (gal)	D-fructose	D-ribose
CHO	CHO	CHO	CH_2OH	CHO
H–C–OH	HO–C–H	H–C–OH	C=O	H–C–OH
HO–C–H	HO–C–H	HO–C–H	HO–C–H	H–C–OH
H–C–OH	H–C–OH	HO–C–H	H–C–OH	H–C–OH
H–C–OH	H–C–OH	H–C–OH	H–C–OH	CH_2OH
CH_2OH	CH_2OH	CH_2OH	CH_2OH	
aldoses	aldoses	aldoses	ketoses	

The mirror image of D-Glc is L-Glc. For common sugars, the prefix D and L refer to the cenre of asymmetry most remote from the aldehyde or ketone. By convention, all chiral centres are related to D- glyceraldehyde, so sugar isomers related to D-glyceraldehyde at their last asymmetric cenre are D sugars.

ISOMERS

Sugars can exists as either configurational isomers (interconverted only by breaking covalent bonds) and conformational isomers.

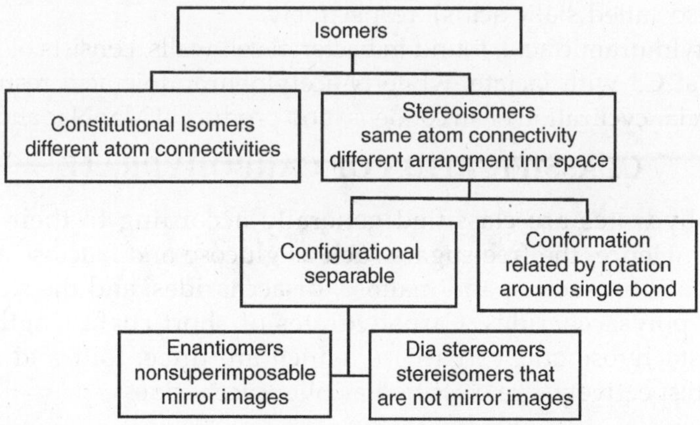

Fig. Different Types of Isomers

The configurational isomers include enantiomers (stereoisomers that are mirror images of each other), diastereomers (stereoisomers that are not mirror images), epimers (diastereomers that differ at one stereocentre) and anomers (a special form of stereoisomer, diastereomer and epimer that differ only in the configuration around the carbon which was attacked in the intramolecular nucleophilic attack to produces the a and b isomers).

Conformational Isomers

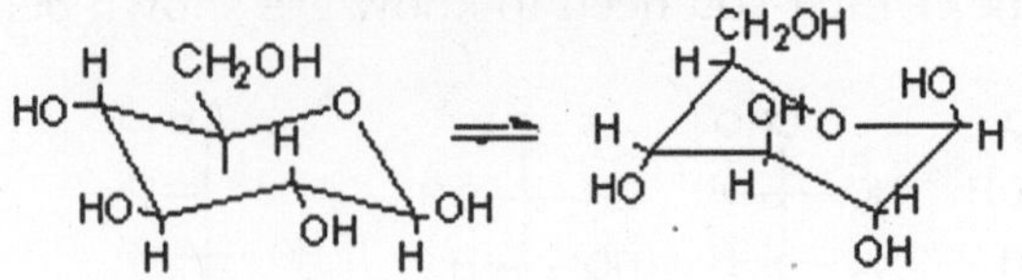

β-D-glucopyranose: chairnforms, boat form also is conformer

Monosaccharide Derivatives

Many derivatives of monosaccharides are found in nature.
These include:

- Oxidized forms in which the aldehyde and/or alcohol functional groups are oxidized to carboxylic acids
- Phosporylated forms in which phosphate is added by ATP to form phosphoester derivatives
- Amine derivatives such as glucosamine or galactosamine
- Acetylated amine derivatives such as N-Acetyl-GlcNAc (GlcNAc) or GalNAc
- Lactone forms (intramolecular esters) in which an OH group attacks a carbony C that was previously oxidized to a carboxylic acid
- Condensation products of sugar derivatives with lactate $(CH_3CHOHCO_2^-)$ and pyruvate, $(CH_3COCO_2^-)$, both from the glycolytic pathway, to form muramic acid and neuraminic acids, (also called sialic acids), respectively.

N-acetylmuramic acid, found in bacterial cell walls, consists of GlcNAc in ether link at C3 with lactate, while N-acetylneuraminic acid results from a intramolecular cyclization of an condensation product of ManNac and pyruvate.

CLASSIFICATION OF CARBOHYDRATE

Carbohydrates are classified generally according to their degree of complexity. Hence, the free sugars such as glucose and fructose are termed monosaccharides; sucrose and maltose, disaccharides; and the starches and celluloses, polysaccharides. Carbohydrates of short chain lengths such as raffinose, stachyose and verbascose, which are three, four and five sugar polymers respectively, are classified as oligosaccharides.

PENTOSES

Pentoses are five-carbon sugars seldom found in the free state in nature. In plants they occur in polymeric forms and are collectively known as pentosans. Thus, xylose and arabinose are the constituents of pentosans present in plant fibres and vegetable gums, respectively. As the sugar moieties in nucleic acids and riboflavin, ribose and deoxyribose are indispensable constituents of the life process. D-ribose has the following chemical structure.

HEXOSES

The hexoses comprise a large group of sugars. Principal among these are: glucose, fructose, galactose and mannose. While glucose and fructose are found free in nature, galactose and mannose occur only in combined form. The hexoses are divided into aldoses and ketoses according to whether they possess aldehydic or ketonic groups. Thus, glucose is an aldo sugar and fructose is a keto sugar.

The presence of aymmetric centres in all sugars with three or more carbon atoms gives rise to stereoisomers. Galactose and mannose are stereoisomers of glucose which, theoretically, is only one of 16 stereoisomers. Because the ketohexoses have only three asymmetric centres, fructose is one of eight stereoisomers. A general phenomenon, known as mutarotation, is observed in a variety of pentoses and hexoses as well as in certain disaccharides. For example, it has been established that two isomers of D-glucose exist, hence requiring an additional asymmetric centre in this sugar. It became apparent that D-glucose and most other sugars have cyclic structures.

The position of the hydroxyl group in relation to the ring oxygen characterizes this additional configurations modification. By convention, the positioning of the hydroxyl group on carbon atom 1 on the same side of the structure as the oxygen ring -modification; and, the positioning of the same hydroxyl group on the opposite side of the ring oxygen b-modification.

Carbohydrases, which catalyse the hydrolysis of glycosidic linkages of simple glycosides, oligosaccharides and polysaccharides often exhibit specificity with regard to substrate configuration. The specificity for enzyme hydrolysis of certain oligosaccharides helps to explain the poor utilization of this class of carbohydrates in fish nutrition.

Sugars containing the aldo or the keto group are capable of reducing copper in alkaline solutions to produce the brick-red colouration of cuprous ions. These sugars are called reducing sugars and the reaction, although not specific for reducing sugars, has use for both qualitative and quantitative determinations.

Glucose is widely distributed in small amounts in fruits, plant juices and honey. It is commercially produced by the acid or enzyme hydrolysis of grain and root starches. Glucose is of special interest in nutrition because it is the end-product of carbohydrate digestion in all non-ruminant animals including fish. Fructose is the only important ketohexose and is found in the free state alongside glucose in ripening fruits and honey. Combined with glucose it forms sucrose.

Fructose is somewhat sweeter than sucrose and is produced in increasing quantities commercially as a sweetener. Galactose occurs in milk in combination with glucose. It is also present in oligo-saccharides of plant origin, in combination with both glucose and fructose. Mannose is present in some plant polysaccharides collectively termed mannans.

DISACCHARIDES

Disaccharides are condensation products of two molecules of monosaccharides. Sucrose is the predominant disaccharide occurring in the free form and is the principal substance of sugar cane and sugar beet. It is also formed during germination of legume seeds. Other common disaccharides are maltose and lactose. Maltose is a dimer of glucose and lactose is a copolymer of galactose and glucose. The two molecules of glucose in maltose are held together in an a -1,4 glycosidic linkage whereas the two hexose entities of galactose are linked at the b -1,4 position. Glucose and fructose are combined in an a -1,2 linkage in sucrose. The abbreviated name of sucrose is D-Glu-(a, 1® 2)-D-Fru.

OLIGOSACCHARIDES

The Oligosaccharides raffinose, stachyose and verbascose are present in significant quantities in legume seeds. Raffinose, which is the most widespread among the three, consists of one molecule of glucose linked to a molecule of sucrose at the α-1, 6 position.

Its abbreviated chemical name is α-D-Gal (1→6) -α - D -Glu - (1 → 2) - β-D-Fru. Further chain elongation at the galactose end with another galactose molecule will yield stachyose. These galactose-galactose linkages are all at the α-l,6 position and digestion of these Oligosaccharides by animals requires a highly specific enzyme not elaborated by the animals themselves but by certain bacteria present in the animals guts.

The gradual disappearance of oligosaccharides from the cotelydons of legume seeds during germination is part of an intricate process beginning with uptake of water by the seed. This uptake of moisture releases gibberellic acid which in turn activates the DNA in the seed, thereby triggering the life cycle of the plant. The DNA directs the production of α-galactosidase which is required for the hydrolysis of these Oligosaccharides. Any interference of the DNA transcription process blocks enzyme production and will be evidenced by continued senescence of the seed and persistence of oligosaccharides in the seed cotelydons.

POLYSACCHARIDES

The polysaccharides represent a large group of complex carbohydrates which are condensation products of undetermined numbers of sugar molecules. The various subgroups are rather ill-defined and there is a lack of agreement on their classification. Most polysaccharides are insoluble in water. Upon hydrolysis with acids or enzymes they eventually yield their constituent monosaccharides.

Starch is a high molecular weight polymer of D-glucose and is the principal reserve carbohydrate in plants. Most starches consist of a mixture of two types of polymers, namely; amylose and amylopectin.

The proportion of amylose and amylopectin is generally one part of amylose and three parts of amylopectin. Enzymes capable of catalyzing the hydrolysis of starch are present in the digestive secretions of animals and fish within their cells. The a-amylases which are found virtually in all living cells cleave the α-D-(1→4) linkages at random and bring about an eventual total conversion of the starch molecule into the reducing sugars. The principal α-amylases of animal origin are those produced in the salivary gland and the pancreas. Starch is insoluble in water and is stained blue by iodine.

Glycogen is the only complex carbohydrate of animal origin. It exists in limited quantities in liver and muscle tissues and acts as a readily available energy source. Dextrins are intermediate compounds resulting from incomplete hydrolysis or digestion of starch.

The presence of α-D-(1→6) linkages in amylopectin and the inability of α-amylase to cleave these bonds give rise to low molecular weight carbohydrate segments called limit dextrins. These residues are acted upon primarily by acidophilic bacteria in the digestive tract. Cellulose is made up of long chains of glucose units held together by β-D-(1→4) linkages. The enzymes which cleave these linkages are not ordinarily present in the digestive secretions of animals and fish although some species of shellfish are believed to elabourate cellulase, the enzyme which catalyzes the hydrolysis of cellulose.

Cellulase producing micro-organisms present in the gut of herbivorous animals and fish impart to their host animals the ability to utilize as food the otherwise indigestible cellulose. Other complex polysaccharides in common occurrence are the hemicelluloses and pentosans. Hemicellulose represents a group of carbohydrates including araban, xylan, certain hexosans and polyuronides. These substances are generally less resistant to chemical treatment and undergo some degree of enzymatic hydrolysis during normal digestive processes. Pentosans are polymers of either xylose or arabinose as constituents of plant structural material and vegetable gums, respectively.

Arbohydrates are classified generally according to their degree of complexity. Hence, the free sugars such as glucose and fructose are termed monosaccharides; sucrose and maltose, disaccharides; and the starches and celluloses, polysaccharides. Carbohydrates of short chain lengths such as raffinose, stachyose and verbascose, which are three, four and five sugar polymers respectively, are classified as oligosaccharides.

DISEASES OF CARBOHYDRATE METABOLISM

DIABETES MELLITUS

Diabetes mellitus often referred to as diabetes—is a condition in which the body either does not produce enough, or does not properly respond to, insulin, a hormone produced in the pancreas. Insulin enables cells to absorb glucose in order to turn it into energy. This causes glucose to accumulate in the blood (hyperglycemia), leading to various potential complications.

Many Types of Diabetes are Recognized

The principal three are:

- *Type* 1: Results from the body's failure to produce insulin. It is estimated that 5–10 per cent of Americans who are diagnosed with diabetes have type 1 diabetes. Presently most persons with type 1 diabetes take insulin injections.
- *Type* 2: Results from insulin resistance, a condition in which cells fail to use insulin properly, sometimes combined with absolute insulin deficiency. Most Americans who are diagnosed with diabetes have type 2 diabetes.
- *Gestational diabetes*: Pregnant women who have never had diabetes before but who have high blood sugar (glucose) levels during pregnancy are said to have gestational diabetes. Gestational diabetes affects about 4 per cent of all pregnant women. It may precede development of type 2 (or rarely type 1) DM.

Other forms of diabetes mellitus are categorized separately from these. Examples include congenital diabetes due to genetic defects of insulin secretion, cystic fibrosis-related diabetes, steroid diabetes induced by high doses of glucocorticoids, and several forms of monogenic diabetes.

All forms of diabetes have been treatable since insulin became medically available in 1921, but a cure is difficult. Pancreas transplants have been tried with limited success in type 1 DM; gastric bypass surgery has been successful in many with morbid obesity and type 2 DM; and gestational diabetes usually resolves after delivery. Diabetes without proper treatments can cause many complications. Acute complications include hypoglycemia, diabetic ketoacidosis, or nonketotic hyperosmolar coma. Serious long-term complications include cardiovascular disease, chronic renal failure, retinal damage. Adequate treatment of diabetes is thus important, as well as blood pressure control and lifestyle factors such as smoking cesation and maintaining a healthy body weight. As of 2000 at least 171 million people worldwide suffer from diabetes, or 2.8 per cent of the population.

CLASSIFICATION

Most cases of diabetes mellitus fall into the three broad categories of type 1 or type 2 and gestational diabetes. A few other types are described.

The term diabetes, without qualification, usually refers to diabetes mellitus, which roughly translates to excessive sweet urine (known as "glycosuria"). Several rare conditions are also named diabetes. The most common of these is diabetes insipidus in which large amounts of urine are produced (polyuria), which is not sweet (insipidus meaning "without taste" in Latin).

The term "type 1 diabetes" has replaced several former terms, including childhood-onset diabetes, juvenile diabetes, and insulin-dependent diabetes mellitus (IDDM). Likewise, the term "type 2 diabetes" has replaced several former terms, including adult-onset diabetes, obesity-related diabetes, and non-insulin-dependent diabetes mellitus (NIDDM).

Beyond these two types, there is no agreed-upon standard nomenclature. Various sources have defined "type 3 diabetes" as: gestational diabetes, insulin-resistant type 1 diabetes (or "double diabetes"), type 2 diabetes which has progressed to require injected insulin, and latent autoimmune diabetes of adults (or LADA or "type 1.5" diabetes.)

Type 1 Diabetes

Type 1 diabetes mellitus is characterized by loss of the insulin-producing beta cells of the islets of Langerhans in the pancreas leading to insulin deficiency. This type of diabetes can be further classified as immune-mediated or idiopathic. The majority of type 1 diabetes is of the immune-mediated nature, where beta cell loss is a T-cell mediated autoimmune attack.

There is no known preventive measure against type 1 diabetes, which causes approximately 10 per cent of diabetes mellitus cases in North America and Europe. Most affected people are otherwise healthy and of a healthy weight when onset occurs. Sensitivity and responsiveness to insulin are usually normal, especially in the early stages. Type 1 diabetes can affect children or adults but was traditionally termed "juvenile diabetes" because it represents a majority of the diabetes cases in children.

Type 2 Diabetes

Type 2 diabetes mellitus is characterized by insulin resistance which may be combined with relatively reduced insulin secretion.

The defective responsiveness of body tissues to insulin is believed to involve the insulin receptor. However, the specific defects are not known. Diabetes mellitus due to a known defect are classified separately. Type 2 diabetes is the most common type. In the early stage of type 2 diabetes, the predominant abnormality is reduced insulin sensitivity. At this stage hyperglycemia can be reversed by a variety of measures and medications that improve insulin sensitivity or reduce glucose production by the liver. As the disease progresses, the impairment of insulin secretion occurs, and therapeutic replacement of insulin often becomes necessary.

Gestational Diabetes

Gestational diabetes mellitus (GDM) resembles type 2 diabetes in several respects, involving a combination of relatively inadequate insulin secretion and responsiveness. It occurs in about 2 per cent–5 per cent of all pregnancies and may improve or disappear after delivery. Gestational diabetes is fully treatable but requires careful medical supervision throughout the pregnancy. About 20 per cent–50 per cent of affected women develop type 2 diabetes later in life. Even though it may be transient, untreated gestational diabetes can damage the health of the fetus or mother. Risks to the baby include macrosomia (high birth weight), congenital cardiac and central nervous system anomalies, and skeletal muscle malformations. Increased fetal insulin may inhibit fetal surfactant production and cause respiratory distress syndrome.

Hyperbilirubinemia may result from red blood cell destruction. In severe cases, perinatal death may occur, most commonly as a result of poor placental perfusion due to vascular impairment. Labour induction may be indicated with decreased placental function. A cesarean section may be performed if there is marked fetal distress or an increased risk of injury associated with macrosomia, such as shoulder dystocia. A 2008 study completed in the U.S. found that more American women are entering pregnancy with preexisting diabetes. In fact the rate of diabetes in expectant mothers has more than doubled in the past 6 years. This is particularly problematic as diabetes raises the risk of complications during pregnancy, as well as increasing the potential that the children of diabetic mothers will also become diabetic in the future.

Other Types

Pre-diabetes indicates a condition that occurs when a person's blood glucose levels are higher than normal but not high enough for a diagnosis of type 2 diabetes. Many people destined to develop type 2 diabetes spend many years in a state of pre-diabetes which has been termed "America's largest health care epidemic,".

Some cases of diabetes are caused by the body's tissue receptors not responding to insulin (even when insulin levels are normal, which is what separates it from type 2 diabetes); this form is very uncommon.

Genetic mutations (autosomal or mitochondrial) can lead to defects in beta cell function. Abnormal insulin action may also have been genetically determined in some cases. Any disease that causes extensive damage to the pancreas may lead to diabetes (for example, chronic pancreatitis and cystic fibrosis). Diseases associated with excessive secretion of insulin-antagonistic hormones can cause diabetes (which is typically resolved once the hormone excess is removed). Many drugs impair insulin secretion and some toxins damage pancreatic beta cells.

The ICD-10 (1992) diagnostic entity, *malnutrition-related diabetes mellitus* (MRDM or MMDM, ICD-10 code E12), was deprecated by the World Health Organization when the current taxonomy was introduced in 1999.

SIGNS AND SYMPTOMS

The classical symptoms of DM are polyuria (frequent urination), polydipsia (increased thirst) and polyphagia (increased hunger). Symptoms may develop quite rapidly (weeks or months) in type 1 diabetes, particularly in children. However, in type 2 diabetes symptoms usually develop much more slowly and may be subtle or completely absent. Type 1 diabetes may also cause a rapid yet significant weight loss (despite normal or even increased eating) and irreducible mental fatigue. All of these symptoms except weight loss can also manifest in type 2 diabetes in patients whose diabetes is poorly controlled, although unexplained weight loss may be experienced at the onset of the disease. Final diagnosis is made by measuring the blood glucose concentration.

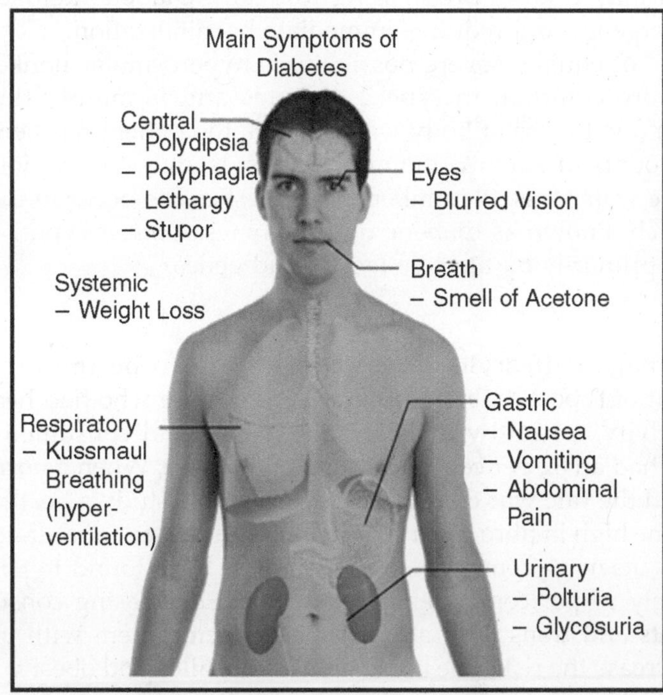

Fig. Overview of the Most Significant Symptoms of Diabetes.

When the glucose concentration in the blood is raised beyond its renal threshold (about 10 mmol/L, although this may be altered in certain conditions, such as pregnancy), reabsorption of glucose in the proximal renal tubuli is incomplete, and part of the glucose remains in the urine (glycosuria).

This increases the osmotic pressure of the urine and inhibits reabsorption of water by the kidney, resulting in increased urine production (polyuria) and increased fluid loss. Lost blood volume will be replaced osmotically from water held in body cells and other body compartments, causing dehydration and increased thirst. Prolonged high blood glucose causes glucose absorption, which leads to changes in the shape of the lenses of the eyes, resulting in vision changes; sustained sensible glucose control usually returns the lens to its original shape. Blurred vision is a common complaint leading to a diabetes diagnosis; type 1 should always be suspected in cases of rapid vision change, whereas with type 2 change is generally more gradual, but should still be suspected. Patients (usually with type 1 diabetes) may also initially present with diabetic ketoacidosis (DKA), an extreme state of metabolic dysregulation characterized by the smell of acetone on the patient's breath; a rapid, deep breathing known as Kussmaul breathing; polyuria; nausea; vomiting and abdominal pain; and any of many altered states of consciousness or arousal (such as hostility and mania or, equally, confusion and lethargy). In severe DKA, coma may follow, progressing to death. Diabetic ketoacidosis is a medical emergency and requires immediate hospitalization.

A rarer but equally severe possibility is hyperosmolar nonketotic state, which is more common in type 2 diabetes and is mainly the result of dehydration due to loss of body water. Often, the patient has been drinking extreme amounts of sugar-containing drinks, leading to a vicious circle in regard to the water loss. A number of skin rashes can occur in diabetes that are collectively known as diabetic dermadromes. *Causes*: Type 2 diabetes is determined primarily by lifestyle factors and genes.

Lifestyle

A number of lifestyle factors are known to be important to the development of type 2 diabtetes. In one study, those who had high levels of physical activity, a healthy diet, did not smoke, and consumed alcohol in moderation had an 82 per cent lower rate of diabetes. When a normal weight was included the rate was 89 per cent lower. In this study a healthy diet was defined as one high in fibre, with a high polyunsaturated to saturated fat ratio, and a lower mean glycemic index. Obesity has been found to contribute to approximately 55 per cent type 2 diabetes, and decreasing consumption of saturated fats and trans fatty acids while replacing them with unsaturated fats may decrease the risk. The increased rate of childhood obesity in between the 1960s and 2000s is believed to have lead to the increase in type 2 diabetes in children and adolescents. Environmental toxins may contribute to recent increases in the rate of type 2 diabetes. A positive correlation has been found between the concentration in the urine of bisphenol A, a constituent of some plastics, and the incidence of type 2 diabetes.

Medical Conditions

Subclinical Cushing's syndrome (cortisol excess) may be associated with DM type 2. The percentage of subclinical Cushing's syndrome in the diabetic population is about 9 per cent. Diabetic patients with a pituitary microadenoma can improve insulin sensitivity by removal of these microadenomas. Hypogonadism is often associated with cortisol excess, and testosterone deficiency is also associated with diabetes mellitus type 2, even if the exact mechanism by which testosterone improve insulin resistance is still not known.

Genetics

Both type 1 and type 2 diabetes are partly inherited. Type 1 diabetes may be triggered by certain infections, with some evidence pointing at Coxsackie B4 virus. There is a genetic element in individual susceptibility to some of these triggers which has been traced to particular HLA genotypes (*i.e.*, the genetic "self" identifiers relied upon by the immune system).

However, even in those who have inherited the susceptibility, type 1 diabetes mellitus seems to require an environmental trigger.

There is a stronger inheritance pattern for type 2 diabetes. Those with first-degree relatives with type 2 have a much higher risk of developing type 2, increasing with the number of those relatives. Concordance among monozygotic twins is close to 100 per cent, and about 25 per cent of those with the disease have a family history of diabetes.

Genes significantly associated with developing type 2 diabetes, include TCF7L2, PPARG, FTO, KCNJ11, NOTCH2, WFS1, CDKAL1, IGF2BP2, SLC30A8, JAZF1, and HHEX. KCNJ11 (potassium inwardly rectifying channel, subfamily J, member 11), encodes the islet ATP-sensitive potassium channel Kir6.2, and TCF7L2 (transcription factor 7-like 2) regulates proglucagon gene expression and thus the production of glucagon-like peptide-1.

Moreover, obesity (which is an independent risk factor for type 2 diabetes) is strongly inherited. Monogenic forms, *e.g.*, MODY, constitute 1-5 per cent of all cases. Various hereditary conditions may feature diabetes, for example myotonic dystrophy and Friedreich's ataxia. Wolfram's syndrome is an autosomal recessive neurodegenerative disorder that first becomes evident in childhood. It consists of diabetes insipidus, diabetes mellitus, optic atrophy, and deafness, hence the acronym DIDMOAD.

Gene expression promoted by a diet of fat and glucose as well as high levels of inflammation related cytokines found in the obese results in cells that "produce fewer and smaller mitochondria than is normal," and are thus prone to insulin resistance.

PATHOPHYSIOLOGY

Insulin production is more or less constant within the beta cells, irrespective of blood glucose levels. It is stored within vacuoles pending release, via exocytosis, which is primarily triggered by food, chiefly food containing absorbable glucose.

The chief trigger is a rise in blood glucose levels after eating. Insulin is the principal hormone that regulates uptake of glucose from the blood into most cells (primarily muscle and fat cells, but not central nervous system cells). Therefore deficiency of insulin or the insensitivity of its receptors plays a central role in all forms of diabetes mellitus.

Most of the carbohydrates in food are converted within a few hours to the monosaccharide glucose, the principal carbohydrate found in blood and used by the body as fuel. The most significant exceptions are fructose, most disaccharides (except sucrose and in some people lactose), and all more complex polysaccharides, with the outstanding exception of starch.

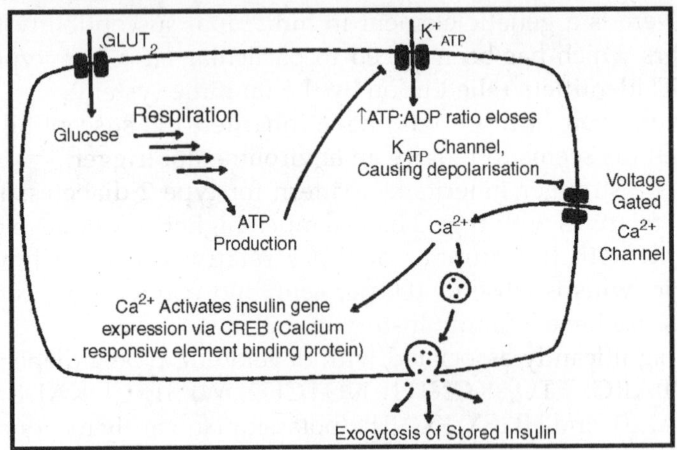

Fig. Mechanism of Insulin Release in Normal Pancreatic Beta Cells.

Insulin is released into the blood by beta cells (β-cells), found in the Islets of Langerhans in the pancreas, in response to rising levels of blood glucose, typically after eating. Insulin is used by about two-thirds of the body's cells to absorb glucose from the blood for use as fuel, for conversion to other needed molecules, or for storage.

Insulin is also the principal control signal for conversion of glucose to glycogen for internal storage in liver and muscle cells. Lowered glucose levels result both in the reduced release of insulin from the beta cells and in the reverse conversion of glycogen to glucose when glucose levels fall. This is mainly controlled by the hormone glucagon which acts in an opposite manner to insulin. Glucose thus recovered by the liver re-enters the bloodstream; muscle cells lack the necessary export mechanism.

Higher insulin levels increase some anabolic ("building up") processes such as cell growth and duplication, protein synthesis, and fat storage. Insulin (or its lack) is the principal signal in converting many of the bidirectional processes of metabolism from a catabolic to an anabolic direction, and vice versa. In particular, a low insulin level is the trigger for entering or leaving ketosis (the fat burning metabolic phase).

If the amount of insulin available is insufficient, if cells respond poorly to the effects of insulin (insulin insensitivity or resistance), or if the insulin itself is defective, then glucose will not be absorbed properly by those body cells that require it nor will it be stored appropriately in the liver and muscles. The net effect is persistent high levels of blood glucose, poor protein synthesis, and other metabolic derangements, such as acidosis.

DIAGNOSIS

1999 WHO Diabetes criteria		
Condition	**2 hour glucose**	**Fasting glucose**
	mmol/l(mg/dl)	mmol/l(mg/dl)
Normal	<7.8 (<140)	<6.1 (<110)
Impaired fasting glycaemia	<7.8 (<140)	≥ 6.1(≥110) & <7.0(<126)
Impaired glucose tolerance	≥7.8 (≥140)	<7.0 (<126)
Diabetes mellitus	≥11.1 (≥200)	≥7.0 (≥126)

Diabetes mellitus is characterized by recurrent or persistent hyperglycemia, and is diagnosed by demonstrating any one of the following:

- Fasting plasma glucose level at or above 7.0 mmol/L (126 mg/dL).
- Plasma glucose at or above 11.1 mmol/L (200 mg/dL) two hours after a 75 g oral glucose load as in a glucose tolerance test.
- Symptoms of hyperglycemia and casual plasma glucose at or above 11.1 mmol/L (200 mg/dL).

About a quarter of people with new type 1 diabetes have developed some degree of diabetic ketoacidosis (a type of metabolic acidosis which is caused by high concentrations of ketone bodies, formed by the breakdown of fatty acids and the deamination of amino acids) by the time the diabetes is recognized. The diagnosis of other types of diabetes is usually made in other ways. These include ordinary health screening; detection of hyperglycemia during other medical investigations; and secondary symptoms such as vision changes or unexplainable fatigue.

Diabetes is often detected when a person suffers a problem that is frequently caused by diabetes, such as a heart attack, stroke, neuropathy, poor wound healing or a foot ulcer, certain eye problems, certain fungal infections, or delivering a baby with macrosomia or hypoglycemia.

A positive result, in the absence of unequivocal hyperglycemia, should be confirmed by a repeat of any of the above-listed methods on a different day. Most physicians prefer to measure a fasting glucose level because of the ease of measurement and the considerable time commitment of formal glucose tolerance testing, which takes two hours to complete and offers no prognostic advantage over the fasting test. According to the current definition, two fasting glucose measurements above 126 mg/dL (7.0 mmol/L) is considered diagnostic for diabetes mellitus. Patients with fasting glucose levels from 100 to 125 mg/dL (6.1 and 7.0 mmol/L) are considered to have impaired fasting glucose. Patients with plasma glucose at or above 140 mg/dL or 7.8 mmol/L, but not over 200, two hours after a 75 g oral glucose load are considered to have impaired glucose tolerance.

Of these two pre-diabetic states, the latter in particular is a major risk factor for progression to full-blown diabetes mellitus as well as cardiovascular disease.

While not used for diagnosis, an elevated level of glucose irreversibly bound to hemoglobin (termed glycated hemoglobin or *HbA1c*) of 6.0 per cent or higher (the 2003 revised U.S. standard) is considered abnormal by most labs; HbA1c is primarily used as a treatment-tracking test reflecting average blood glucose levels over the preceding 90 days (approximately) which is the average lifetime of red blood cells which contain hemoglobin in most patients. However, some physicians may order this test at the time of diagnosis to track changes over time. The current recommended goal for HbA1c in patients with diabetes is 6.5 per cent.

SCREENING

Diabetes screening is recommended for many people at various stages of life, and for those with any of several risk factors. The screening test varies according to circumstances and local policy, and may be a random blood glucose test, a fasting blood glucose test, a blood glucose test two hours after 75 g of glucose, or an even more formal glucose tolerance test. Many health care providers recommend universal screening for adults at age 40 or 50, and often periodically thereafter.

Earlier screening is typically recommended for those with risk factors such as obesity, family history of diabetes, high-risk ethnicity.

Many medical conditions are associated with diabetes and warrant screening. A partial list includes: subclinical Cushing's syndrome, testosterone deficiency, high blood pressure, elevated cholesterol levels, coronary artery disease, past gestational diabetes, polycystic ovary syndrome, chronic pancreatitis, fatty liver, hemochromatosis, cystic fibrosis, several mitochondrial neuropathies and myopathies, myotonic dystrophy, Friedreich's ataxia, some of the inherited forms of neonatal hyperinsulinism.

The risk of diabetes is higher with chronic use of several medications, including long term corticosteroids, some chemotherapy agents (especially L-asparaginase), as well as some of the antipsychotics and mood stabilizers (especially phenothiazines and some atypical antipsychotics).

People with a confirmed diagnosis of diabetes are tested routinely for complications. This includes yearly urine testing for microalbuminuria and examination of the retina of the eye for retinopathy.

PREVENTION

Type 1

Type 1 diabetes risk is known to depend upon a genetic predisposition based on HLA types (particularly types DR3 and DR4), an unknown environmental trigger (suspected to be an infection, although none has proven definitive in all cases), and an uncontrolled autoimmune response that attacks the insulin producing beta cells. Some research has suggested that breastfeeding decreased the risk in later life; various other nutritional risk factors are being studied, but no firm evidence has been found. Giving children 2000 IU of Vitamin D during their first year of life is associated with reduced risk of type 1 diabetes, though the causal relationship is obscure. Children with antibodies to beta cell proteins (ie at early stages of an immune reaction to them) but no overt diabetes, and treated with vitamin B-3 (niacin), had less than half the diabetes onset incidence in a 7-year time span as did the general population, and an even lower incidence relative to those with antibodies as above, but who received no vitamin B3.

Type 2

Lifestyle

Type 2 diabetes risk can be reduced in many cases by making changes in diet and increasing physical activity. The American Diabetes Association (ADA) recommends maintaining a healthy weight, getting at least 2½ hours of exercise per week (several brisk sustained walks appear sufficient), having a modest fat intake, and eating sufficient fibre (*e.g.*, from whole grains).

The ADA does not recommend alcohol consumption as a preventive, but it is interesting to note that moderate alcohol intake may reduce the risk (though heavy consumption absolutely and clearly increases damage to bodily systems significantly); a similarly confused connection between low dose alcohol consumption and heart disease is termed the French Paradox. There is inadequate evidence that eating foods of low glycemic index is clinically helpful despite recommendations and suggested diets emphasizing this approach.

Diets that are very low in saturated fats reduce the risk of becoming insulin resistant and diabetic. Study group participants whose "physical activity level and dietary, smoking, and alcohol habits were all in the low-risk group had an 82 per cent lower incidence of diabetes."

In another study of dietary practice and incidence of diabetes, "foods rich in vegetable oils, including non-hydrogenated margarines, nuts, and seeds, should replace foods rich in saturated fats from meats and fat-rich dairy products.

Consumption of partially hydrogenated fats should be minimized." There are numerous studies which suggest connections between some aspects of Type II diabetes with ingestion of certain foods or with some drugs. Breastfeeding may also be associated with the prevention of type 2 of the disease in mothers.

Medications

Some studies have shown delayed progression to diabetes in predisposed patients through prophylactic use of metformin, rosiglitazone, or valsartan. In patients on hydroxychloroquine for rheumatoid arthritis, incidence of diabetes was reduced by 77 per cent though causal mechanisms are unclear. Lifestyle interventions are however more effective than metformin at preventing diabetes regardless of weightloss.

MANAGEMENT

Diabetes mellitus is a chronic disease which is difficult to cure. Management concentrates on keeping blood sugar levels as close to normal ("euglycemia") as possible without presenting undue patient danger.

This can usually be with close dietary management, exercise, and use of appropriate medications (insulin only in the case of type 1 diabetes mellitus. Oral medications may be used in the case of type 2 diabetes, as well as insulin).

Lifestyle modifications

There are roles for patient education, dietetic support, sensible exercise, with the goal of keeping both short-term and long-term blood glucose levels within acceptable bounds. In addition, given the associated higher risks of cardiovascular disease, lifestyle modifications are recommended to control blood pressure in patients with hypertension, cholesterol in those with dyslipidmia, as well as exercising more, smoking less or ideally not at all, consuming a recommended diet. Patients with foot problems are also recommended to wear diabetic socks, and possibly diabetic shoes.

Medications

In countries using a general practitioner system, such as the United Kingdom, care may take place mainly outside hospitals, with hospital-based specialist care used only in case of complications, difficult blood sugar control, or research projects. In other circumstances, general practitioners and specialists share care of a patient in a team approach.

Optometrists, podiatrists/chiropodists, dietitians, physio-therapists, nursing specialists (*e.g.*, DSNs (Diabetic Specialist Nurse)), nurse practitioners, or Certified Diabetes Educators, may jointly provide multidisciplinary expertise. In countries where patients must provide for their own health care (*e.g.* in the US, and in much of the undeveloped world). Peer support links people living with diabetes. Within peer support, people with a common illness share knowledge and experience that others, including many health workers, do not have. Peer support is frequent, ongoing, accessible and flexible and can take many forms-phone calls, text messaging, group meetings, home visits, and even grocery shopping. It complements and enhances other health care services by creating the emotional, social and practical assistance necessary for managing disease and staying healthy.

PROGNOSIS

Patient education, understanding, and participation is vital since the complications of diabetes are far less common and less severe in people who have well-managed blood sugar levels. Wider health problems may accelerate the deleterious effects of diabetes.

These include smoking, elevated cholesterol levels, obesity, high blood pressure, and lack of regular exercise. According to one study, women with high blood pressure (hypertension) were three times more likely to develop type 2 diabetes as compared with women with optimal BP after adjusting for various factors such as age, ethnicity, smoking, alcohol intake, body mass index (BMI), exercise, family history of diabetes, etc. The study was conducted by researchers from the Brigham and Women's Hospital, Harvard Medical School and the Harvard School of Public Health, USA, who followed over 38,000 female health professionals for ten years.

Except in the case of type 1 diabetes, which always requires insulin replacement, the way type 2 diabetes is managed may change with age. Insulin production decreases because of age-related impairment of pancreatic beta cells. Additionally, insulin resistance increases because of the loss of lean tissue and the accumulation of fat, particularly intra-abdominal fat, and the decreased tissue sensitivity to insulin.

Glucose tolerance progressively declines with age, leading to a high prevalence of type 2 diabetes and postchallenge hyperglycemia in the older population. Age-related glucose intolerance in humans is often accompanied by insulin resistance, but circulating insulin levels are similar to those of younger people. Treatment goals for older patients with diabetes vary with the individual, and take into account health status, as well as life expectancy, level of dependence, and willingness to adhere to a treatment regimen.

EPIDEMIOLOGY

In 2000, according to the World Health Organization, at least 171 million people worldwide suffer from diabetes, or 2.8 per cent of the population. Its incidence is increasing rapidly, and it is estimated that by the year 2030, this number will almost double. Diabetes mellitus occurs throughout the world, but is more common (especially type 2) in the more developed countries. The greatest increase in prevalence is, however, expected to occur in Asia and Africa, where most patients will probably be found by 2030.

The increase in incidence of diabetes in developing countries follows the trend of urbanization and lifestyle changes, perhaps most importantly a "Western-style" diet. This has suggested an environmental (*i.e.,* dietary) effect, but there is little understanding of the mechanism(s) at present, though there is much speculation, some of it most compellingly presented.

For at least 20 years, diabetes rates in North America have been increasing substantially. In 2008 there were about 24 million people with diabetes in the United States alone, from those 5.7 million people remain undiagnosed. Other 57 million people are estimated to have pre-diabetes. The Centers for Disease Control has termed the change an epidemic. The National Diabetes Information Clearinghouse estimates that diabetes costs $132 billion in the

United States alone every year. About 5 per cent-10 per cent of diabetes cases in North America are type 1, with the rest being type 2.

The fraction of type 1 in other parts of the world differs. Most of this difference is not currently understood. The American Diabetes Association cite the 2003 assessment of the National Centre for Chronic Disease Prevention and Health Promotion (Centers for Disease Control and Prevention) that 1 in 3 Americans born after 2000 will develop diabetes in their lifetime.

According to the American Diabetes Association, approximately 18.3 per cent (8.6 million) of Americans age 60 and older have diabetes. Diabetes mellitus prevalence increases with age, and the numbers of older persons with diabetes are expected to grow as the elderly population increases in number. The National Health and Nutrition Examination Survey (NHANES III) demonstrated that, in the population over 65 years old, 18 per cent to 20 per cent have diabetes, with 40 per cent having either diabetes or its precursor form of impaired glucose tolerance. Indigenous populations in first world countries have a higher prevalence and increasing incidence of diabetes than their corresponding non-indigenous populations. In Australia the age-standardised prevalence of self-reported diabetes in Indigenous Australians is almost 4 times that of non-indigenous Australians. Preventative community health programmes such as Sugar Man (diabetes education) are showing some success in tackling this problem.

History

The term *diabetes* was coined by Aretaeus of Cappadocia. " The verb *diabeinein* meant "to stride, walk, or stand with legs asunder"; hence, its derivative *diabetes* meant "one that straddles," or specifically "a compass, siphon." The sense "siphon" gave rise to the use of *diabçtçs* as the name for a disease involving the discharge of excessive amounts of urine.

Diabetes is first recorded in English, in the form diabete, in a medical text written around 1425. In 1675, Thomas Willis added the word *mellitus,* from the Latin meaning "honey", a reference to the sweet taste of the urine. This sweet taste had been noticed in urine by the ancient Greeks, Chinese, Egyptians, Indians, and Persians. In 1776, Matthew Dobson confirmed that the sweet taste was because of an excess of a kind of sugar in the urine and blood of people with diabetes.

Diabetes mellitus appears to have been a death sentence in the ancient era. Hippocrates makes no mention of it, which may indicate that he felt the disease was incurable. Aretaeus did attempt to treat it but could not give a good prognosis; he commented that "life (with diabetes) is short, disgusting and painful." Sushruta (6th century BCE) identified diabetes and classified it as *Medhumeha*. He further identified it with obesity and sedentary lifestyle, advising exercises to help "cure" it. The ancient Indians tested for diabetes by observing whether ants were attracted to a person's urine, and called the ailment "sweet urine disease" (Madhumeha). The Korean, Chinese, and Japanese words for diabetes are based on the same ideographs which mean "sugar urine disease".

In medieval Persia, Avicenna (980–1037) provided a detailed account on diabetes mellitus in *The Canon of Medicine,* "describing the abnormal appetite and the collapse of sexual functions," and he documented the sweet taste of diabetic urine. Like Aretaeus before him, Avicenna recognized a primary and secondary diabetes. He also described diabetic gangrene, and treated diabetes using a mixture of lupine, trigonella (fenugreek), and zedoary seed, which produces a considerable reduction in the excretion of sugar, a treatment which is still prescribed in modern times. Avicenna also "described diabetes insipidus very precisely for the first time", though it was later Johann Peter Frank (1745–1821) who first differentiated between diabetes mellitus and diabetes insipidus.

Although diabetes has been recognized since antiquity, and treatments of various efficacy have been known in various regions since the Middle Ages, and in legend for much longer, pathogenesis of diabetes has only been understood experimentally since about 1900.

The discovery of a role for the pancreas in diabetes is generally ascribed to Joseph von Mering and Oskar Minkowski, who in 1889 found that dogs whose pancreas was removed developed all the signs and symptoms of diabetes and died shortly afterwards. In 1910, Sir Edward Albert Sharpey-Schafer suggested that people with diabetes were deficient in a single chemical that was normally produced by the pancreas—he proposed calling this substance *insulin,* from the Latin *insula,* meaning island, in reference to the insulin-producing islets of Langerhans in the pancreas.

The endocrine role of the pancreas in metabolism, and indeed the existence of insulin, was not further clarified until 1921, when Sir Frederick Grant Banting and Charles Herbert Best repeated the work of Von Mering and Minkowski, and went further to demonstrate they could reverse induced diabetes in dogs by giving them an extract from the pancreatic islets of Langerhans of healthy dogs. Banting, Best, and colleagues (especially the chemist Collip) went on to purify the hormone insulin from bovine pancreases at the University of Toronto. This led to the availability of an effective treatment—insulin injections—and the first patient was treated in 1922.

For this, Banting and laboratory director MacLeod received the Nobel Prize in Physiology or Medicine in 1923; both shared their Prize money with others in the team who were not recognized, in particular Best and Collip.

Banting and Best made the patent available without charge and did not attempt to control commercial production. Insulin production and therapy rapidly spread around the world, largely as a result of this decision. Banting is honoured by World Diabetes Day which is held on his birthday, November 14.

The distinction between what is now known as type 1 diabetes and type 2 diabetes was first clearly made by Sir Harold Percival (Harry) Himsworth, and published in January 1936. Despite the availability of treatment, diabetes has remained a major cause of death. For instance, statistics reveal that the cause-specific mortality rate during 1927 amounted to about 47.7 per 100,000 population in Malta.

Other landmark discoveries include:

- Identification of the first of the sulfonylureas in 1942
- Reintroduction of the use of biguanides for Type 2 diabetes in the late 1950s. The initial phenformin was withdrawn worldwide (in the U.S. in 1977) due to its potential for sometimes fatal lactic acidosis and metformin was first marketed in France in 1979, but not until 1994 in the US.
- The determination of the amino acid sequence of insulin (by Sir Frederick Sanger, for which he received a Nobel Prize)
- The radioimmunoassay for insulin, as discovered by Rosalyn Yalow and Solomon Berson (gaining Yalow the 1977 Nobel Prize in Physiology or Medicine)
- The three-dimensional structure of insulin (PDB 2INS)
- Dr Gerald Reaven's identification of the constellation of symptoms now called metabolic syndrome in 1988
- Demonstration that intensive glycemic control in type 1 diabetes reduces chronic side effects more as glucose levels approach 'normal' in a large longitudinal study, and also in type 2 diabetics in other large studies
- Identification of the first thiazolidinedione as an effective insulin sensitizer during the 1990s
- In 1980, U.S. biotech company Genentech developed human insulin. The insulin is isolated from genetically altered bacteria (the bacteria contain the human gene for synthesizing human insulin), which produce large quantities of insulin. Scientists then purify the insulin and distribute it to pharmacies for use by diabetes patients.

Society and Culture

The 1990 "St Vincent Declaration" was the result of international efforts to improve the care accorded to those with diabetes. Doing so is important both in terms of quality of life and life expectancy but also economically-expenses due to diabetes have been shown to be a major drain on health-and productivity-related resources for health care systems and governments.

Several countries established more and less successful national diabetes programmes to improve treatment of the disease. A study shows that diabetic patients with neuropathic symptoms such as numbness or tingling in feet or hands are twice as likely to be unemployed as those without the symptoms.

LACTOSE INTOLERANCE

Lactose intolerance is the inability to metabolize lactose, because of a lack of the required enzyme lactase in the digestive system. It is estimated that 75 per cent of adults worldwide show some decrease in lactase activity during adulthood. The frequency of decreased lactase activity ranges from as little as 5 per cent in northern Europe, up to 71 per cent for Sicily, to more than 90 per cent in some African and Asian countries.

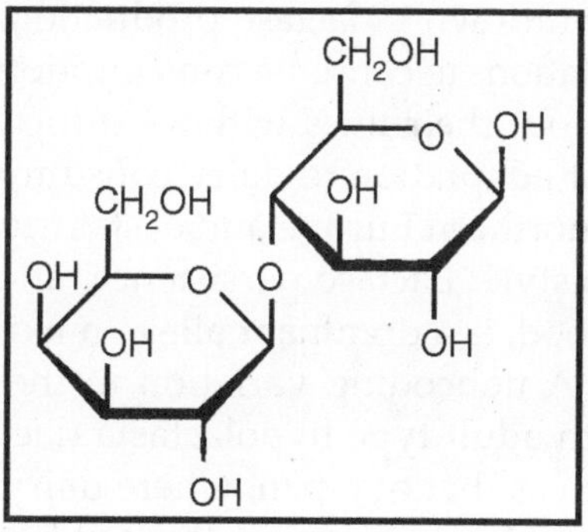

Fig. Lactose (disaccharide of β-D-galactose & β-D-glucose) is Normally Split by Lactase.

Symptoms

Abdominal bloating cramping and diarrhea.

It may be congenital or may begin in childhood, adolescence, or young adulthood

Classification

There are three major types of lactose intolerance:

- *Primary lactose intolerance*: Environmentally induced when weaning a child in non–dairy consuming societies. This is found in many Asian and African cultures, where industrialized and commercial dairy products are uncommon.
- *Secondary lactose intolerance*: Environmentally induced, resulting from certain gastrointestinal diseases, including exposure to intestinal parasites such as Giardia lamblia. In such cases the production of lactase may be permanently disrupted. A very common cause of temporary lactose intolerance is gastroenteritis, particularly when the gastroenteritis is caused by rotavirus. Another form of temporary lactose intolerance is lactose overload in infants.
- *Congenital lactase deficiency*: A genetic disorder which prevents enzymatic production of lactase. Present at birth, and diagnosed in early infancy.

Lactase Biology

The normal mammalian condition is for the young of a species to experience reduced lactase production at the end of the weaning period (a species-specific length of time). In humans, in non-dairy consuming societies, lactase production usually drops about 90 per cent during the first four years of life, although the exact drop over time varies widely.

However, certain human populations have a mutation on chromosome 2 which eliminates the shutdown in lactase production, making it possible for members of these populations to continue consumption of fresh milk and other dairy products throughout their lives without difficulty. This appears to be an evolutionarily recent adaptation to dairy consumption, and has occurred independently in both northern Europe and east Africa in populations with a historically pastoral lifestyle. Lactase persistence, allowing lactose digestion to continue into adulthood, is a dominant allele, making lactose intolerance a recessive genetic trait. A noncoding variation in the MCM6 gene has been strongly associated with adult type hypolactasia (lactose intolerance).

Some cultures, such as that of Japan, where dairy consumption has been on the increase, demonstrate a lower prevalence of lactose intolerance in spite of a genetic predisposition. Pathological lactose intolerance can be caused by coeliac disease, which damages the villi in the small intestine that produce lactase. This lactose intolerance is temporary. Lactose intolerance associated with coeliac disease ceases after the patient has been on a gluten-free diet long enough for the villi to recover. Certain people who report problems with consuming lactose are not actually lactose intolerant. In a study of 323 Sicilian adults, Carroccio et al. (1998) found only 4 per cent were both lactose intolerant and lactose maldigesters, while 32.2 per cent were lactose maldigesters but did not test as lactose intolerant. However, Burgio et al. (1984) found that 72 per cent of 100 Sicilians were lactose intolerant in their study and 106 of 208 northern Italians (*i.e.*, 51 per cent) were lactose intolerant.

Diagnosis

To assess lactose intolerance, the intestinal function is challenged by ingesting more dairy than can be readily digested. Clinical symptoms typically appear within 30 minutes but may take up to 2 hours, depending on other foods and activities. Substantial variability of the clinical response (symptoms of nausea, cramping, bloating, diarrhea, and flatulence) is to be expected, as the extent and severity of lactose intolerance varies between individuals.

When considering the need for confirmation, it is important to distinguish lactose intolerance from milk allergy, which is an abnormal immune response (usually) to milk proteins. Since lactose intolerance is the normal state for most adults on a worldwide scale and is not considered a disease condition, a medical diagnosis is not normally required. However, if confirmation is necessary, three tests are available.

Hydrogen Breath Test

In a hydrogen breath test, after an overnight fast, 50 grams of lactose (in a solution with water) is swallowed. If the lactose cannot be digested, enteric bacteria metabolize it and produce hydrogen. This, along with methane, can be detected in the patient's breath by a clinical gas chromato-graph or a compact solid state detector. The test takes about 2 to 3 hours. A medical condition with similar symptoms is fructose malabsorption.

In conjunction, measuring the blood glucose level every 10 - 15 minutes after ingestion will show a "flat curve" in individuals with lactose malabsorption, while the lactase persistent will have a significant "top", with an elevation of typically 50 to 100 per cent within 1 - 2 hours. However, given the need for frequent blood draws, this approach has been largely supplanted by breath testing.

Stool Acidity Test

This test can be used to diagnose lactose intolerance in small infants, for whom other forms of testing are risky or impractical.

Intestinal Biopsy

An intestinal biopsy can confirm lactose intolerance following discovery of elevated hydrogen in the hydrogen breath test. However, given the invasive nature of this test, and the need for a highly specialized laboratory to measure lactase enzymes or mRNA in the biopsy tissue, this approach is used almost exclusively in clinical research.

History of Diagnosis

The ancient Greek physician Hippocrates (460-370 B.C.) first noted gastrointestinal upset and skin problems in some who consumed milk; patients experiencing the former symptom may likely have been suffering from lactose intolerance. However, it was only in the last few decades that the syndrome was more widely described by modern medical science.

The condition was first recognized in the 1950s and 1960s when various organizations like the United Nations began to engage in systematic famine-relief efforts in countries outside Europe for the first time. Holzel et al. (1959) and Durand (1959) produced two of the earliest studies of lactose intolerance. As anecdotes of embarrassing dairy-induced discomfort increased, the First World donor countries could no longer ascribe the reports to spoilage in transit or inappropriate food preparation by the Third World recipients.

Because the first nations to industrialize and develop modern scientific medicine were dominated by people of European descent, adult dairy consumption was long taken for granted. Westerners for some time did not recognize that the majority of the human ethno-genetic groups could not consume dairy products during adulthood.

Although there had been regular contact between Europeans and non-Europeans throughout history, the notion that large-scale medical studies should be representative of the ethnic diversity of the human populations (as well as all genders and ages) did not become well-established until after the American Civil Rights Movement.

Since then, the relationship between lactase and lactose has been thoroughly investigated in food science due to the growing market for dairy products among non-Europeans. Originally it was hypothesised that gut bacteria such as E. coli produced the lactase enzyme needed to cleave lactose into its constituent monosaccharides, and thus become metabolisable and digestible by humans. Some form of human-bacteria symbiosis was proposed as a means of producing lactase in the human digestive tract. By the early 1970s, genetics and protein analysis techniques revealed this to be untrue; humans produce their own lactase enzyme natively in intestine cells.

Nomenclature

According to Heyman, approximately 70 per cent of the global population cannot tolerate lactose in adulthood. Thus, some argue that the terminology should be reversed — lactose intolerance should be seen as the norm, and the minority groups should be labeled as having *lactase persistence.* A counter-argument to this is that the cultures that don't generally consume unmodified milk products have little need to discuss their intolerance to it, leaving the cultures for which lactose intolerance is a significant dietary issue to define its terminology.

History of Genetic Prevalence

Lactose intolerance has been studied as an aid in understanding ancient diets and population movement in prehistoric societies. Milking an animal vastly increases the calories that may be extracted from the animal, as compared to the consumption of its meat alone. It is not surprising, then, that consuming milk products became an important part of the agricultural way of life in the Neolithic. It is believed that most of the milk was used to make mature cheeses, which are mostly lactose free.

Roman authors recorded that the people of northern Europe, particularly Britain and Germany, drank unprocessed milk (as opposed to the Romans who made cheese). This corresponds very closely with modern European distributions of lactose intolerance, where the people of Britain, Germany and Scandinavia have a good tolerance, and those of southern Europe, especially Italy, have a poorer tolerance.

In east Asia, historical sources also attest that the Chinese did not consume milk, whereas the nomads that lived on the borders did. Again, this reflects modern distributions of intolerance. China is particularly notable as a place of poor tolerance, whereas in Mongolia and the Asian steppes horse milk is drunk regularly. This tolerance is thought to be advantageous, as the nomads do not settle down long enough to process mature cheese. Given that their prime source of income is generated through horses, to ignore their milk as a source of calories would be greatly detrimental. The nomads also make an alcoholic beverage, called Kumis, from horse milk, although the fermentation process reduces the amount of lactose present.

The African Fulani have a nomadic origin and their culture once completely revolved around cow, goat, and sheep herding. Dairy products were once a large source of nutrition for them. As might be expected if lactase persistence evolved in response to dairy product consumption, they are particularly tolerant to lactose (about 77 per cent of the population). Many Fulani live in Guinea-Conakry, Burkina Faso, Mali, Nigeria, Niger, Cameroon, and Chad. There is some debate on exactly where and when genetic mutation(s) occurred, although a recent study suggests that the genetic change that enabled early Europeans to drink milk without getting sick has appeared in dairying farmers who lived around 7,500 years ago in a region between the central Balkans and central Europe.

Some have argued earlier for separate mutation events in Sweden (which has one of the lowest levels of lactose intolerance in the world) and the Arabian Peninsula around 4000 BC. However, others argue for a single mutation event in the Middle East at about 4500 BC, which then subsequently radiated. Some sources suggest a third and more recent mutation in the East African Tutsi.

Whatever the precise origin in time and place, most modern Northern Europeans and people of India, as well as people of European or Indian ancestry, show the effects of this mutation (that is, they are able to safely consume milk products all their lives), while most modern East Asians, sub-Saharan Africans and native peoples of America and the Pacific Islands do not (making them lactose intolerant as adults). The Maasai ability to consume dairy without exhibiting symptoms may be due to a different genetic mutation, or it may be due to the fact that they curdle their milk before they consume it, removing the lactose.

Managing Lactose Intolerance

For persons living in societies where the diet contains relatively little dairy, lactose intolerance is not considered a condition that requires treatment. However, those living among societies that are largely lactose-tolerant may find lactose intolerance troublesome.

Although there are still no methodologies to reinstate lactase production, some individuals have reported their intolerance to vary over time (depending on health status and pregnancy). Lactose intolerance is not usually an all-or-nothing condition: the reduction in lactase production—and hence, the amount of lactose that can be tolerated—varies from person to person. Since lactose intolerance poses no further threat to a person's health, managing the condition consists of minimizing the occurrence and severity of symptoms. Berdanier and Hargrove recognise 4 general principles:

- Avoidance of dietary lactose;
- Substitution to maintain nutrient intake;
- Regulation of calcium intake;
- Use of enzyme substitute.

Avoiding Lactose-containing Products

Since each individual's tolerance to lactose varies, according to the US National Institute of Health, "Dietary control of lactose intolerance depends on people learning through trial and error how much lactose they can handle." Label reading is essential, as commercial terminology varies according to language and region. Lactose is present in two large food categories: conventional dairy products, and as a food additive (in dairy and non dairy products).

Dairy Products

Lactose is a water-soluble molecule. Therefore fat percentage and the curdling process have an impact on which foods may be tolerated. After the curdling process, lactose is found in the water portion (along with whey and casein) but is not found in the fat portion.

Dairy products which are "fat reduced" or "fat free" generally have a slightly higher lactose percentage. Additionally, low fat dairy foods also often have various dairy derivatives such as milk solids added to them to enhance sweetness, increasing the lactose content.

- *Milk*: Human milk has the highest lactose percentage at around 9 per cent. Unprocessed cow milk has 4.7 per cent lactose. Unprocessed milk from other bovids contains similar lactose percentages (goat milk 4.1 per cent, buffalo 4.86 per cent, yak 4.93 per cent, sheep milk 4.6 per cent)
- *Butter*: The butter-making process separates the majority of milk's water components from the fat components. Lactose, being a water soluble molecule, will still be present in small quantities in the butter unless it is also fermented to produce cultured butter.
- *Yogurt:*Yogurt and kefir. People can be more tolerant of traditionally made yogurt than milk, because it contains lactase enzyme produced by the bacterial cultures used to make the yogurt. However, many commercial brands contain milk solids, increasing the lactose content.
- *Cheeses*: Traditionally made hard cheese (such as Swiss cheese) and soft ripened cheeses may create less reaction than the equivalent amount of milk because of the processes involved. Fermentation and higher fat content contribute to lesser amounts of lactose.

Traditionally made Swiss or Cheddar might contain 10 per cent of the lactose found in whole milk. In addition, the traditional aging methods of cheese (over 2 years) reduces their lactose content to practically nothing. Commercial cheese brands, however, are generally manufactured by modern processes that do not have the same lactose reducing properties, and as no regulations mandate what qualifies as an "aged" cheese, this description does not provide any indication of whether the process used significantly reduced lactose.

- Sour cream and ice cream, like yogurt, if made the traditional way, may be tolerable, but most modern brands add milk solids. Consult labels.

Examples of Lactose Levels in Foods

As scientific consensus has not been reached concerning lactose percentage analysis methods (non-hydrated form or the mono-hydrated form), and considering that dairy content varies greatly according to labeling practices, geography and manufacturing processes, lactose numbers may not be very reliable.

The following are examples of lactose levels in foods which commonly set off symptoms. These quantities are to be treated as guidelines only.

Dairy product	Lactose Content
Yogurt, plain, low-fat, 240 mL	5 g
Milk, reduced fat, 240 mL	11 g
Swiss cheese, 28 g	1 g
Ice cream, 120 mL	6 g
Cottage cheese, 120 mL	2–3 g

Lactose in Non-dairy Products

Lactose (also present when labels state lactoserum, whey, milk solids, modified milk ingredients, etc.) is a commercial food additive used for its texture, flavour and adhesive qualities, and is found in foods such as processed meats (sausages/hot dogs, sliced meats), gravy stock powder, margarines sliced breads, breakfast cereals, potato chips, dried fruit, processed foods, medications, pre-prepared meals, meal replacement (powders and bars), and protein supplements (powders and bars).

Kosher products labeled *pareve* are free of milk. However, if a "D" (for "Dairy") is present next to the circled "K", "U", or other hechsher, the food likely contains milk solids (although it may also simply indicate that the product was produced on equipment shared with other products containing milk derivatives).

Alternative Products

Plant based milks and derivatives are inherently lactose free: soy milk, rice milk, almond milk, hazelnut milk, oat milk, hemp milk, peanut milk, horchata (which can be made with dairy milk, consult ingredients).

The dairy industry has created low-lactose or lactose-free products to replace regular dairy. Lactose-free milk can be produced by passing milk over lactase enzyme bound to an inert carrier; once the molecule is cleaved, there are no lactose ill-effects. A form is available with reduced amounts of lactose (typically 30 per cent of normal), and alternatively with nearly 0 per cent.

Finland, where approximately 17 per cent of the Finnish-speaking population has hypolactasia, has had "HYLA" (acronym for *hydrolysed lactose*) products available for many years. These low-lactose level cow's milk products, ranging from ice cream to cheese, use a Valio patented chromatographic separation method to remove lactose. The ultra-pasteurization process, combined with aseptic packaging, ensures a long shelf-life. Recently, the range of low-lactose products available in Finland has been augmented with milk and other dairy products (such as ice cream, butter, and buttermilk) that contain no lactose at all. The remaining about 20 per cent of lactose in HYLA products is taken care of enzymatically. These typically cost slightly more than equivalent products containing lactose. Valio also markets these products in Sweden and in Estonia.

In the UK, where an estimated 15 per cent of the population are affected by lactose intolerance, Lactofree produces milk, cheese, and yogurt products which contain only 0.03 per cent lactose. Alternatively, a bacterium such as *L. acidophilus* may be added, which affects the lactose in milk the same way it affects the lactose in yogurt.

Lactase Supplementation

When lactose avoidance is not possible, or on occasions when a person chooses to consume such items, then enzymatic lactase supplements may be used. Lactase enzymes similar to those produced in the small intestines of humans are produced industrially by fungi of the genus *Aspergillus*. The enzyme, β-galactosidase, is available in tablet form in a variety of doses, in many countries without a prescription. It functions well only in high-acid environments, such as that found in the human gut due to the addition of gastric juices from the stomach.

Unfortunately, too much acid can denature it, and it therefore should not be taken on an empty stomach. Also, the enzyme is ineffective if it does not reach the small intestine by the time the problematic food does. Lactose-sensitive individuals can experiment with both timing and dosage to fit their particular needs. But supplements such as these may not be able to provide the accurate amount of lactase needed to adequately digest the lactose contained in dairy products, which may lead to symptoms similar to the existing lactose intolerance. While essentially the same process as normal intestinal lactose digestion, direct treatment of milk employs a different variety of industrially produced lactase. This enzyme, produced by yeast from the genus *Kluyveromyces*, takes much longer to act, must be thoroughly mixed throughout the product, and is destroyed by even mildly acidic environments.

It therefore has been much less popular as a consumer product (sold, where available, as a liquid) than the *Aspergillus* -produced tablets, despite its predictable effectiveness. Its main use is in producing the lactose-free or lactose-reduced dairy products sold in supermarkets. Enzymatic lactase supplementation may have an advantage over avoiding dairy products, in that alternative provision does not need to be made to provide sufficient calcium intake, especially in children.

Rehabituation to Dairy Products

For healthy individuals with secondary lactose intolerance, it may be possible in some cases for the bacteria in the large intestine to adapt to an altered diet and break down small quantities of lactose more effectively by habitually consuming small amounts of dairy products several times a day over a period of time. Reintroducing dairy in this way to people who have an underlying or chronic illness, however, is not recommended, as certain illnesses damage the intestinal tract in a way which prevents the lactase enzyme from being expressed. Some studies indicate that environmental factors (more specifically, the consumption of lactose) may "play a more important role than genetic factors in the etio-pathogenesis of milk intolerance", but some other publications suggest that lactase production does not seem to be induced by dairy/lactose consumption.

Nutritional Concerns

Primary Lactose Intolerance

Populations where primary lactose intolerance is the norm have demonstrated similar health levels to westerners, or better health.

Secondary Lactose Intolerance

While secondary lactose intolerance does not inherently affect an individual's nutritional needs, according to accepted medical doctrines in western European and North American countries dairy is an essential part of a healthy diet. Dairy products are relatively good and accessible sources of calcium and potassium and many countries mandate that milk be fortified with vitamin A and vitamin D.

Consequently, in dairy-consuming societies, dairy is often a main source of these nutrients and, for lacto-vegetarians, a main source of vitamin B12. Individuals who reduce or eliminate consumption of dairy must obtain these nutrients elsewhere. However, Asian populations for whom dairy is not part of their food culture do not present decreased health and sometimes present above average health, as in Japan.

Plant based milk substitutes are not naturally rich in calcium, potassium, or vitamins A or D (and, like most non-animal products, contain no vitamin B12). However, prominent brands are often voluntarily fortified with many of these nutrients. An increasing number of calcium-fortified breakfast foods - such as orange juice, bread, and dry cereal - have been appearing on supermarket shelves. Many fruits and vegetables are rich in potassium and vitamin A; animal products like meat and eggs are rich in vitamin B12, and the human body itself produces some vitamin D from exposure to direct sunlight. Finally, a dietitian or physician may recommend a vitamin or mineral supplement to make up for any remaining nutritional shortfall. Lactose-reduced dairy products have the same nutritional content as their full-lactose counterparts, but their taste and appearance may differ slightly.

Most infants with gastroenteritis due to rotavirus do not develop lactose intolerance, so these infants do not benefit from being put on a lactose-free diet unless symptoms of lactose intolerance are severe and persistent.

Congenital Lactase Deficiency

Congenital lactase deficiency, or CLD, is an autosomal recessive disorder which prevents the expression of lactase. Before the 20th century, infants with this disease rarely survived. As substitute and lactose-free infant formulas later became available, nursing infants affected with CLD could now have their normal nutritional needs met. Beyond infancy, individuals with CLD usually have the same nutritional concerns as those affected by secondary lactose intolerance.

CARBOHYDRATES MONOSACCHARIDES

- (Monosaccharoses) Dioses ($C_2H_4O_2$) - Glycolose.
- Moses ($C_3H_6O_3$). Aldoses - Glycerose. Ketose - Dioxyacetone.
- Tetroses ($C_4H_8O_4$).
- Aldoses - Erythrose,3 Threose.3 Ketose - Erythrulose.2
- Pentoses ($C_6H_{10}O_5$).
- Aldoses - Arabinose,2 Xylose,2 Ribose,2 Lyxose.8 Ketoses
- - Araboketose3, Xyloketose (ketoxylose).2 [Methyl pentoses ($C_6H_{10}O_5$) - Rhamnose,2 Fucose2].

Some writers use the term polysaccharides to include all carbohydrates other than monosaccharides. Mathews applies it to all carbohydrates more complex than the disaccharides.

Names of a few of the most important carbohydrates are printed in small capitals. Separate mention of the d, l, and dl forms of the various sugars is omitted, since in the study of food and nutrition we are practically concerned only with that one of the three forms which is found in or derived from natural products.

- Hexoses ($C_6H_{12}O_6$).
- Aldoses - Glucose,1 Mannose,2 Galactose,2 Gulose,3 Idose,3Talose,3.
- Allose,2 Altrose.3 Ketoses - Fructose,1 Sorbose,2 Tagatose.3.
- Heptoses ($C_7H_{14}O_7$). Aldose - Mannoheptose.1 Ketose - Sedoheptose.1.

Note: No attempt is here made to summarize the occurrence of any but the tetroses, pentoses, hexoses and heptoses. Glycolose and the trioses if formed in nature are probably too reactive to accumulate sufficiently for identification.

Disaccharides (Disaccharoses)

Dihexoses (Hexobioses) - ($C_{12}H_{22}O_{11}$). Anhydride of glucose + fructose-Sucrose. Anhydrides of glucose + galactose - Lactose, Melibiose. Anhydrides of glucose + glucose- Maltose, Isomaltose, Trehalose, Turanose.

Trisaccharides (Trisaccharoses)

Trihezoses ($C_{18}H_{22}O_{16}$). Anhydride of glucose + galactose +fructose - Raffinose. Anhydride of glucose + glucose + glucose - Melezitose. Anhydride of fructose + fructose + fructose - Secalose.

Tetrasaccharides (Tetrasaccharoses)

Tetrahexoses ($C_{24}H_{42}O_{21}$). Anhydrides of 2 galactose + glucose + fructose - Stachyose, Lupeose.

Polysaccharides (Polysaccharoses)

Pentosans (chief constituents of gums and mucilages). Anhydrides of xylose - Xylans. Anhydrides of arabinose - Arabans.

Hexosans

Anhydrides of glucose-Starch, Cellulose, Glycogen, Dextrin (and other"dextrans"). Anhydrides of mannose - Mannans. Anhydrides of galactose-Galactans (pectins). Anhydrides of fructose - Inulin (and other"levulans").

- Occurs free in nature.
- Not yet found free in nature (or only in small amounts) but obtained by hydrolysis or fermentation of natural product.
- Known only (with certainty) as a Labouratory product.

SIMPLE BIOCHEMICAL TESTS

Materials

- Benedict's solution (not"quantitative")

- Iodine solution (I/KI)
- Waterbaths set at about 100°C
- Boiling tubes Racks
- 1 per cent glucose soln
- 1 per cent fructose soln
- per cent sucrose soln
- other sugars?
- 1 per cent starch sol solution X

HCl, NaOH, litmus/universal indicator paper filter funnels, papers, flasks? measuring cylinders (10, 25, 50, 100 ml) plastic pipettes glass pipettes (1, 5, 10 ml?), pipette pumps

Objectives

Familiarisation with Benedict's test - for reducing sugars modification of Benedict's test for non-reducing sugars iodine test dilution techniques sensitivity of above tests.

Procedure

See also Roberts Students' manual p. 42 You will be expected to write up and hand in a concise account of your techniques and the results. There is no need to use large amounts of test substances or reagents.

SAFETY

Irritant Solutions Wear Safety Glasses And Lab Coats Hot Water And Steam Take Care: Keep Working Area Clear:

- Carry out Benedict's test on the named sugar solutions and starch as provided.
- Use fixed (noted!) quantities of test substance and reagents.
- Do not contaminate stock solutions shared by the group.
- Note results - colour and intensity - separately from conclusions.
  - *Refinement*: Filter the contents of the tube in which you have been testing glucose (using paper previously labelled with pencil). Consider the colour of the filtrate, and accordingly modify the quantities of reagents used. Repeat filtration process. Explain your conclusions in your write-up.
- Dilute the sugar solutions (say ×10, ×100 etc), and carry out Benedict's test on suitable samples (*i.e.* amounts same as before).
- Explain the dilution technique you used in your write-up.
- What does this tell you about the sensitivity of the test?
- Using sucrose, carry out the test for non reducing sugars. Having obtained a negative result from ordinary benedict's test, a second test is performed after hydrolysing it to reducing sugars (acid hydrolysis, followed by neutralisation), then performing Benedict's test:-

Add... drops/ml HCl - leave for... minutes at room temperature (... °C) or... mins at 100 °C - add... drops/ml NaOH (same strength) - check neutrality with litmus (colour......) or universal indicator paper (colour.....).

- Test solution X using a combination of the above tests.

GLYCOPROTEIN STRUCTURE AND STEROID HORMONE FUNCTION

Glycoproteins play an important part in hormone function. The action of hormones depends on the initial binding of the hormone to a protein receptor molecule. In many cases this molecule is a glycoprotein. Many hormones bind to receptors in the cell membrane; these hormones never actually enter the cell.

Steroids, on the other hand, bind to an intracellular protein receptor. There is still controversy over whether the steroid hormone receptor is found in the nucleus or the cytosol. However, it is clear that after the steroid binds, the hormone-receptor complex moves to the nucleus. The hormone binding domain of the receptor is found at the C-terminus. The amino acid sequence of this region is highly diverse; it is not conserved from one protein to the next.

Binding of the hormone stimulates a conformational change in the hormone receptor. This change allows the hormone-receptor complex to bind to the DNA. The DNA binding domain is highly conserved and is found within the central core of the protein. The central core contains a very basic amino acid sequence. In the cellular environment, these bases will tend to pick up a hydrogen and become positively charged. The positive charge will attract the hormone-receptor complex to the negatively charged DNA. Binding of the complex to the DNA stimulates transcription (Evans 890-891 and Mathews 808-809).

SIMILARITIES BETWEEN GLYCOPROTEINS, CYTOCHROMES AND METALLOENZYMES

Cytochromes have features that are quite similar to transferrins, a particular class of glycoproteins. First of all, cytochromes make up some of the integral proteins found in membranes, as do glycoproteins. Second of all, the main function of cytochromes is the binding of ion cations. Via oxidoreduction with these cations of 2+state, cytochromes bind iron to a porphyrin prosthetic group, forming a heme group (Mathews). The function is to transport cations in and out of the cell. Transferrin is an iron binding protein as well.

It interacts with iron cations of the 2+state which also bind to a porphyrin group, Protoporphyrin IX, to form a heme group. Transferrins carry ions to many different locations in the body including in and out of cells (Gottschalk).

Another group of molecules that serve a function related to glycoproteins are the metalloenzymes. These enzymes, as their name implies, bind to metal ions via the same type of heme groups just discussed (Gottschalk). One class of glycoprotein enzymes, the oxidoreductases, contain an enzyme called chloroperoxidase. Like transferrin, chloroperoxidase contains protophyrin IX and forms the heme group with iron 3+ (Mathews). Eicosanoids, are also related to glycoproteins because they recognize the receptors on cells which are glycoproteins.

INHIBITION OF ASPARAGINE LINKED GLYCOPROTEINS BY ANTIBIOTICS

Synthesis of N-linked glycoproteins can be inhibited with the use of antibiotics (function of antibiotics). Examples are monensin, mevinolin and tunicamycin. Monensin, a monocovalent ionophore, inhibits the transport of proteins through the endoplasmic reticulum- Golgi complex (general virology 855). Mevinolin on the other hand, is an inhibitor of mevalonate synthesis. Mevalonate is the precursor of dolichol and other isoprenoid lipids (cellular physiology 284). Finally, as mentioned above, asparagine linked glycosylation involves the first step of addition of GlcNAc to dolichol diphosphate. This reaction is catalyzed by the enzyme UDP-GlcNAc:dolichol phosphate N-acetylglucosamine-1-phosphate transferase or GPT. Tunicamycin is a specific inhibitor of GPT, therefore it can inhibit the synthesis of all N-linked glycoproteins (biological chemistry 8895). To give just an example of the effectiveness of tunicamycin, when it was injected to rats' hearts, it killed 12, 24, 50 and 60 h after the injection (histochemistry 607). Another connection could be seen with the function of the ribosomes, since the polypeptide chain of all glycoproteins.

PEPTIDOGLYCAN

Peptidoglycan, also known as murein, is a polymer consisting of sugars and amino acids that forms a mesh-like layer outside the plasma membrane of eubacteria. The sugar component consists of alternating residues of β-(1,4) linked N-acetylglucosamine and N-acetylmuramic acid residues. Attached to the N-acetylmuramic acid is a peptide chain of three to five amino acids. The peptide chain can be cross-linked to the peptide chain of another strand forming the 3D mesh-like layer. Some Archaea have a similar layer of pseudopeptidoglycan. Peptidoglycan serves a structural role in the bacterial cell wall, giving structural strength, as well as counteracting the osmotic pressure of the cytoplasm. A common misconception is that peptidoglycan gives the cell its shape; however, whereas peptidoglycan helps maintain the structure of the cell, it is actually the MreB protein that facilitates cell shape. Peptidoglycan is also involved in binary fission during bacterial cell reproduction.

The peptidoglycan layer is substantially thicker in Gram-positive bacteria (20 to 80 nm) than in Gram-negative bacteria (7 to 8 nm), with the attachment of the S-layer. Peptidoglycan forms around 90 per cent of the dry weight of Gram-positive bacteria but only 10 per cent of Gram-negative strains. In Gram-positive strains, it is important in attachment roles and sterotyping purposes. For both Gram-positive and Gram-negative bacteria, particles of approximately 2 nm can pass through the peptidoglycan.

ANTIBIOTIC INHIBITION

Some antibacterial drugs such as penicillin interfere with the production of peptidoglycan by binding to bacterial enzymes known as penicillin-binding proteins or transpeptidases. Penicillin-binding proteins form the bonds between oligopeptide crosslinks in peptidoglycan. For a bacterial cell to reproduce through binary fission, more than a million peptidoglycan subunits (NAM-NAG+oligopeptide) must be attached to existing subunits. Mutations in transpeptidases that lead to reduced interactions with an antibiotic are a significant source of emerging antibiotic resistance.

Considered the human body's *own antibiotic,* lysozymes found in tears work by breaking the â-(1,4)-glycosidic bonds in peptidoglycan (see below) and thereby destroying many bacterial cells. Antibiotics such as penicillin commonly target bacterial cell wall formation (of which peptidoglycan is an important component) because animal cells do not have cell walls.

Structure

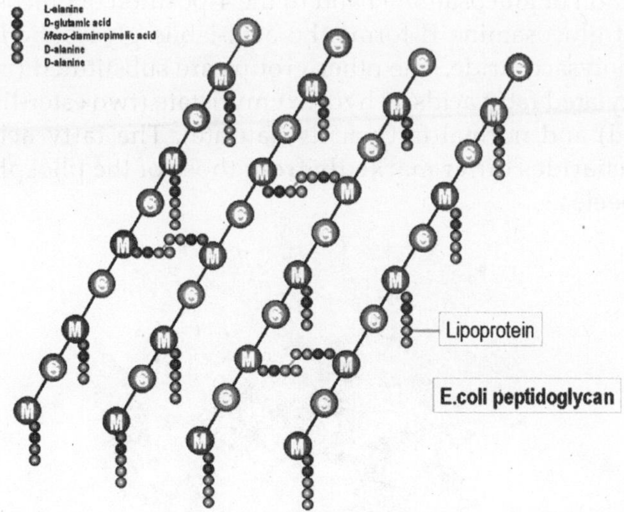

Fig. Structure of Peptidoglycan

The peptidoglycan layer in the bacterial cell wall is a crystal lattice structure formed from linear chains of two alternating amino sugars, namely

N-acetylglucosamine (GlcNAc or NAG) and *N*-acetylmuramic acid (MurNAc or NAM). The alternating sugars are connected by a â-(1,4)-glycosidic bond.

Each MurNAc is attached to a short (4- to 5-residue) amino acid chain, normally containing D-alanine, D-glutamic acid and mesodiaminopimelic acid. These three amino acids do not occur in proteins and are thought to help protect against attacks by most peptidases. Cross-linking between amino acids in different linear amino sugar chains by an enzyme known as transpeptidase result in a 3-dimensional structure that is strong and rigid. The specific amino acid sequence and molecular structure vary with the bacterial species.

LIPOPOLYSACCHARIDES

LIPOPOLYSACCHARIDES

These complex compounds (LPS) are the endotoxic O-antigens found in the cell walls of Gram-negative bacteria (S-lipopolysaccharides) but are also present in one fungus, they may cause several pathophysiological symptoms, such as fever, diarrhea, blood pressure decrease, septic shock and death. A lipid part (Lipid A) forms a complex with a core polysaccharide part through a glycosidic linkage. The core part is linked to a third external region of a highly immunogenic and variable O-chain polysaccharide or O-antigen made up of repeating oligosaccharide units. The latter region of the LPS molecule is responsible for bacterial serological strain specificity. Lipid A consists of a backbone of b-1,6-glucosaminyl-glucosamine with two phosphoester groups in the 1-position of glucosamine I and in the 4-position of glucosamine II. The 3-position of glucosamine II forms the acid-labile glycosidic linkage to the long-chain polysaccharide. The other groups are substituted (in Escherichia) with hydroxylated fatty acids as hydroxymyristate (two ester-linked and two amide-linked) and normal fatty acids (laurate). The fatty acid profiles of lipopolysaccharides differ markedly from those of the phospholipids from the same species.

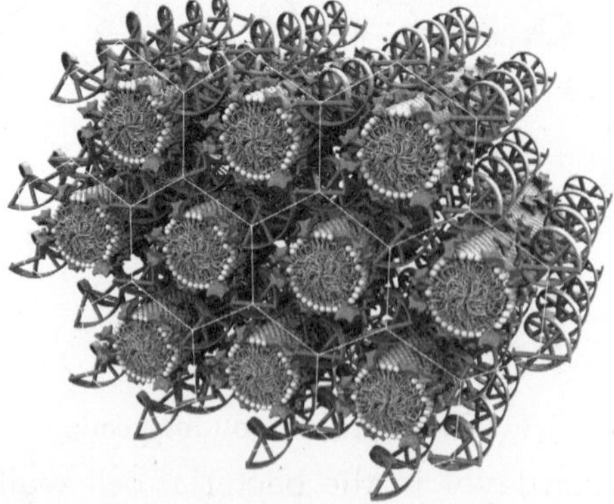

Fig. Structure of Lipid A of *Escherichia*

Although the hydroxylated fatty acids are generally 3-hydroxyalkanoic acids, 2-hydroxyalkanoic acids are also common. Exceptionally, some bacteria seem to lack hydroxy acids such as some species of *Gardnerella, Thermus, Thiobacillus, Bacteroides, Sorangium, Achromobacter* and *Brucella*. In most organisms, the major hydroxy acids are straight-chain, even-carbon compounds (from 12 to 18 carbon atoms). Odd-carbon acids are minor components (13 to 17 carbon atoms) in some enterobacteria, some *Vibrio*, some Pseudomonads and *Bordetella*.

They are also found in *Pectinatus, Veillonella* and *Selenomonas*. Branched-chain hydroxy fatty acids are also present in *Pseudomonas, Moraxella, Desulfovibrio, Capnocytophaga, Flavobacterium, Legionella* species (3-OH i-11:0 up to3-OH i-17:0). Few unsaturated hydroxy fatty acids have been described, but 2-OH-18:1 is present in some *Pseudomonas* and *Azospirillum* species.

Lipopolysaccharides are the major toxins of Gram-negative bacteria (endotoxin). The only Gram-positive bacteria that possesses LPS is *Listeria monocytogenes*, an infective agent in milk. While the polysaccharide is non-toxic, the lipid A part is responsible for the toxic activities of these bacteria which result in septic shock.

FOOD LABELLING LAWS

Producers of processed food are not currently required by law to mark foods containing "fructose in excess of glucose." This can cause some surprises and pitfalls for fructose malabsorbers.

Note that foods (such as bread) marked "gluten-free" are usually suitable for fructose malabsorbers, though sufferers need to be careful of gluten-free foods that contain dried fruit or high fructose corn syrup or fructose itself in sugar form.

However, fructose malabsorbers do not need to avoid gluten as do those with celiac disease. Many fructose malabsorbers can eat breads made from rye and corn flour. However, these may contain wheat unless marked "wheat-free" (or "gluten-free") (Note, rye bread is NOT gluten-free).

Although often assumed to be an acceptable alternative to wheat, spelt flour is not suitable for sufferers of fructose malabsorption, just as it is not appropriate for those with wheat-allergies or celiac disease (while bread made from spelt sprouts may be tolerated by the latter).

However, some fructose malabsorbers do not have difficulty with fructans from wheat products while they may have problems with foods that contain excess free fructose. Note that there are many breads on the market that advertise themselves with phrases like: "No High Fructose Corn Syrup".

Many bakeries now produce special breads with a high-inulin content, where inulin is a replacement in the baking process for all three: high fructose corn syrup, flour, and fat. Because of the caloric reduction, lower fat content, dramatic fibre increase, and prebiotic tendencies of the replacement inulin, these breads are considered a healthier alternative to traditionally prepared leavening breads. Inulin breads may even highlight the absence of high fructose corn syrup in the bread. However significant these health advances may be, sufferers of fructose malabsorption will likely find no difference between these new breads and traditionally prepared breads in the areas of flatulence or stool, because inulin is a fructan.

4

Enzyme Biotechnology

As major gene products, enzymes in the CNS are likely candidates for heritable variation. The range of enzymes active in neural tissue is wide and provides a wealth of material for pharmacological study. At a time when research on Behavioural control by various neurotransmitter systems is so predominant, it is perhaps easy to overlook other kinds of enzyme systems that could markedly influence drug response. Not the least of these are those biochemical pathways involved in energy metabolism.

Small perturbations in cerebral energy metabolism can have potentially major effects on a wide variety of Behaviours. The example provided by hypoglycemia and its effects on cognitive and affective characters is clear. The brain is a high energy utilization organ. All specialized functions of neural tissue depend on the more basic biochemistry of energy sources. Although we are not aware of genetic studies of enzymes involved in such energy metabolism and pharmacological alterations of activity, the caveat is relevant in the context of the distinction between direct and indirect drug effects on Behaviour.

A drug such as alcohol may very well modify neurotransmitter activity and some related Behaviour by actions on synthesis, degradation, receptor binding characteristics, or channel alteration. Nu merous other "side" effects might also occur. For example, ethanol consumption produces a marked alteration in general metabolism, producing among other things a general acidosis and increasing blood a-hydroxybutyrate levels. As â hydroxybutyrate is a usable energy source for brain tissue, marked changes in cerebral energy metabolism are likely to follow.

Such "indirect" effects on neural function, and perhaps Behaviour, would certainly appear subject to the same sort of heritable variation as other biochemical systems. Thus the degree of Behavioural modification by indirect as well as direct action of the pharmacological agent might also be influenced by genetic factors. While hoping not to be redundant, we believe the conclusion is obivous. That is, gene action may affect pharamcological responsiveness in a variety of ways other than via the specific neural system controlling the phenotype of interest. Keeping the foregoing caution in mind, numerous informative avenues of research on neuro-transmitter system regulation by genetic variables are available.

The most well researched of these is probably the area of catecholamine neurotransmitters. Norepinephrine and dopamine systems are subject to functional alteration by numerous pharmacological agents; many of these agents are of use in a clinical setting precisely because of their Behavioural effects. These functional alterations can occur as a result of inhibition or induction of the enzymes responsible for synthesis and degradation of these catecholamine transmitters. Early work by Ciaranello, Barchas, Kessler, and Barchas first established the fact of strain differences in brain activity of tyrosine hydroxylase (TH), the rate limiting enzyme in catecholamine biosynthesis.

The same report also demonstrated marked strain differences in adrenal phenylethanolamine-N methyl transferase (PNMT), the enzyme responsible for epinephrine production from norepinephrine. We here describe further examination of these mouse strain differences in TH activity in a context that serves as an example for pharmacogenetic study as well as neural development. In midbrain of CBA/J mice, the activity of TH is 20per cent less than in BALB/cJ mice. A comparable difference in immunocyto chemically detectable dopamine neurons in midbrain suggests that this difference in TH activity is due to more cells with the enzyme, rather than to greater specific activity per cell. A correspondingly more dense innervation of the striatum, by what are presumably the axons of these midbrain nigral cells, occurs in the BALB/c mice.

It is likely that this difference is related in a general way to the fact that the BALB/c brain is simply a larger one. Nonetheless, this represents a clear example of genetic control over neuronal organization and neurochemical characteristics of a Behaviourally important system. The differences in density of striatal innervation apparently predispose a differential response to two agents with dopaminergic activity, amphetamine and apomorphine.

BALB/c mice are more responsive to the locomotor stimulant effects of amphetamine and less sensitive to the stereotypy producing effects of apomorphine, a dopamine receptor agonist. Interestingly, these charac teristics follow a developmental time course that lags slightly behind the appearance of the TH activity difference. Apparently both pre and postsynaptic functions of the dopamine system are different between the two strains as reflected in the actions of the two drugs. A more detailed understanding of the genetic control over dopamine cell proliferation, growth, and effects on target tissue in this model system will answer many questions not only of pharmacogenetic interest, but also of basic importance to developmental neurobiology.

In this example of TH activity differences, the enzyme activity itself is probably not the focal point of either gene or pharmacological action. As noted, a more basic developmental issue is central. Yet the enzyme is a clear marker for a dramatic genetic influence. In other work, TH is seen to be more directly involved. Gamma-butyrolactone (GBL), through its metabolite gamma-hydroxy butyric acid (GHB), produces an inhibition of impulse transmission in dopamine neurons. A compensatory activation of TH occurs in these cells and an accumulation of dopamine occurs. In the LS and SS mice, GLB has hypnotic effects that differ in the same direction as ethanol effects. Of interest here is the fact that dopamine accumulation is 50per cent greater in the LS mice after GBL treatment.

Presumably the activation of TH was greater in LS mice. This activation occurs via kinetic rather than molecular genetic induction and apparently reflects a differential effectiveness of GHB at a subcellular site in LS and SS dopamine neurons. Although several studies of neurotransmitter systems' enzymes exist, there is yet to be a complete characterization of all biosynthetic and degradatory enzymes, transmitter levels and turnover rate, and receptor kinetics for any single system where genetic variation is also of interest. The difficulties of such a massive undertaking are obvious, but the benefits accruing to all three fields, neuropharmacology of drug action, neurochemical organization and dynamics, and mechanisms of genetic action are equally impressive. The field must surely progress along these lines.

CLASSIFICATION OF ENZYMES

Based on catalyzed reactions, the nomenclature committee of the International Union of Biochemistry and Molecular Biology (IUBMB) recommended the following classification.

1. Oxidoreductases catalyze a variety of oxidation-reduction reactions. Common names include dehydrogenase, oxidase, reductase and catalase.
2. Transferases catalyze transfers of groups (acetyl, methyl, phosphate, etc.). Common names include acetyltransferase, methylase, protein kinase and polymerase. The first three subclasses play major roles in the regulation of cellular processes. The polymerase is essential for the synthesis of DNA and RNA.
3. Hydrolases catalyze hydrolysis reactions where a molecule is split into two or more smaller molecules by the addition of water. Common examples are given below.

Proteases splits protein molecules. Examples: HIV protease and caspase. HIV protease is essential for HIV replication. Caspase plays a major role in apoptosis.

Apoptosis

Apoptosis is a tightly regulated form of cell death, also called the programmed cell death. Morphologically, it is characterized by chromatin condensation and cell shrinkage in the early stage. Then the nucleus and cytoplasm fragment, forming membrane-bound apoptotic bodies which can be engulfed by phagocytes. In contrast, cells undergo another form of cell death, necrosis, swell and rupture. The released intracellular contents can damage surrounding cells and often cause inflammation.

Apoptosis is an important process during normal development. It is also involved in aging and various diseases such as cancer, AIDS, Alzheimer's disease and Parkinson's disease.

THE MECHANISM

During apoptosis, the cell is killed by a class of proteases called caspases. More than 10 caspases (web link) have been identified. Some of them (*e.g.*, caspase 8 and 10) are involved in the initiation of apoptosis, others (caspase 3, 6, and 7) execute the death order by destroying essential proteins in the cell. The apoptotic process can be summarized as follows:

- Activation of initiating caspases by specific signals.
- Activation of executing caspases by the initiating caspases which can cleave inactive caspases at specific sites.
- Degradation of essential cellular proteins by the executing caspases with their protease activity.
  - Death receptors: Fas/CD95, DR4/DR5, DR3, and TNFR (Tumor Necrosis Factor Receptor).
  - Adaptors: FADD (Fas-associated death domain protein) and TRADD (TNFR-associated death domain protein).
  - Activation: Binding of death ligands (FasL/CD95L, TRAIL/APO-2L, APO-3L and TNF) induces trimerization of their receptors, which then recruit adaptors and activate caspases.

    Note: TRADD is involved only in the coupling between caspases and DR3 or TNFR. This adaptor can also recruit other proteins (web link) to inhibit apoptosis through the NF-kB pathway.

As shown in the above figure, a variety of death ligands (FasL/CD95L, TRAIL, APO-3L and TNF) can induce apoptosis. It is natural to see if they can kill cancer cells without affecting normal cells. TNF was first investigated in the 1980s for cancer therapy, but with disappointing results. Then CD95L (FasL) was tested in the 1990s. The results were still not satisfactory. Recently, TRAIL has been demonstrated to be highly selective for transformed cells, with minimal effects on normal cells. It could be an effective drug for both cancer and AIDS.

Nucleases splits nucleic acids (DNA and RNA). Based on the substrate type, they are divided into RNase and DNase. RNase catalyzes the hydrolysis of RNA and DNase acts on DNA. They may also be divided into exonuclease and endonuclease. The exonuclease progressively splits off single nucleotides from one end of DNA or RNA. The endonuclease splits DNA or RNA at internal sites.

Phosphatase catalyzes dephosphorylation (removal of phosphate groups). Example: calcineurin.The immuno suppressive drugs FK506 and Cyclosporin A are the inhibitors of calcineurin.

NFAT, Calcineurin and Immunosuppression

Organ transplantation often elicits immune responses to reject the foreign object. The most widely used drugs to deal with such graft reaction are FK506 and cyclosporin A (CsA). Their direct target is a phosphatase known as calcineurin. CsA combines with a cellular protein called cyclophilin to inhibit calcineurin; FK506 combines with FKBP (FK506 binding protein) to inhibit the phosphatase. How could inhibition of a phosphatase suppress the immune response? The answer lies in the transport of NFAT (nuclear factor of activated T cells) which is a transcription factor regulating IL-2.

NFAT contains both nuclear localization sequence (NLS) and nuclear export sequence (NES), but one of them is buried in the protein interior, inaccessible to importin and exportin. Whether NLS or NES is masked depends on the phosphorylation state of specific serine residues in the regulatory domain. Phosphorylation of these serine residues exposes NES whereas dephosphorylation exposes NLS. In resting cells, the NLS of the cytoplasmic NFAT is masked due to phosphorylation on these serine residues. In stimulated cells, an increase of intracellular calcium ions activates calcineurin, which then dephosphorylates the masking residues. Consequently, NLS is exposed and NFAT can be carried into the nucleus by the importin a/b. Inside the nucleus, NFAT may be re-phosphorylated by protein kinase, exposing its NES and be exported by the exportin Crm1.

- The structure of the FK506-FKBP-calcineurin complex. PDB ID = 1TCO.
- In the normal process, calcineurin may dephosphorylate NFAT, exposing its NLS so that NFAT can move into the nucleus to stimulate the expression of IL-2, inducing a cascade of immune responses. If calcineurin is inhibited, the immune response will be suppressed.
- Lyases catalyze the cleavage of C-C, C-O, C-S and C-N bonds by means other than hydrolysis or oxidation. Common names include decarboxylase and aldolase.
- Isomerases catalyze atomic rearrangements within a molecule. Examples include rotamase, protein disulfide isomerase (PDI), epimerase and racemase.
- Ligases catalyze the reaction which joins two molecules. Examples include peptide synthase, aminoacyl-tRNA synthetase, DNA ligase and RNA ligase.

The IUBMB committee also defines subclasses and sub-subclasses. Each enzyme is assigned an EC (Enzyme Commission) number. For example, the EC number of catalase is EC1.11.1.6. The first digit indicates that the enzyme belongs to oxidoreductase (class 1). Subsequent digits represent subclasses and sub-subclasses.

CATALYTIC MECHANISMS OF ENZYMES

From the energetic point of view, the reason why an enzyme can accelerate a reaction is because it can lower the energy barrier (activation energy) separating the substrate and the reaction product. For example, the covalent bond energy between two atoms ranges from 50 to 200 kcal/mol, which is far greater than the thermal energy (0.6 kcal/mol) at room temperature. Therefore, the covalent bond is unlikely to break in the absence of external interactions. Enzymes can provide a proper environment to lower the energy barrier.

The exact mechanism of lowering the energy barrier depends on individual systems. RNase A is a very interesting example. This enzyme can cleave an RNA molecule through hydrolysis reaction, but has no effect on DNA.

ENZYMES AS BIOLOGICAL CATALYSTS

In cells and organisms most reactions are catalyzed by enzymes, which are regenerated during the course of a reaction. These biological catalysts are physiologically important because they speed up the rates of reactions that would otherwise be too slow to support life. Enzymes increase reaction rates, sometimes by as much as one million-fold, but more typically by about one thousand fold. Catalysts speed up the forward and reverse reactions proportionately so that, although the magnitude of the rate constants of the forward and reverse reactions is are increased, the ratio of the rate constants remains the same in the presence or absence of enzyme. Since the equilibrium constant is equal to a ratio of rate constants, it is apparent that enzymes and other catalysts have no effect on the equilibrium constant of the reactions they catalyze.

Enzymes increase reaction rates by decreasing the amount of energy required to form a complex of reactants that is competent to produce reaction products. This complex is known as the activated state or transition state complex for the reaction. Enzymes and other catalysts accelerate reactions by lowering the energy of the transition state. The free energy required to form an activated complex is much lower in the catalyzed reaction. The amount of energy required to achieve the transition state is lowered; consequently, at any instant a greater proportion of the molecules in the population can achieve the transition state. The result is that the reaction rate is increased.

Michaelis-Menten Kinetics

In typical enzyme-catalyzed reactions, reactant and product concentrations are usually hundreds or thousands of times greater than the enzyme concentration. Consequently, each enzyme molecule catalyzes the conversion to product of many reactant molecules. In biochemical reactions, reactants are commonly known as substrates. The catalytic event that converts substrate to product involves the formation of a transition state, and it occurs most easily at a specific binding site on the enzyme. This site, called the catalytic site of the enzyme, has been evolutionarily structured to provide specific, high-affinity binding of substrate(s) and to provide an environment that favours the catalytic events. The complex that forms, when substrate(s) and enzyme combine, is called the enzyme substrate (ES) complex. Reaction products arise when the ES complex breaks down releasing free enzyme.

Between the binding of substrate to enzyme, and the reappearance of free enzyme and product, a series of complex events must take place. At a minimum an ES complex must be formed; this complex must pass to the transition state (ES*); and the transition state complex must advance to an enzyme product complex (EP). The latter is finally competent to dissociate to product and free enzyme.

The series of events can be shown thus:

$$E + S \leftrightarrow ES \leftrightarrow ES^* \leftrightarrow EP \leftrightarrow E + P$$

The kinetics of simple reactions like that above were first characterized by biochemists Michaelis and Menten. The concepts underlying their analysis of enzyme kinetics continue to provide the cornerstone for understanding metabolism today, and for the development and clinical use of drugs aimed at selectively altering rate constants and interfering with the progress of disease states. The Michaelis-Menten equation is a quantitative description of the relationship among the rate of an enzyme- catalyzed reaction [v_1], the concentration of substrate [S] and two constants, V_{max} and K_m (which are set by the particular equation). The symbols used in the Michaelis-Menten equation refer to the reaction rate [v_1], maximum reaction rate (V_{max}), substrate concentration [S] and the Michaelis-Menten constant (K_m).

$$V_1 = \frac{V_{max}[s]}{\{K_m + [S]\}}$$

The Michaelis-Menten equation can be used to demonstrate that at the substrate concentration that produces exactly half of the maximum reaction rate, *i.e.* ½ V_{max}, the substrate concentration is numerically equal to K_m. This fact provides a simple yet powerful bioanalytical tool that has been used to characterize both normal and altered enzymes, such as those that produce the symptoms of genetic diseases. Rearranging the Michaelis-Menten equation leads to:

$$K_m = [S]\left\{\left[\frac{V_{max}}{V_1}\right] - 1\right\}$$

From this equation it should be apparent that when the substrate concentration is half that required to support the maximum rate of reaction, the observed rate, v_1, will, be equal to V_{max} divided by 2; in other words, v_1 = [$V_{max}/2$]. At this substrate concentration V_{max}/v_1 will be exactly equal to 2, with the result that:

$$[S](1) = K_m$$

The latter is an algebraic statement of the fact that, for enzymes of the Michaelis-Menten type, when the observed reaction rate is half of the maximum possible reaction rate, the substrate concentration is numerically equal to the Michaelis-Menten constant. In this derivation, the units of K_m are those used to specify the concentration of S, usually Molarity.

The Michaelis-Menten equation has the same form as the equation for a rectangular hyperbola; graphical analysis of reaction rate (v) versus substrate concentration [S] produces a hyperbolic rate plot.

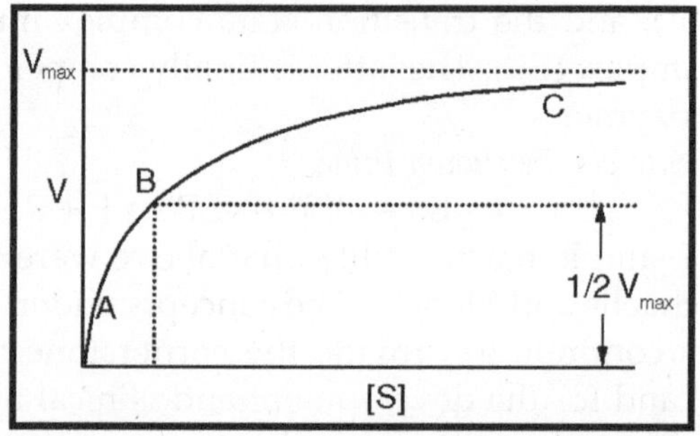

Fig. Plot of Substrate Concentration Versus Reaction Velocity

The key features of the plot are marked by points A, B and C. At high substrate concentrations the rate represented by point C the rate of the reaction is almost equal to V_{max}, and the difference in rate at nearby concentrations of substrate is almost negligible. If the Michaelis-Menten plot is extrapolated to infinitely high substrate concentrations, the extrapolated rate is equal to V_{max}. When the reaction rate becomes independent of substrate concentration, or nearly so, the rate is said to be zero order. (Note that the reaction is zero order only with respect to this substrate. If the reaction has two substrates, it may or may not be zero order with respect to the second substrate).

The very small differences in reaction velocity at substrate concentrations around point C (near V_{max}) reflect the fact that at these concentrations almost all of the enzyme molecules are bound to substrate and the rate is virtually independent of substrate, hence zero order. At lower substrate concentrations, such as at points A and B, the lower reaction velocities indicate that at any moment only a portion of the enzyme molecules are bound to the substrate. In fact, at the substrate concentration denoted by point B, exactly half the enzyme molecules are in an ES complex at any instant and the rate is exactly one half of V_{max}. At substrate concentrations near point A the rate appears to be directly proportional to substrate concentration, and the reaction rate is said to be first order.

Inhibition of Enzyme Catalyzed Reactions

To avoid dealing with curvilinear plots of enzyme catalyzed reactions, biochemists Lineweaver and Burk introduced an analysis of enzyme kinetics based on the following rearrangement of the Michaelis-Menten equation:

$$\frac{1}{V}=\left[\frac{K_m(1)}{V_{max}(S)}+\frac{1}{V_{max}}\right]$$

Plots of 1/v versus 1/[S] yield straight lines having a slope of K_m/V_{max} and an intercept on the ordinate at $1/V_{max}$.

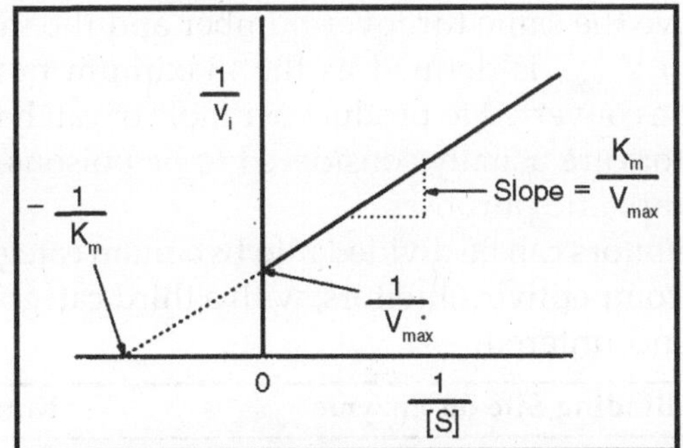

Fig. A Lineweaver-Burk Plot

An alternative linear transformation of the Michaelis-Menten equation is the Eadie-Hofstee transformation:

$$\frac{V}{[S]}=-V\left[\frac{1}{K_m}+\right]+\left[\frac{V_{max}}{K_m}\right]$$

and when v/[S] is plotted on the y-axis versus v on the x-axis, the result is a linear plot with a slope of $-1/K_m$ and the value V_{max}/K_m as the intercept on the y-axis and V_{max} as the intercept on the x-axis.

Both the Lineweaver-Burk and Eadie-Hofstee transformation of the Michaelis-Menton equation are useful in the analysis of enzyme inhibition. Well- known examples of such therapy include the use of methotrexate in cancer chemotherapy to semi-selectively inhibit DNA synthesis of malignant cells, the use of aspirin to inhibit the synthesis of prostaglandins which are at least partly responsible for the aches and pains of arthritis, and the use of sulfa drugs to inhibit the folic acid synthesis that is essential for the metabolism and growth of disease-causing bacteria. In addition, many poisons, such as cyanide, carbon monoxide and polychlorinated biphenols (PCBs) produce their life-threatening effects by means of enzyme inhibition.

Enzyme inhibitors fall into two broad classes: those causing irreversible inactivation of enzymes and those whose inhibitory effects can be reversed. Inhibitors of the first class usually cause an inactivating, covalent modification of enzyme structure. Cyanide is a classic example of an irreversible enzyme inhibitor: by covalently binding mitochondrial cytochrome oxidase, it inhibits all the reactions associated with electron transport.

The kinetic effect of irreversible inhibitors is to decrease the concentration of active enzyme, thus decreasing the maximum possible concentration of ES complex. Since the limiting enzyme reaction rate is often k_2[ES], it is clear that under these circumstances the reduction of enzyme concentration will lead to decreased reaction rates. Note that when enzymes in cells are only partially inhibited by irreversible inhibitors, the remaining unmodified enzyme molecules are not distinguishable from those in untreated cells; in particular, they have the same turnover number and the same K_m. Turnover number, related to V_{max}, is defined as the maximum number of moles of substrate that can be converted to product per mole of catalytic site per second. Irreversible inhibitors are usually considered to be poisons and are generally unsuitable for therapeutic purposes.

Reversible inhibitors can be divided into two main categories; competitive inhibitors and noncompetitive inhibitors, with a third category, uncompetitive inhibitors, rarely encountered.

Inhibitor Type	Binding Site on Enzyme	Kinetic effect
Competitive Inhibitor	Specifically at the catalytic site, where it competes with substrate for binding in a dynamic equilibrium-like process. Inhibition is reversible by substrate.	V_{max} is unchanged; K_m, as defined by [S] required for ½ maximal activity, is increased.
Noncompetitive Inhibitor	Binds E or ES complex other than at the catalytic site. Substrate binding unaltered, but ESI complex cannot form products. Inhibition cannot be reversed by substrate.	K_m appears unaltered; V_{max} is decreased proportionately to inhibitor concentration.
Uncompetitive Inhibitor	Binds only to ES complexes at locations other than the catalytic site. Substrate binding modifies enzyme structure, making inhibitor- binding site available. Inhibition cannot be reversed by substrate.	Apparent V_{max} decreased; K_m, as defined by [S] required for ½ maximal activity, is decreased.

The hallmark of all the reversible inhibitors is that when the inhibitor concentration drops, enzyme activity is regenerated. Usually these inhibitors bind to enzymes by non-covalent forces and the inhibitor maintains a reversible equilibrium with the enzyme. The equilibrium constant for the dissociation of enzyme inhibitor complexes is known as K_i:

$$K_i = \frac{[E][I]}{[EI]}$$

The importance of K_I is that in all enzyme reactions where substrate, inhibitor and enzyme interact, the normal K_m and or V_{max} for substrate enzyme interaction appear to be altered. These changes are a consequence of the influence of K_i on the overall rate equation for the reaction. The effects of K_i are best observed in Lineweaver-Burk plots.

Probably the best known reversible inhibitors are competitive inhibitors, which always bind at the catalytic or active site of the enzyme. Most drugs that alter enzyme activity are of this type. Competitive inhibitors are especially attractive as clinical modulators of enzyme activity because they offer two routes for the reversal of enzyme inhibition, while other reversible inhibitors offer only one. First, as with all kinds of reversible inhibitors, a decreasing concentration of the inhibitor reverses the equilibrium regenerating active free enzyme. Second, since substrate and competitive inhibitors both bind at the same site they compete with one another for binding Raising the concentration of substrate (S), while holding the concentration of inhibitor constant, provides the second route for reversal of competitive inhibition. The greater the proportion of substrate, the greater the proportion of enzyme present in competent ES complexes. As noted earlier, high concentrations of substrate can displace virtually all competitive inhibitor bound to active sites. Thus, it is apparent that V_{max} should be unchanged by competitive inhibitors. This characteristic of competitive inhibitors is reflected in the identical vertical-axis intercepts of Lineweaver-Burk plots, with and without inhibitor.

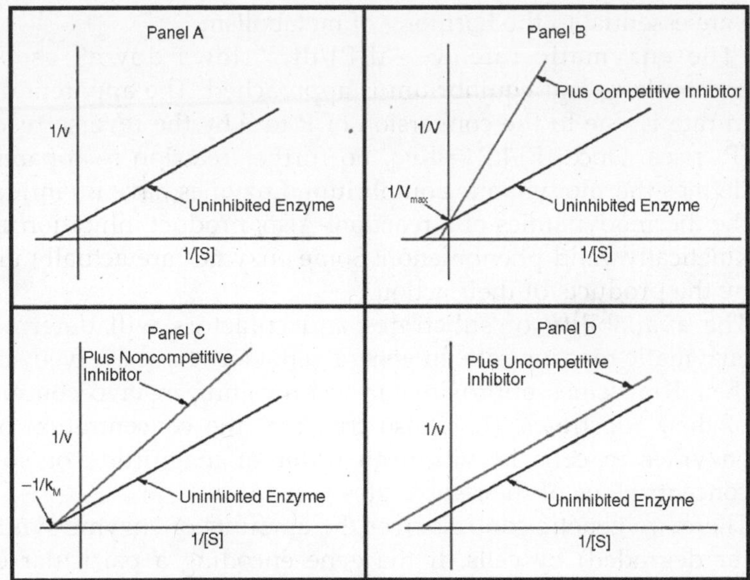

Fig. Lineweaver-Burk Plots of Inhibited Enzymes

Since attaining V_{max} requires appreciably higher substrate concentrations in the presence of competitive inhibitor, K_m (the substrate concentration at half maximal velocity) is also higher, as demonstrated by the differing negative intercepts on the horizontal axis in panel B.

Analogously, panel C illustrates that noncompetitive inhibitors appear to have no effect on the intercept at the x-axis implying that noncompetitive inhibitors have no effect on the K_m of the enzymes they inhibit. Since noncompetitive inhibitors do not interfere in the equilibration of enzyme, substrate and ES complexes, the K_m's of Michaelis-Menten type enzymes are not expected to be affected by noncompetitive inhibitors, as demonstrated by x-axis intercepts in panel C. However, because complexes that contain inhibitor (ESI) are incapable of progressing to reaction products, the effect of a noncompetitive inhibitor is to reduce the concentration of ES complexes that can advance to product. Since $V_{max} = k_2[E_{total}]$, and the concentration of competent E_{total} is diminished by the amount of ESI formed, noncompetitive inhibitors are expected to decrease V_{max}, as illustrated by the y-axis intercepts in panel C.

A corresponding analysis of uncompetitive inhibition leads to the expectation that these inhibitors should change the apparent values of K_m as well as V_{max}. Changing both constants leads to double reciprocal plots, in which intercepts on the x and y axes are proportionately changed; this leads to the production of parallel lines in inhibited and uninhibited reactions.

CONTROLS OVER ENZYMATIC ACTIVITY

The activity displayed by enzymes is affected by a variety of factors, some of which are essential to the harmony of metabolism.

1. The enzymatic rate, v = d[P]/dt, "slows down" as product accumulates and equilibrium is approached. The apparent decrease in rate is due to the conversion of P to S by the reverse reaction as [P] rises. Once [P]/[S] = Keq, no further reaction is apparent. Keq defines thermodynamic equilibrium. Enzymes have no influence on the thermodynamics of a reaction. Also, product inhibition can be a kinetically valid phenomenon: Some enzymes are actually inhibited by the products of their action.
2. The availability of substrates and cofactors will determine the enzymatic reaction rate. In general, enzymes have evolved such that their Km values approximate the prevailing in vivo concentration of their substrates. (It is also true that the concentration of some enzymes in cells is within an order of magnitude or so of the concentrations of their substrates.)
3. There are genetic controls over the amounts of enzyme synthesized (or degraded) by cells. If the gene encoding a particular enzyme protein is turned on or off, changes in the amount of enzyme activity

soon follow. Induction, which is the activation of enzyme synthesis, and repression, which is the shutdown of enzyme synthesis, are important mechanisms for the regulation of metabolism. By controlling the amount of an enzyme that is present at any moment, cells can either activate or terminate various metabolic routes. Genetic controls over enzyme levels have a response time ranging from minutes in rapidly dividing bacteria to hours (or longer) in higher eukaryotes.

4. Enzymes can be regulated by covalent modification, the reversible covalent attachment of a chemical group. For example, a fully active enzyme can be converted into an inactive form simply by the covalent attachment of a functional group, such as a phosphoryl moiety (Figure). Alternatively, some enzymes exist in an inactive state unless specifically converted into the active form through covalent addition of a functional group. Covalent modification reactions are catalyzed by special converter enzymes, which are themselves subject to metabolic regulation. Although covalent modification represents a stable alteration of the enzyme, a different converter enzyme operates to remove the modification, so that when the conditions that favoured modification of the enzyme are no longer present, the process can be reversed, restoring the enzyme to its unmodified state. Many examples of covalent modification at important metabolic junctions will be encountered in our discussions of metabolic pathways. Because covalent modification events are enzyme-catalyzed, they occur very quickly, with response times of seconds or even less for significant changes in metabolic activity. The 1992 Nobel Prize in physiology or medicine was awarded to Edmond Fischer and Edwin Krebs for their pioneering studies of reversible protein phosphorylation as an important means of cellular regulation.

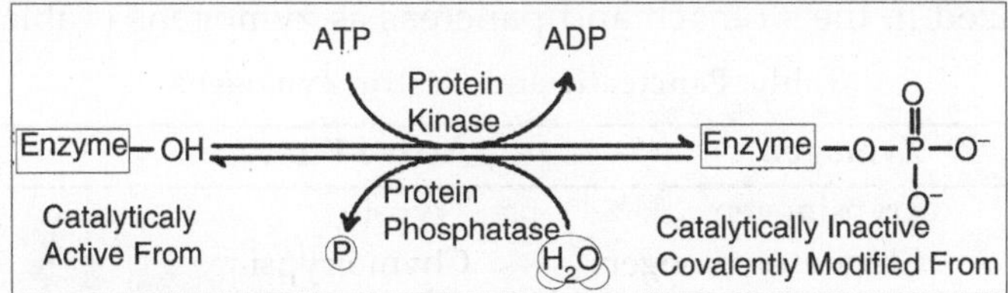

Fig. Enzymes Regulated by Covalent Modification are Called Interconvertible Enzymes.

The enzymes (protein kinase and protein phosphatase, in the example shown here) catalyzing the conversion of the interconvertible enzyme between its two forms are called converter enzymes..

Enzymatic activity can also be activated or inhibited through noncovalent interaction of the enzyme with small molecules (metabolites) other than the substrate. This form of control is termed allosteric regulation, because the activator or inhibitor binds to the enzyme at a site other than (allo means "other") the active site. Further, such allosteric regulators, or effector molecules, are often quite different sterically from the substrate.

Because this form of regulation results simply from reversible binding of regulatory ligands to the enzyme, the cellular response time can be virtually instantaneous. Specialized controls: Enzyme regulation is an important matter to cells, and evolution has provided a variety of additional options, including zymogens, isozymes, and modulator proteins.

ZYMOGENS

Most proteins become fully active as their synthesis is completed and they spontaneously fold into their native, three-dimensional conformations. Some proteins, however, are synthesized as inactive precursors, called zymogens or proenzymes, that only acquire full activity upon specific proteolytic cleavage of one or several of their peptide bonds.

Unlike allosteric regulation or covalent modification, zymogen activation by specific proteolysis is an irreversible process. Activation of enzymes and other physiologically important proteins by specific proteolysis is a strategy frequently exploited by biological systems to switch on processes at the appropriate time and place.

Insulin

Some protein hormones are synthesized in the form of inactive precursor molecules, from which the active hormone is derived by proteolysis. For instance, insulin, an important metabolic regulator, is generated by proteolytic excision of a specific peptide from proinsulin.

PROTEOLYTIC ENZYMES OF THE DIGESTIVE TRACT

Enzymes of the digestive tract that serve to hydrolyze dietary proteins are synthesized in the stomach and pancreas as zymogens (Table).

Table. Pancreatic and Gastric Zymogens

Origin	Zymogen	Active Protease
Pancreas	Trypsinogen	Trypsin
Pancreas	Chymotrypsinogen	Chymotrypsin
Pancreas	Procarboxypeptidase	Carboxypeptidase
Pancreas	Proelastase	Elastase
Stomach	Pepsinogen	Pepsin

Only upon proteolytic activation are these enzymes able to form a catalytically active substrate-binding site. The activation of chymotrypsinogen is an interesting example (Figure). Chymotrypsino-gen is a 245-residue polypeptide chain cross-linked by five disulfide bonds. Chymotrypsinogen is converted to an enzymatically active form called π-chymotrypsin when trypsin cleaves the peptide bond joining Arg15 and Ile16.

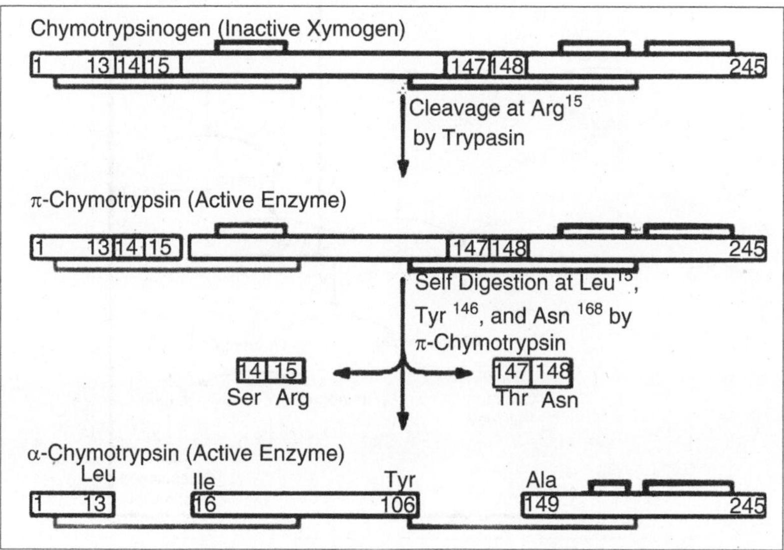

Fig. The Proteolytic Activation of Chymotrypsinogen.

The enzymatically active π-chymotrypsin acts upon other π-chymotrypsin molecules, excising two dipeptides, Ser^{14}-Arg^{15} and Thr^{147}-Asn^{148}. The end product of this processing pathway is the mature protease a -chymotrypsin, in which the three peptide chains, A (residues 1 through 13), B (residues 16 through 146), and C (residues 149 through 245), remain together because they are linked by two disulfide bonds, one from A to B, and one from B to C. It is interesting to note that the transformation of inactive chymotrypsinogen to active π-chymotrypsin requires the cleavage of just one particular peptide bond.

BLOOD CLOTTING

The formation of blood clots is the result of a series of zymogen activations (Figure). The amplification achieved by this cascade of enzymatic activations allows blood clotting to occur rapidly in response to injury. Seven of the clotting factors in their active form are serine proteases: kallikrein, XIIa, XIa, IXa, VIIa, Xa, andthrombin. Two routes to blood clot formation exist. The intrinsic pathway is instigated when the blood comes into physical contact with abnormal surfaces caused by injury; the extrinsic pathway is initiated by factors released from injured tissues. The pathways merge at Factor X and culminate in clot formation. The intrinsic and extrinsic pathways converge at Factor X, and the final common pathway involves the activation of thrombin and its conversion of fibrinogen into fibrin, which aggregates into ordered filamentous arrays that become cross-linked to form the clot.

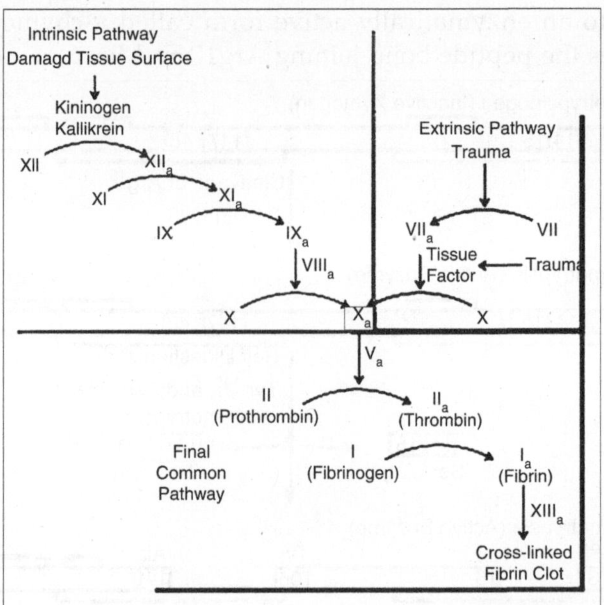

Fig. The Cascade of Activation Steps Leading to Blood Clotting.

Thrombin excises peptides rich in negative charge from fibrinogen, converting it to fibrin, a molecule with a different surface charge distribution. Fibrin readily aggregates into ordered fibrous arrays that are subsequently stabilized by covalent cross-links. Thrombin specifically cleaves Arg-Gly peptide bonds and is homologous to trypsin, which is also a serine protease (recall that trypsin acts only at Arg and Lys residues).

Protein Kinases

Protein kinases are converter enzymes that catalyze the ATP-dependent phosphorylation of serine, threonine, and/or tyrosine hydroxyl groups in target proteins (table). Phosphorylation introduces a bulky group bearing two negative charges, causing conformational changes that alter the target protein's function. (Unlike a phosphoryl group, no amino acid side chain can provide two negative charges.) Protein kinases represent a protein superfamily whose members are widely diverse in terms of size, subunit structure, and subcellular localization. Nevertheless, all share a common catalytic mechanism based on a conserved catalytic core/kinase domain of approximately 260 amino acid residues (see figure). Protein kinases are classified as Ser/Thr- and/or Tyr-specific and are subclassified in terms of the allosteric activators they require and the consensus amino acid sequence within the target protein that is recognized by the kinase. For example, cAMP-dependent protein kinase (PKA) phosphorylates proteins having Ser or Thr residues within an R(R/K)X (S*/T*) target consensus sequence (* denotes the residue that becomes phosphorylated). That is, PKA phosphorylates Ser or Thr residues that occur in an Arg-(Arg or Lys)-(any amino acid)-(Ser or Thr) sequence segment (table). Targeting of protein kinases to particular consensus sequence elements within proteins creates a means to regulate these kinases byintrasteric control. Intrasteric control occurs when a regulatory subunit (or protein domain) has a pseudosubstrate sequence that mimics the target sequence but lacks a OH-bearing side chain at the right place. For example, the cAMP-binding regulatory subunits of PKA (R subunits in Figure) possess the pseudosubstrate sequence RRGA*I and this sequence binds to the active site of PKA catalytic subunits, blocking their activity. This pseudosubstrate sequence has an alanine residue where serine occurs in the PKA target sequence; Ala is sterically similar to serine but lacks a phosphorylatable OH-group. When these PKA regulatory subunits bind cAMP, they undergo a conformational change and dissociate from the catalytic (C) subunits, and the active site of PKA is free to bind and phosphorylate its targets.

The abundance of many protein kinases in cells is an indication of the great importance of protein phosphorylation in cellular regulation. Exactly 113 protein kinase genes have been recognized in yeast, and it is estimated that the human genome encodes more than 1000 different protein kinases. Tyrosine kinases(protein kinases that phosphorylate Tyr residues) occur only in multicellular organisms (yeast has no tyrosine kinases). Tyrosine kinases are components of signaling pathways involved in cell - cell communication.

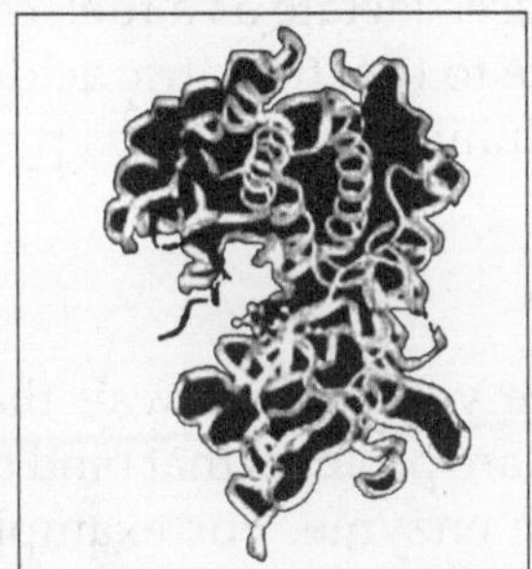

Cyclic AMP-dependent protein kinase is shown complexed with a pseudosubstrate peptide. This complex also includes ATP and two Mn2+ ions bound at the active site.

ISOZYMES

A number of enzymes exist in more than one quaternary form, differing in their relative proportions of structurally equivalent but catalytically distinct polypeptide subunits. A classic example is mammalian lactate dehydrogenase (LDH), which exists as five different isozymes, depending on the tetrameric association of two different subunits, A and B: A_4, A_3B, A_2B_2, AB_3, and B_4 (Figure). The kinetic properties of the various LDH isozymes differ in terms of their relative affinities for the various substrates and their sensitivity to inhibition by product. Different tissues express different isozyme forms, as appropriate to their particular metabolic needs. By regulating the relative amounts of A and B subunits they synthesize, the cells of various tissues control which isozymic forms are likely to assemble, and, thus, which kinetic parameters prevail.

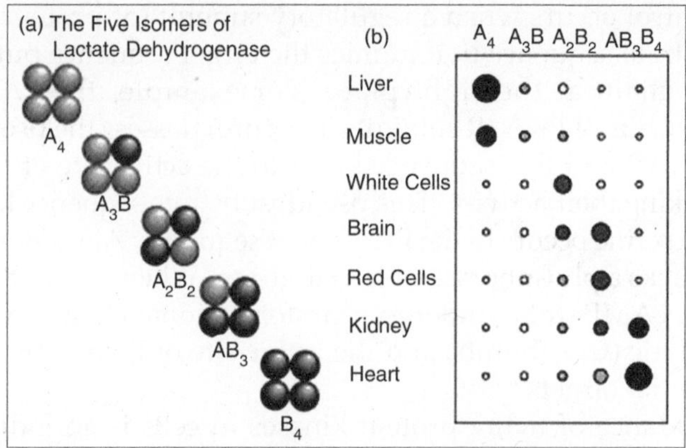

Fig. The Isozymes of Lactate Dehydrogenase (LDH).

Active muscle tissue becomes anaerobic and produces pyruvate from glucose via glycolysis. It needs LDH to regenerate NAD^+ from NADH so glycolysis can continue. The lactate produced is released into the blood. The muscle LDH isozyme (A_4) works best in the NAD^+-regenerating direction. Heart tissue is aerobic and uses lactate as a fuel, converting it to pyruvate via LDH and using the pyruvate to fuel the citric acid cycle to obtain energy. The heart LDH isozyme (B_4) is inhibited by excess pyruvate so the fuel won't be wasted.

MODULATOR PROTEINS

Modulator proteins are yet another way that cells mediate metabolic activity. Modulator proteins are proteins that bind to enzymes, and by binding, influence the activity of the enzyme. For example, some enzymes, such as cAMP-dependent protein kinase, exist as dimers of catalytic subunits and regulatory subunits. These regulatory subunits are modulator proteins that suppress the activity of the catalytic subunits. Dissociation of the regulatory subunits (modulator proteins) activates the catalytic subunits; reassociation once again suppresses activity.

Phosphoprotein phosphatase inhibitor-1 (PPI-1) is another example of a modulator protein. When PPI-1 is phosphorylated on one of its serine residues, it binds to phosphoprotein phosphatase (Figure), inhibiting its phosphatase activity.

The result is an increased phosphorylation of the interconvertible enzyme targeted by the protein kinase/phosphoprotein phosphatase cycle (Figure). We will meet other important representatives of this class as the processes of metabolism unfold in subsequent chapters. For now, let us focus our attention on the fascinating kinetics of allosteric enzymes.

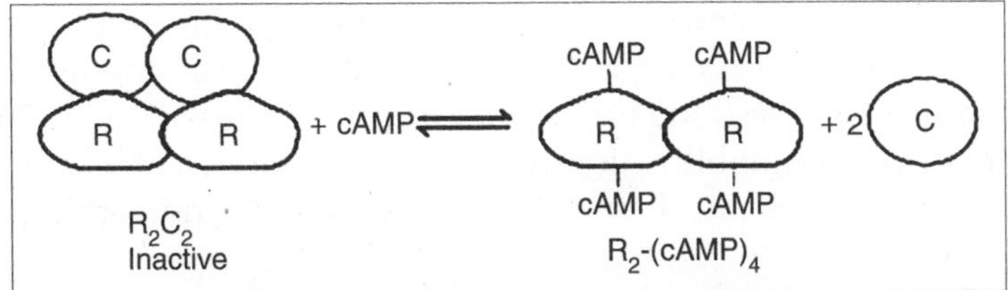

Fig. Cyclic AM π- dependent protein kinase (also known as PKA) is a 150- to 170-kD R_2C_2 tetramer in mammalian cells.

The two R (regulatory) subunits bind cAMP (KD = 3 × 10-8 M); cAMP binding releases the R subunits from the C (catalytic) subunits. C subunits are enzymatically active as monomers.

ENZYME IMMOBILIZATION TECHNOLOGY

The term *immobilized enzyme* was adopted in 1971 at the first Enzyme Engineering Conference. It describes enzymes physically confined at or localized in a certain region of space with retention of catalytic activity and which can be used repeatedly and continuously. The immobilization of biocatalysts (not only enzymes but also other bioactive molecules such as growth factors and hormones, cellular organelles, microbial cells, and plant and animal cells) is attracting worldwide attention in biotechnology applications. In general, immobilized biocatalysts are more stable and easier to handle compared with their free counterparts.

At present, applications of immobilized biocatalysts include the production of useful compounds by stereospecific or regiospecific bioconversion, the production of energy by biological processes, the selective treatment of specific pollutants to solve environmental problems, continuous analyses of compounds with a high sensitivity and specificity, and medical uses such as new types of drugs for enzyme therapy or artificial organs. Immobilized enzymes are already being used in medical applications for clinical diagnosis and also for intra- and extracorporeal enzyme therapy. Applications in clinical analysis are mainly related to biosensors, which have been used to detect the presence of various organic compounds for many years. For example, glucose oxidase and catalase have been used to measure blood glucose concentration, and cholesterol oxidase and cholesterol esterase to determine cholesterol levels. In addition, enzymes can be immobilized on different prosthetic devices or used extracorporeally (*e.g.*, artificial heart, artificial lung, artificial kidney, equipment for hemodialysis and specific blood purification) as surface modifiers in order to increase the biocompatibility of these devices and to prevent blood clotting.

Methods for Immobilizing Enzymes in Polymeric Carriers

Various methods have been developed for the immobilization of biocatalysts, which are being used extensively today. A wide range of support materials has also been employed for enzyme immobilization. The support type can be classified according to their chemical composition, such as organic or inorganic supports, and the former can be further classified into natural or synthetic matrices. Immobilization techniques can be divided into different categories: physical, chemical, enzymatic, and genetic engineering methods.

Adsorption

The adsorption of an enzyme onto a support or film material is the simplest method of obtaining an immobilized enzyme. Basically, the enzyme is attached to the support material by noncovalent linkages and does not require any preactivation step of the support. The interactions formed between the enzyme and the support material will be dependent on the existing surface chemistry of the support and on the type of amino acids exposed at the surface of the enzyme molecule. Enzyme immobilization by adsorption involves, normally, weak interactions between the support and the enzyme such as ionic or hydrophobic interactions, hydrogen bonding, and van der Waals forces. Most of the support materials available have sufficient surface-charge properties suitable for immobilization by adsorption. They include inorganic carriers (ceramic, alumina, activated carbon, kaolinite, bentonite, porous glass), organic synthetic carriers (nylon, polystyrene), and natural organic carriers (chitosan, dextran, gelatin, cellulose, starch).

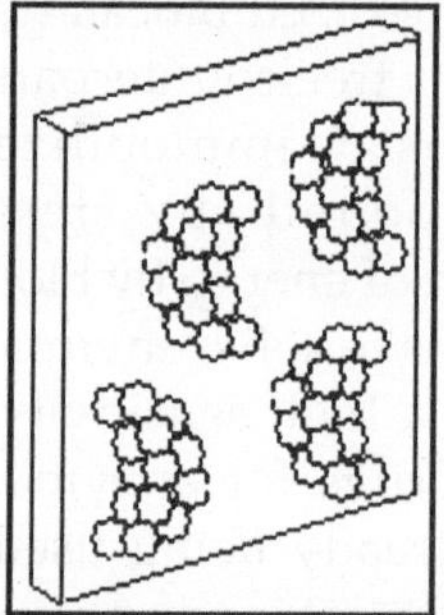

Fig. Biocatalysts Bound to a Carrier by Adsorption.

The method consists of simply mixing an aqueous solution of enzyme with the support material for a period of time, after which the excess enzyme is washed away from the immobilized enzyme on the support. The procedure requires strict control of the pH and ionic strength, because these can alter the charges of the enzyme and the support and therefore affect the level of adsorption. A simple shift in pH can cancel ionic interactions and promote the release of the enzyme from the support.

The main advantages of adsorption are the method simplicity, the little effect on the conformation/activity of the biocatalyst, and the possibility of regenerating inactive enzyme by addition of fresh enzyme. The main disadvantage is the desorption of the biocatalyst from the support due to the weak interactions established. The enzyme desorption can easily occur by changes in the environment medium such as pH, temperature, solvent, and ionic strength or in the case of extended reactions.

Ionic Binding

Immobilization via ionic binding is based, mainly, on ionic binding of enzyme molecules or active molecule to solid supports containing ionic charges. In this method, the amount of enzyme bound to the carrier and the activity after immobilization depends on the nature of the carrier. How the enzyme is bound to the carrier. In some cases, physical adsorption may also take place. The main difference between ionic binding and physical adsorption is the strength of the interaction, which is much stronger for ionic binding, although less strong than covalent binding. The preparation of immobilized enzymes using ionic binding is based on the same procedure as described for physical adsorption.

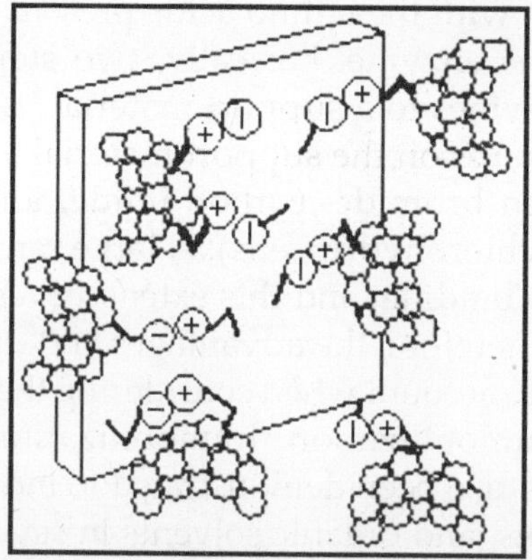

Fig. Biocatalysts Bound to a Carrier by Ionic Binding.

The ionic nature of the binding forces between the enzyme and the support also depends on pH variations, support charge, enzyme concentrations, and temperature. The supports used for ionic binding may be based on polysaccharide derivatives (*e.g.*, diethylamino-ethylcellulose, dextran, chitosan, carboxymethyl-cellulose), synthetic polymers (*e.g.*, polystyrene derivatives, polyethylene vinylalcohol), and inorganic materials (*e.g.*, ambertite, alumina, silicates, bentonite, sepiolite, silica gel, etc.). The immobilization by ionic binding has the advantage that changes in the enzyme conformation only occur in a small extent, resulting in immobilized enzymes with high enzymatic activities. The main disadvantage is the possible interference of other ions, and special attention should be paid in maintaining the correct ionic strength and pH conditions in order to prevent their easy detachment.

Covalent Binding by Chemical Coupling

The covalent binding method is based on the binding of enzymes, or other active molecules, to a support or matrix by means of covalent bonds. The bond is normally formed between a functional group present on the support surface and amino acid residues on the surface of the enzyme. Those which are most often involved in covalent binding are the amino (NH_2) group of lysine or arginine, carboxyl (CO_2H) group of aspartic acid or glutamic acid, hydroxyl (OH) group of serine or threonine, and sulphydryl (SH) group of cysteine. There are many reaction procedures for joining an enzyme to a material with a covalent bond (diazotation, amino bond, Schiff's base formation, amidation reactions, thiol-disulfide, peptide bond, and alkylation reactions). The connection between the support and the biocatalyst can be achieved either by direct linkage between the components or via an intercalated link of different length, the so-called spacer or harm. The advantage of using a spacer molecule is that it gives a greater degree of mobility to the coupled biocatalysts so that its activity can, under certain circumstances, be higher than if it is bound directly to the support. It is important to choose a method that will not involve the reaction with the amino acids present in the active site, since this could inactivate the enzyme. Basically, two steps are involved in the covalent binding of enzymes to a support material.

First, functional groups on the support material are activated by specific reagents (*e.g.*, cyanogen bromide, carbodiimide, aminoalkylethoxysilane, isothiocyanate, and epichlorohydrin, etc.). A large range of support materials is available for covalent binding, and this extensive range reflects the fact that no ideal matrix exists. Therefore, the advantages and disadvantages of a given matrix must be taken into account when considering the appropriate procedure for a given enzyme immobilization. Immobilization of enzymes through covalent attachment has also been demonstrated to induce higher resistance to temperature, denaturants, and organic solvents in several cases.

The extent of these improvements may depend on other conditions of the system, *e.g.*, the nature of the enzyme, type of support, and the method of immobilization. Many factors may influence the selection of the support, and some of the more important are its cost and availability, the binding capacity (amount of enzyme bound per given weight of matrix), hydrophilicity (the ability to incorporate water into the matrix and stability of matrix), structural rigidity, and durability during applications.

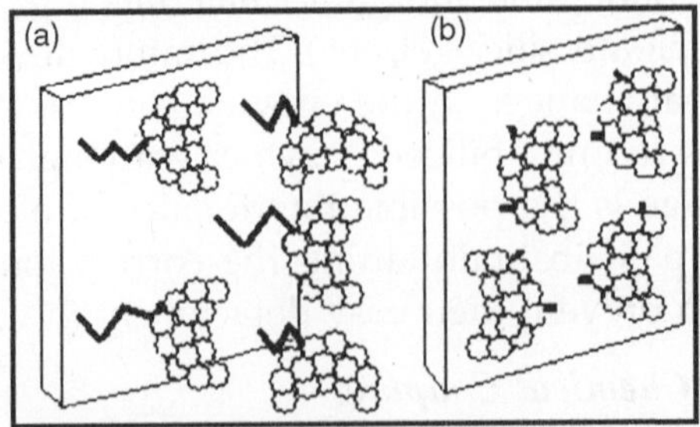

Fig. Covalent Bond Between the Biocatalysts and a Carrier with (a) and without Spacer (b).

Natural polymers, which are very hydrophilic, are popular support materials for enzyme immobilization since the residues in these polymers contain hydroxyl groups, which are ideal functional groups for participating in covalent bonds. A frequently encountered disadvantage of immobilization by covalent binding is that it places great stress on the enzyme. The necessary harshness of the immobilization procedure nearly always leads to considerable changes in conformation and a resultant loss of catalytic activity.

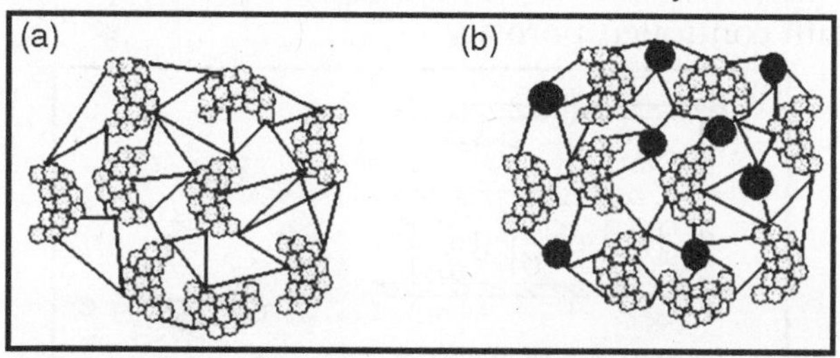

Fig. Biocatalysts Immobilized by Means of Crosslinking (a) and Co-Crosslinking with Inert Molecules Incorporated (b).

Crosslinking

The crosslinking method is based on the formation of covalent bonds between the enzyme or active molecules, by means of bi- or multifunctional reagents. The individual biocatalytic units (enzymes, organelles, whole cells) are joined to one another with the help of bi- or multifunctional reagents (*e.g.*, glutardialdehyde, glutaraldehyde, glyoxal, diisocyanates, hexamethylene diisocyanate, toluene diisocyanate, etc.). Enzyme crosslinking involves normally the amino groups of the lysine but, in occasional cases, the sulfhydryl groups of cysteine, phenolic OH groups of tyrosine, or the imidazol group of histidine can also be used for binding. How the biocatalysts can be linked by a simple crosslinking process and also by co-crosslinking, in which inert molecules are incorporated in the high-polymer network in order to improve the mechanical and enzymatic immobilized preparation. The advantages and disadvantages of a given matrix must be taken into account when considering the appropriate procedure for a given enzyme immobilization. One advantage is the simplicity of the process.

The main disadvantages are the fragility of the particles produced in some cases and diffusion limitations. Since crosslinking and co-crosslinking usually involve covalent bonds, immobilized biocatalysts in this way frequently undergo changes in the conformation with a resultant loss of activity. The isomerization of glucose process is a very important example of the industrial application using biocatalysts crosslinked with glutaraldehyde. Some of the immobilized preparations used in these large-scale processes are produced simply by glutaraldehyde treatment of bacterial cell masses that have formed fine particles.

Entrapment and Encapsulation

The entrapment method for immobilization consists of the physical trapping of the active components into a film, gel, fibre, coating, or microencapsulation. This method can be achieved by mixing an enzyme or active molecule with a polymer and then crosslinking the polymer to form a lattice structure that traps the enzyme. Microencapsulated enzymes are formed by enclosing enzymes solution within spherical semipermeable polymer membranes with controlled porosity.

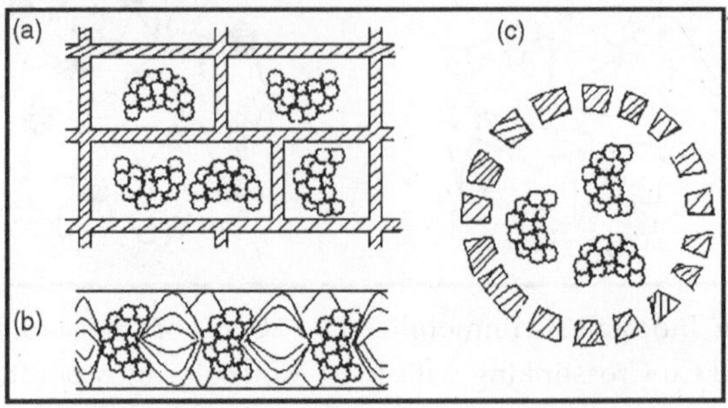

Fig. Enzyme Encapsulation in a Matrix (a), Fibre (b), or Capsule (c).

While the encapsulation of dyes, drugs, and other chemicals has been known for some time, it was not until the mid-1960s that such a method was first applied to enzymes. Since that first report, a number of other enzymes have been successfully immobilized via microencapsulation, using a number of different materials and methods to prepare the microcapsules. The advantages of this immobilization method are the extremely large surface area between the substrate and the enzyme, within a relatively small volume, and the real possibility of simultaneous immobilization. The major disadvantages of this method include the occasional inactivation of enzyme during microencapsulation and the high enzyme concentration required. In addition, to retain the enzyme, the pore size needs to be very low and these systems tend to be very diffusion limited.

ENZYME INHIBITION

If the velocity of an enzymatic reaction is decreased or inhibited, the kinetics of the reaction obviously have been perturbed. Systematic perturbations are a basic tool of experimental scientists; much can be learned about the normal workings of any system by inducing changes in it and then observing the effects of the change. The study of enzyme inhibition has contributed significantly to our understanding of enzymes.

REVERSIBLE VERSUS IRREVERSIBLE INHIBITION

Enzyme inhibitors are classified in several ways. The inhibitor may interact either reversibly or irreversibly with the enzyme. Reversible inhibitors interact with the enzyme through noncovalent association/dissociation reactions. In contrast, irreversible inhibitors usually cause stable, covalent alterations in the enzyme. That is, the consequence of irreversible inhibition is a decrease in the concentration of active enzyme.

REVERSIBLE INHIBITION

Reversible inhibitors fall into two major categories: competitive and noncompetitive (although other more unusual and rare categories are known). Competitive inhibitors are characterized by the fact that the substrate and inhibitor compete for the same binding site on the enzyme, the so–called active site or S–binding site. Thus, increasing the concentration of S favours the likelihood of S binding to the enzyme instead of the inhibitor, I. That is, high [S] can overcome the effects of I. The other major type, noncompetitive inhibition, cannot be overcome by increasing [S]. The two types can be distinguished by the particular patterns obtained when the kinetic data are analysed in linear plots, such as double–reciprocal (Lineweaver–Burk) plots. A general formulation for common inhibitor interactions in our simple enzyme kinetic model would include

$$E + I \rightleftharpoons EI \text{ and/or } I + ES \rightleftharpoons IES$$

That is, we consider here reversible combinations of the inhibitor with E and/or ES.

Competitive Inhibition

Consider the following system,

$$E + S \underset{k_{-1}}{\overset{k_1}{\rightleftharpoons}} ES \xrightarrow{k_2} E + P$$

$$E + I \underset{k_{-3}}{\overset{k_5}{\rightleftharpoons}} EI$$

where an inhibitor, I, binds *reversibly* to the enzyme at the same site as S. S–binding and I–binding are mutually exclusive, *competitive* processes. Formation of the ternary complex, EIS, where both S and I are bound, is physically impossible. This condition leads us to anticipate that S and I must share a high degree of structural similarity because they bind at the same site on the enzyme. Also notice that, in our model, EI does not react to give rise to E + P. That is, I is not changed by interaction with E. The rate of the product–forming reaction is $v = k_2[ES]$.

It is revealing to compare the equation for the uninhibited case, Equation (the Michaelis–Menten equation) with Equation for the rate of the enzymatic reaction in the presence of a fixed concentration of the competitive inhibitor,

$$v = \frac{V_{max}[S]}{[S] + K_m}$$

$$v = \frac{V_{max}[S]}{[S] + K_m\left(1 + \frac{[I]}{K_r}\right)}$$

The K_m term in the denominator in the inhibited case is increased by the factor $(1 + [I]/K_I)$; thus, v is less in the presence of the inhibitor, as expected. Clearly, in the absence of I, the two equations are identical. Figure shows a Lineweaver–Burk plot of competitive inhibition. Several features of competitive inhibition are evident. First, at a given [I], v decreases ($1/v$ increases). When [S] becomes infinite, $v = V_{max}$ and is unaffected by I because all of the enzyme is in the ES form. Note that the value of the $-x$–intercept decreases as [I] increases. This $-x$–intercept is often termed the *apparent* K_m (or K_{mapp}) because it is the K_m apparent under these conditions. The diagnostic criterion for competitive inhibition is that V_{max} is unaffected by I; that is, all lines share a common y–intercept. This criterion is also the best experimental indication of binding at the same site by two substances. Competitive inhibitors resemble S structurally.

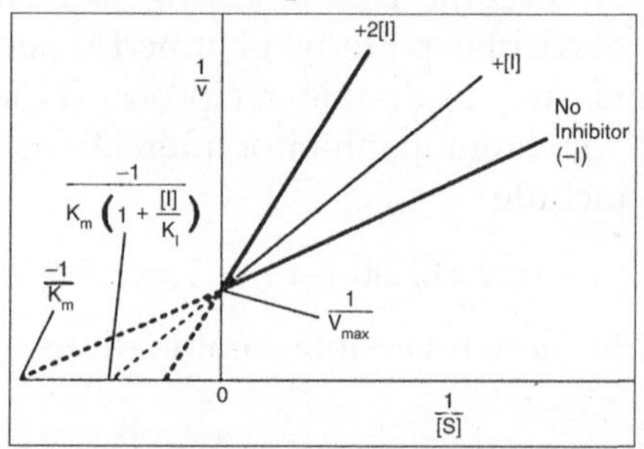

Fig. Lineweaver–Burk Plot of Competitive Inhibition, Showing Lines for No I, [I], and 2[I]

Note that when [S] is infinitely large (1/[S]50), Vmax is the same, whether I is present or not. In the presence of I, the negative x–intercept $521/K_m(11[I]/K_I)$.

Succinate Dehydrogenase

The enzyme *succinate dehydrogenase* (SDH) is competitively inhibited by malonate. Figure shows the structures of succinate and malonate. The structural similarity between them is obvious and is the basis of malonate's ability to mimic succinate and bind at the active site of SDH. However, unlike succinate, which is oxidized by SDH to form fumarate, malonate cannot lose two hydrogens; consequently, it is unreactive.

Substrate		Product	Competitive Inhibitor
COO^- – CH_2 – CH_2 – COO^-	SDH → (2H)	COO^- – CH = HC – COO^-	COO^- – CH_2 – COO^-
Succinate		Fumarate	Malonate

Fig. Structures of Succinate, the Substrate of Succinate Dehydrogenase (SDH), and Malonate, the Competitive Inhibitor

Fumarate (the product of SDH action on succinate) is also shown.

Noncompetitive Inhibition

Noncompetitive inhibitors interact with both E and ES (or with S and ES, but this is a rare and specialized case). Obviously, then, the inhibitor is not binding to the same site as S, and the inhibition cannot be overcome by raising [S].

There are two types of noncompetitive inhibition:

1. Pure and
2. Mixed.

Pure Noncompetitive Inhibition

In this situation, the binding of I by E has no effect on the binding of S by E. That is, S and I bind at different sites on E, and binding of I does not affect binding of S. Consider the system,

$$E + I \underset{}{\overset{K_I}{\rightleftharpoons}} EI \quad ES + I \underset{}{\overset{K_I'}{\rightleftharpoons}} IES$$

Pure noncompetitive inhibition occurs if $K_I = K_I'$. This situation is relatively uncommon; the Lineweaver–Burk plot for such an instance is given in Figure. Note that K_m is unchanged by I (the x–intercept remains the same, with or without I). Note also that V_{max} decreases. A similar pattern is seen if the amount of enzyme in the experiment is decreased. Thus, it is as if I lowered [E].

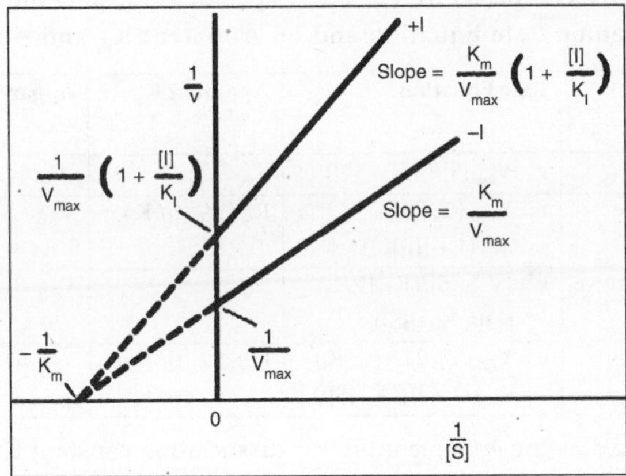

Fig. Lineweaver–Burk Plot of Pure Noncompetitive Inhibition

Note that I does not alter K_m but that it decreases V_{max}. In the presence of I, the y–intercept is equal to $(1/V_{max})(1+I/K_I)$.

Mixed Noncompetitive Inhibition

In this situation, the binding of I by E influences the binding of S by E.

Either the binding sites for I and S are near one another or conformational changes in E caused by I affect S binding. In this case, K_I and K_I', as defined previously, are not equal. Both K_m and V_{max} are altered by the presence of I, and K_m/V_{max} is not constant (Figure).

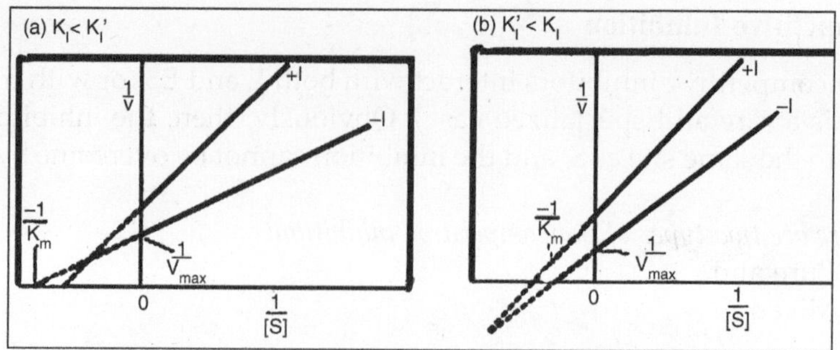

Fig. Lineweaver–Burk Plot of Mixed Noncompetitive Inhibition

Note that both intercepts and the slope change in the presence of I.
(a) When K_I is less than K_I';
(b) When K_I is greater than K_I'.

This inhibitory pattern is commonly encountered. A reasonable explanation is that the inhibitor is binding at a site distinct from the active site, yet is influencing the binding of S at the active site. Presumably, these effects are transmitted via alterations in the protein's conformation. Table includes the rate equations and apparent K_m and V_{max} values for both types of noncompetitive inhibition.

Table. The Effect of Various Types of Inhibitors on the Michaelis–Menten Rate Equation and on Apparent K_m and Apparent V_{max}

Inhibition Type	Rate Equation	Apparent K_m	Apparent V_{max}
None.	$v = V_{max}[S]/(K_m + [S])$	K_m	V_{max}
Competitive	$v = V_{max}[S]/([S] + K_m(1 + [I]/K_I))$	$K_m(1 + [I]/K_I)$	V_{max}
Noncompetitive	$v = (V_{max}[S]/(1 + [I]/K_I))/(K_m + [S])$	K_m	$V_{max}/(1 + [I]/K_I)$
Mixed	$v = V_{max}[S]/((1 + [I]/K_I) K_m 1(1 + [I]/K_I'[S]))$	$K_m(1 + [I]/K_I)/(1 + [I]/K_I')$	$V_{max}/(1 + [I]/K_I')$

- K_I is defined as the enzyme:inhibitor dissociation constant $K_I = [E][I]/[EI]$; K_I' is defined as the enzyme substrate complex: inhibitor dissociation constant $K_I' = [ES][I]/[ESI]$

IRREVERSIBLE INHIBITION

If the inhibitor combines irreversibly with the enzyme—for example, by covalent attachment—the kinetic pattern seen is like that of noncompetitive inhibition, because the net effect is a loss of active enzyme. Usually, this type of inhibition can be distinguished from the noncompetitive, reversible inhibition case since the reaction of I with E (and/or ES) is not instantaneous. Instead, there is a *time–dependent decrease in enzymatic activity* as E + I → EI proceeds, and the rate of this inactivation can be followed. Also, unlike reversible inhibitions, dilution or dialysis of the enzyme:inhibitor solution does not dissociate the EI complex and restore enzyme activity.

Suicide Substrates—Mechanism-Based Enzyme Inactivators

Suicide substrates are inhibitory substrate analogs designed so that, via normal catalytic action of the enzyme, a very reactive group is generated. This reactive group then forms a covalent bond with a nearby functional group within the active site of the enzyme, thereby causing irreversible inhibition. Suicide substrates, also called *Trojan horse substrates,* are a type of affinity label. As substrate analogs, they bind with specificity and high affinity to the enzyme active site; in their reactive form, they become covalently bound to the enzyme. This covalent link effectively labels a particular functional group within the active site, identifying the group as a key player in the enzyme's catalytic cycle.

Penicillin—A Suicide Substrate

Several drugs in current medical use are mechanism–based enzyme inactivators. For example, the antibiotic penicillin exerts its effects by covalently reacting with an essential serine residue in the active site of *glycoprotein peptidase,* an enzyme that acts to cross–link the peptidoglycan chains during synthesis of bacterial cell walls (Figure). Once cell wall synthesis is blocked, the bacterial cells are very susceptible to rupture by osmotic lysis, and bacterial growth is halted.

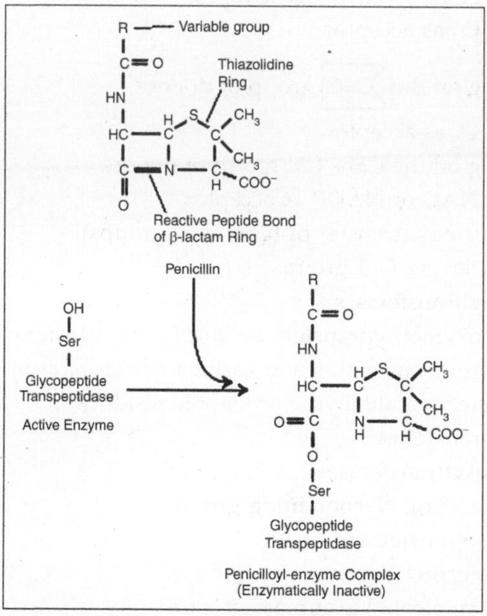

Fig. Penicillin is an Irreversible Inhibitor of the Enzyme *Glycoprotein Peptidase,* which Catalyzes an Essential Step in Bacterial Cell Wall Synthesis

Penicillin consists of a thiazolidine ring fused to a β–lactam ring to which a variable group R is attached. A reactive peptide bond in the β–lactam ring covalently attaches to a serine residue in the active site of the glycopeptide transpeptidase.

ENZYME NOMENCLATURE

Traditionally, enzymes often were named by adding the suffix–*ase* to the name of the substrate upon which they acted, as in *urease* for the urea–hydrolyzing enzyme or *phosphatase* for enzymes hydrolyzing phosphoryl groups from organic phosphate compounds. Other enzymes acquired names bearing little resemblance to their activity, such as the peroxide–decomposing enzyme *catalase* or the proteolytic enzymes (*proteases*) of the digestive tract, *trypsin* and *pepsin*. Because of the confusion that arose from these trivial designations, an International Commission on Enzymes was established in 1956 to create a systematic basis for enzyme nomenclature. Although common names for many enzymes remain in use, all enzymes now are classified and formally named according to the reaction they catalyze. Six classes of reactions are recognized.

Table. Systematic Classification of Enzymes According to the Enzyme Commission

E.C. Number	Systematic Name and Subclasses
1	*Oxidoreductases* (oxidation–reduction reactions)
1.1	Acting on CH—OH group of donors
1.1.1	With NAD or NADP as acceptor
1.1.3	With O_2 as acceptor
1.2	Acting on the $\rangle C{=}O$ group of donors
1.2.3	With O_2 as acceptor
1.3	Acting on the CH—CH group of donors
1.3.1	With NAD or NADP as acceptor
2	*Transferases* (transfer of functional groups)
2.1	Transferring C–1 groups
2.1.1	Methyltransferases
2.1.2	Hydroxymethyltransferases and formyltransferases
2.1.3	Carboxyltransferases and carbamoyltransferases
2.2	Transferring aldehydic or ketonic residues
2.3	Acyltransferases
2.4	Glycosyltransferases
2.6	Transferring N–containing groups
2.6.1	Aminotransferases
2.7	Transferring P–containing groups
2.7.1	With an alcohol group as acceptor
3	*Hydrolases* (hydrolysis reactions)
3.1	Cleaving ester linkage

3.1.1	Carboxylic ester hydrolases
3.1.3	Phosphoric monoester hydrolases
3.1.4	Phosphoric diester hydrolases
4	*Lyases* (addition to double bonds)
4.1	C=C lyases
4.1.1	Carboxy lyases
4.1.2	Aldehyde lyases
4.2	C=O lyases
4.2.1	Hydrolases
4.3	C=N lyases
4.3.1	Ammonia–lyases
5	*Isomerases* (isomerization reactions)
5.1	Racemases and epimerases
5.1.3	Acting on carbohydrates
5.2	*Cis*–trans isomerases
6	*Ligases* (formation of bonds with ATP cleavage)
6.1	Forming C—O bonds
6.1.1	Amino acid–RNA ligases
6.2	Forming C—S bonds
6.3	Forming C—N bonds
6.4	Forming C—C bonds
6.4.1	Carboxylases

Within each class are subclasses, and under each subclass are sub–subclasses within which individual enzymes are listed. Classes, subclasses, sub–subclasses, and individual entries are each numbered, so that a series of four numbers serves to specify a particular enzyme. A systematic name, descriptive of the reaction, is also assigned to each entry. To illustrate, consider the enzyme that catalyzes this reaction:

$$\text{ATP} + \delta\text{–glucose} \rightarrow \text{ADP} + \delta\text{–glucose–6–phosphate}$$

A phosphate group is transferred from ATP to the C–6–OH group of glucose, so the enzyme is a *transferase*. Subclass 7 of transferases is *enzymes transferring phosphorus–containing groups*, and sub–subclass 1 covers those *phosphotransferases with an alcohol group as an acceptor*. Entry 2 in this sub–subclass is ATP: δ–glucose–6–phosphotransferase, and its classification number is 2.7.1.2. In use, this number is written preceded by the letters E.C., denoting the Enzyme Commission. For example, entry 1 in the same sub–subclass is E.C.2.7.1.1, ATP: δ–hexose–6–phosphotransferase, an ATP–dependent enzyme that transfers a phosphate to the 6–OH of hexoses (that is, it is nonspecific regarding its hexose acceptor). These designations can be cumbersome, so in everyday usage, trivial names are employed frequently. The glucose–specific enzyme, E.C.2.7.1.2, is called glucokinase and the nonspecific E.C.2.7.1.1 is known as*hexokinase*. *Kinase* is a trivial term for enzymes that are ATP–dependent phosphotransferases.

COENZYMES

Many enzymes carry out their catalytic function relying solely on their protein structure. Many others require nonprotein components, called cofactors (Table). Cofactors may be metal ions or organic molecules referred to as coenzymes. Cofactors, because they are structurally less complex than proteins, tend to be stable to heat (incubation in a boiling water bath). Typically, proteins are denatured under such conditions. Many coenzymes are vitamins or contain vitamins as part of their structure.

Table. Enzyme Cofactors: Some Metal Lons and Coenzymes and the Enzymes with which they are Associated

Metal Ions and Some Enzymes That Require Them		Coenzymes Serving as Transient Carriers of Specific Atoms or Functional Groups		Representative Enzymes Using Coenzymes
Metal Ion	Enzyme	Coenzyme	Entity Transferred	
Fe^{2+} or Fe^{3+}	Cytochrome oxidase Catalase Peroxidase	Thiamine pyrophosphate (TPP) Flavin adenine dinucleotide (FAD) Nicotinamide adenine dinucleotide (NAD)	Aldehydes Hydrogen atoms Hydride ion (H^-)	Pyruvate dehydrogenase Succinate dehydrogenase Alcohol dehydrogenase
Cu^{2+}	Cytochrome oxidase			
Zn^{2+}	DNA polymerase Carbonic anhydrase	Coenzyme A (CoA) Pyridoxal phosphate (PLP)	Acyl groups Amino groups	Acetyl–CoA carboxylase Aspartate aminotransferase
	Alcohol dehydrogenase	5'–Deoxyadenosylcobalamin (vitamin B_{12})	H atoms and alkyl groups	Methylmalonyl–CoA mutase
Mg^{2+}	Hexokinase Glucose–6–phosphatase	Biotin (biocytin)	CO_2	Propionyl–CoA carboxylase
Mn^{2+}	Arginase	Tetrahydrofolate (THF)	Other one–carbon groups	Thymidylate synthase
K^+	Pyruvate kinase (also requires Mg^{2+})			
Ni^{2+}	Urease			
Mo	Nitrate reductase			
Se	Glutathione peroxidase			

Usually coenzymes are actively involved in the catalytic reaction of the enzyme, often serving as intermediate carriers of functional groups in the conversion of substrates to products. In most cases, a coenzyme is firmly associated with its enzyme, perhaps even by covalent bonds, and it is difficult to separate the two. Such tightly bound coenzymes are referred to as prosthetic groups of the enzyme. The catalytically active complex of protein and prosthetic group is called the holoenzyme. The protein without the prosthetic group is called the apoenzyme; it is catalytically inactive.

ENZYME KINETICS

Kinetics is the branch of science concerned with the rates of chemical reactions. The study of enzyme kinetics addresses the biological roles of enzymatic catalysts and how they accomplish their remarkable feats. In enzyme kinetics, we seek to determine the maximum reaction velocity that the enzyme can attain and its binding affinities for substrates and inhibitors. Coupled with studies on the structure and chemistry of the enzyme, analysis of the enzymatic rate under different reaction conditions yields insights regarding the enzyme's mechanism of catalytic action. Such information is essential to an overall understanding of metabolism. Significantly, this information can be exploited to control and manipulate the course of metabolic events. The science of pharmacology relies on such a strategy. Pharmaceuticals, or drugs, are often special inhibitors specifically targeted at a particular enzyme in order to overcome infection or to alleviate illness. A detailed knowledge of the enzyme's kinetics is indispensable to rational drug design and successful pharmacological intervention.

REVIEW OF CHEMICAL KINETICS

Before beginning a quantitative treatment of enzyme kinetics, it will be fruitful to review briefly some basic principles of chemical kinetics.Chemical kinetics is the study of the rates of chemical reactions. Consider a reaction of overall stoichiometry,

$$A \rightarrow P$$

Although we treat this reaction as a simple, one–step conversion of A to P, it more likely occurs through a sequence of elementary reactions, each of which is a simple molecular process, as in,

$$A \rightarrow I \rightarrow J \rightarrow P$$

where I and J represent intermediates in the reaction. Precise description of all of the elementary reactions in a process is necessary to define the overall reaction mechanism for A → P. Let us assume that A → P *is* an elementary reaction and that it is spontaneous and essentially irreversible. Irreversibility is easily assumed if the rate of P conversion to A is very slow *or* the concentration of P (expressed as [P]) is negligible under the conditions chosen. The velocity, v, or rate, of the reaction A → P is the amount of P formed or the amount of A consumed per unit time, t. That is,

$$v = \frac{d[P]}{dt}$$

or,

$$v = \frac{-d[A]}{dt}$$

The mathematical relationship between reaction rate and concentration of reactant(s) is the rate law. For this simple case, the rate law is,

$$v = \frac{-d[A]}{dt} = k[A]$$

From this expression, it is obvious that the rate is proportional to the concentration of A, and *k* is the proportionality constant, or rate constant. *k* has the units of $(\text{time})^{-1}$, usually sec^{-1}. v is a function of [A] to the first power, or, in the terminology of kinetics, *v* is first–order with respect to A. For an elementary reaction, the order for any reactant is given by its exponent in the rate equation. The number of molecules that must simultaneously interact is defined as the molecularity of the reaction. Thus, the simple elementary reaction of A → P is a first–order reaction. Figure portrays the course of a first–order reaction as a function of time. The rate of decay of a radioactive isotope, like ^{14}C or ^{32}P, is a first–order reaction, as is an intramolecular rearrangement, such as A → P. Both are unimolecular reactions (the molecularity equals).

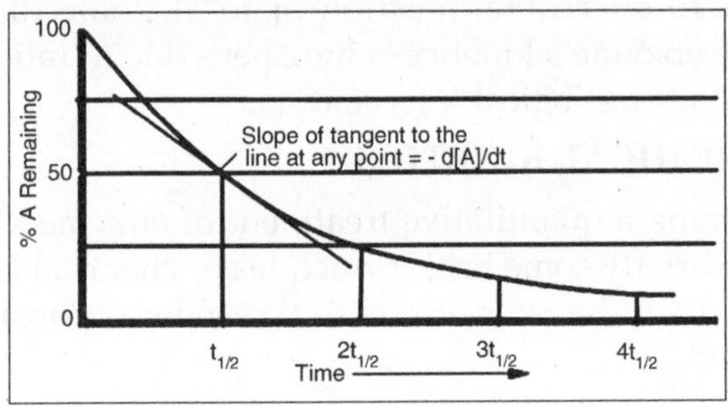

Fig. Plot of the Course of a First–order Reaction

The half–time, $t_{1/2}$, is the time for one–half of the starting amount of A to disappear.

BIOMOLECULAR REACTIONS

Consider the more complex reaction, where two molecules must react to yield products:

$$A + B \rightarrow P + Q$$

Assuming this reaction is an elementary reaction, its molecularity is 2; that is, it is a bimolecular reaction. The velocity of this reaction can be determined from the rate of disappearance of either A or B, or the rate of appearance of P or Q:

$$v = \frac{-d[A]}{dt} = \frac{-d[B]}{dt} = \frac{d[P]}{dt} = \frac{d[Q]}{dt}$$

The rate law is,

$$v = k[A][B]$$

The rate is proportional to the concentrations of both A and B. Because it is proportional to the product of two concentration terms, the reaction is second–order overall, first–order with respect to A and first–order with respect to B. (Were the elementary reaction 2 → P + Q, the rate law would be $v = k[A]^2$, second–order overall and Second–order with respect to A.) Second–order rate constants have the units of (concentration) $^{-1}$ (time) $^{-1}$, as in M^{-1} sec^{-1}.

Molecularities greater than two are rarely found (and greater than three, never). When the overall stoichiometry of a reaction is greater than two (for example, as in A+B+Cn or 2A+Bn), the reaction almost always proceeds via uni– or bimolecular elementary steps, and the overall rate obeys a simple first– or Second–order rate law.

At this point, it may be useful to remind ourselves of an important caveat that is the first principle of kinetics: *Kinetics cannot prove a hypothetical mechanism*. Kinetic experiments can only rule out various alternative hypotheses because they don't fit the data. However, through thoughtful kinetic studies, a process of elimination of alternative hypotheses leads ever closer to the reality.

DECREASING ΔG‡ INCREASES REACTION RATE

We are familiar with two general ways that rates of chemical reactions may be accelerated. First, the temperature can be raised. This will increase the average energy of reactant molecules, which in effect lowers the energy needed to reach the transition state. The rates of many chemical reactions are doubled by a 10°C rise in temperature. Second, the rates of chemical reactions can also be accelerated by catalysts. Catalysts work by lowering the energy of activation rather than by raising the average energy of the reactants. Catalysts accomplish this remarkable feat by combining transiently with the reactants in a way that promotes their entry into the reactive, transition–state condition.

5

Analysis of Amino Acids

One of a class of organic compounds containing the amino (NH_2) and the carboxyl (COOH) group, occurring naturally in plant and animal tissues and forming the chief constituents of protein; many of them are necessary for human and animal growth and nutrition and hence are called Essential Amino Acids. Amino Acids are the building blocks of protein.

Structure of a Typical Amino Acid:

Fig. Anatomy of an Amino Acid

Except for proline and its derivatives, all of the amino acids commonly found in proteins possess this type of structure.

The structure of a single typical amino acid is shown in Figure. Central to this structure is the tetrahedral alpha (α) carbon (C_α), which is covalently linked to both the amino group and the carboxyl group. Also bonded to this α–carbon is a hydrogen and a variable side chain. It is the side chain, the so–called R group, that gives each amino acid its identity. The detailed acid–base properties of amino acids are discussed in the following sections.

It is sufficient for now to realise that, in neutral solution (pH 7), the carboxyl group exists as $–COO^-$ and the amino group as $–NH_3^+$. Because the resulting amino acid contains one positive and one negative charge, it is a neutral molecule called azwitterion.

Amino acids are also *chiral* molecules. With four different groups attached to it, the α–carbon is said to be *asymmetric*. The two possible configurations for the α–carbon constitute nonidentical mirror image isomers or enantiomers.

AMINO ACIDS ARE WEAK POLYPROTIC ACIDS

From a chemical point of view, the common amino acids are all weak polyprotic acids. The ionizable groups are not strongly dissociating ones, and the degree of dissociation thus depends on the pH of the medium. All the amino acids contain at least two dissociable hydrogens.

Consider the acid–base behaviour of glycine, the simplest amino acid. At low pH, both the amino and carboxyl groups are protonated and the molecule has a net positive charge. If the counterion in solution is a chloride ion, this form is referred to as glycine hydrochloride.If the pH is increased, the carboxyl group is the first to dissociate, yielding the neutral zwitterionic species Gly^0 (Figure).

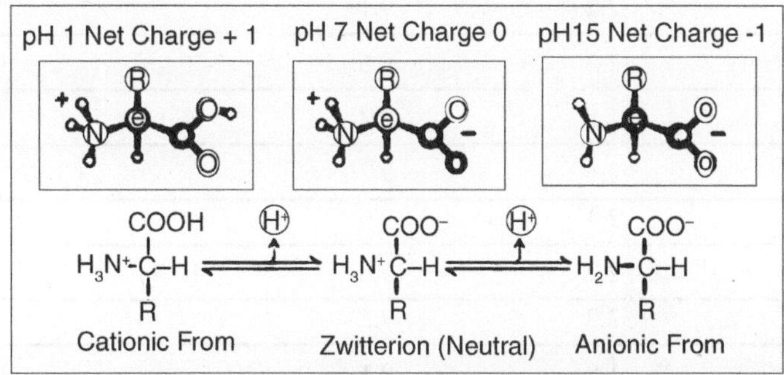

Fig. The Ionic Forms of the Amino Acids, Shown without Consideration of any Ionizations on the Side Chain

The cationic form is the low pH form, and the titration of the cationic species with base yields the zwitterion and finally the anionic form. *(Irving Geis)*

Further increase in pH eventually results in dissociation of the amino group to yield the negatively charged glycinate. If we denote these three forms as Gly^+, Gly^0, and Gly^-, we can write the first dissociation of Gly^+ as,

$$Gly^+ + H_2O \rightleftharpoons Gly^0 + H_3O^+$$

and the dissociation constant K_1 as

$$K = \frac{[Gly^\circ] + [H3O+]}{[Gly^+]}$$

Values for K_1 for the common amino acids are typically 0.4 to 1.0 × 10^{-2} *M*, so that typical values of pK_1 center on values of 2.0 to 2.4. In a similar manner, we can write the second dissociation reaction as

$$Gly^0 + H_2O \rightleftharpoons Gly^- + H_3O^+$$

and the dissociation constant K_2 *as:*

$$K_2 = \frac{\left[Gly^-\right]\left[H_3O^+\right]}{Gly^0}$$

Table. pK_a Values of Common Amino Acids

Amino Acid	α–COOH pK_a	α–NH31 pK_a	R group pK_a
Alanine	2.4	9.7	
Arginine	2.2	9.0	12.5
Asparagine	2.0	8.8	
Aspartic acid	2.1	9.8	3.9
Cysteine	1.7	10.8	8.3
Glutamic acid	2.2	9.7	4.3
Glutamine	2.2	9.1	
Glycine	2.3	9.6	
Histidine	1.8	9.2	6.0
Isoleucine	2.4	9.7	
Leucine	2.4	9.6	
Lysine	2.2	9.0	10.5
Methionine	2.3	9.2	
Phenylalanine	1.8	9.1	
Proline	2.1	10.6	
Serine	2.2	9.2	~ 13.00
Threonine	2.6	10.4	~ 13.00
Tryptophan	2.4	9.4	
Tyrosine	2.2	9.1	10.1
Valine	2.3	9.6	

Typical values for pK_2 are in the range of 9.0 to 9.8. At physiological pH, the α–carboxyl group of a simple amino acid (with no ionizable side chains) is completely dissociated, whereas the a–amino group has not really begun its dissociation. The titration curve for such an amino acid.

Example:

What is the pH of a glycine solution in which the a–NH_3^+ group is one–third dissociated?

Solution:

The appropriate Henderson–Hasselbalch equation is,

$$pH = pK_a + \log_{10} \frac{[Gly^-]}{[Gly^0]}$$

If the α–amino group is one–third dissociated, there is one part Gly^- for every two parts Gly^0. The important pK_a is the pK_a for the amino group. The glycine α–amino group has a pK_a of 9.6. The result is,

$$pH = 9.6 + \log_{10}(1/2)$$
$$pH = 9.3$$

Note that the dissociation constants of both the α–carboxyl and α–amino groups are affected by the presence of the other group. The adjacent a–amino group makes the a–COOH group more acidic (that is, it lowers the pK_a) so that it gives up a proton more readily than simple alkyl carboxylic acids.

Thus, the pK_1 of 2.0 to 2.1 for a–carboxyl groups of amino acids is substantially lower than that of acetic acid ($pK_a = 4.76$), for example. What is the chemical basis for the low pK_a of the a–COOH group of amino acids? The α–NH_3^+ (ammonium) group is strongly electron–withdrawing, and the positive charge of the amino group exerts a strong field effect and stabilizes the carboxylate anion.

Ionization of Side Chains

As we have seen, the side chains of several of the amino acids also contain dissociable groups. Thus, aspartic and glutamic acids contain an additional carboxyl function, and lysine possesses an aliphatic amino function. Histidine contains an ionizable imidazolium proton, and arginine carries a guanidinium function. The β–carboxyl group of aspartic acid and the γ–carboxyl side chain of glutamic acid exhibit pK_a values intermediate to the α–COOH on the one hand and typical aliphatic carboxyl groups on the other hand.

In a similar fashion, the ε–amino group of lysine exhibits a pK_a that is higher than the α–amino group but similar to that for a typical aliphatic amino group. These intermediate values for side–chain pK_a values reflect the slightly diminished effect of the α–carbon dissociable groups that lie several carbons removed from the side–chain functional groups. Figure shows typical titration curves for glutamic acid and lysine, along with the ionic species that predominate at various points in the titration. The only other side–chain groups that exhibit any significant degree of dissociation are the *para*–OH group of tyrosine and the –SH group of cysteine.

The pK_a of the cysteine sulfhydryl is 8.32, so that it is about 12 per cent dissociated at pH 7. The tyrosine *para*–OH group is a very weakly acidic group, with a pK_a of about 10.1. This group is essentially fully protonated and uncharged at pH 7.

ESSENTIAL AMINO ACIDS

For the most part, plants and free-living microorganisms make these. The fact that humans can't synthesize them except by degrading protein has practical consequences. Any single amino acid can be limiting in the diet. Thus, for example, an individual who does not get enough tryptophan in the diet can't compensate by eating more methionine. The only way to get enough tryptophan would be to break down muscle to release enough tryptophan to support life. This means that a person can literally starve to death while getting what appears to be enough calories and protein. In the 1970s, a fad of eating only hydrolyzed liquid collagen and vitamins in an attempt to lose weight appeared. These commercial products were removed from the market after several people died of heart problems. The cause was simple: Collagen is very poor in aromatic amino acids. People were breaking down their muscle, including heart muscle, to provide these amino acids for the synthesis of essential proteins. Eventually they died of cardiac insufficiency. Animal and dairy proteins usually contain a balanced supply of amino acids. That is, they are complete proteins. Plant proteins aren't always complete, so individuals who eat primarily vegetable sources of proteins must eat complementary foods to supply a full set of amino acids. This is the reason for the coexistence of rice and beans in so many cuisines. The combination supplies all the essential amino acids, while neither food does so on its own.

AMINO ACID CATABOLISM

Excess amino acids are degraded, rather than stored, by almost all biological systems. Seed formation in plants and the synthesis of yolk and proteins in eggs constitute the major exceptions. Thus, a "high-protein" diet normally provides little benefit. Most healthy individuals need a relatively small amount of dietary protein, unless they are growing children. A typical "Western diet," with a large meat intake, isn't necessary for health.

The products of amino acid breakdown are of two kinds. Ketone bodies—that is, acetoacetate and hydroxybutyrate—are formed from the catabolism of the branched-chain amino acids, lysine and some aromatic amino acids. Tricarboxylic acid (TCA) cycle intermediates, including pyruvate and glutamate, are formed from most of the other aliphatic and aromatic amino acids. Amino acids whose metabolisms produce ketone bodies such as acetoacetate are called ketogenic; amino acids whose metabolisms produce TCA cycle intermediates are called glucogenic, because TCA cycle intermediates are substrates for gluconeogenesis. Individual amino acids can be exclusively ketogenic, exclusively glucogenic, or both. Only leucine and lysine are considered to be exclusively ketogenic, and some suspicion remains that they may also give rise to TCA cycle intermediates.

NITROGEN SOURCES

In order to synthesize amino acids, a source of nitrogen is needed. In animals glutamate and glutamine play the pivotal roles. The ±"amino group of most of the amino acids comes from the transamination reaction transferring the amino group from glutamate to an ±-ketoacid acceptor. Glutamate is synthesized from ammonia and ±-ketoglutarate by the action of glutamate dehydrogenase.

$$NH_4^+ \ \alpha\text{-ketoglutarate} + NAD(P)H + H^+ \rightleftarrows \text{Glutamate} + NAD(P)^+ + H_2O$$

This reaction occurs in two steps first a Schiff base is formed between the ammonia and the ketone of aketoglutarate. The Schiff base is reduced by hydride transfer from either NADH or NADPH to form glutamate. The importance of the hydride transfer is that it establishes the stereochemistry at the ±-carbon. Glutamate dehyrogenase binds ±-ketoglutarate is such a way that the hydride is transferred to specific face to form only the L-isomer of glutamate.

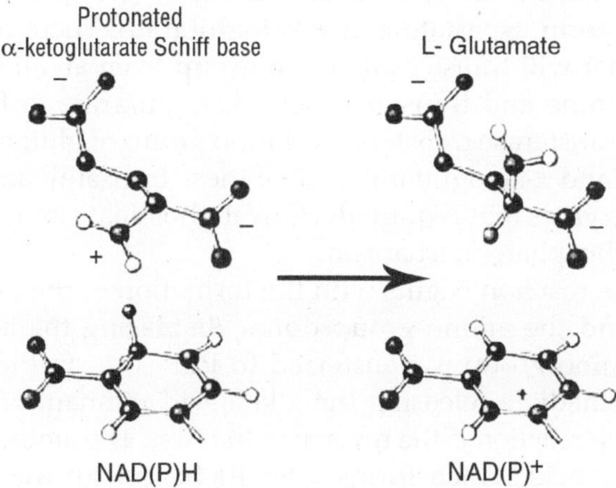

A second ammonium ion is incorporated into glutamate by the action of glutamine synthetase to form glutamine. The regulation of glutamine synthetase plays a crucial role in controlling nitrogen metabolism. The dynamic duo of glutamate dehydrogenase and glutamine synthetase are found in all living organisms.

AMINO ACIDS ARE SYNTHESIZED FROM METABOLITES OF OTHER PATHWAYS

Plants and bacteria can synthesize all 20 of the amino acids. Where as we humans cannot synthesize of them. These nine amino acids must come from our diets and are called essential amino acids. The essential amino acids are: Histidine, Isoleucine, Leucine, Lysine, Methionine, Phenylalanine, Threonine, Tryptophan and Valine.

The other 11 amino acids are called non-essential amino acids they include: Alanine, Arginine, Asparagine, Aspartate, Cysteine, Glutamate, Glutamine, Glycine, Proline, Serine and Tyrosine. These non-essential amino acids are synthesize by simple pathways. Whereas the biosynthesis of the essential amino acids are quite complex.

The major metabolic precursors of the amino acids. The amino acids that are precursors for other amino acids are shown in yellow. The nine essential amino acids are shown in boldface. The carbon skeletons come from intermediates of glycolysis, the pentose phosphate pathway and the citric acid cycle. On the basis of the starting points the 20 amino acids can be group into 6 categories depending on the precursor: oxaloacetate, PEP, ±-ketoglutarate, pyruvate, 3-phosphoglycerate, ribose-5P.

ASPARTATE, GLUTAMATE AND ALANINE BIOSYNTHESIS

Oxaloacetate, pyruvate and ±-ketoglutarate are all ±-ketoacids which are substrates for transamination reactions that we very familiar with by now. Asparatate aminotransferase transfers an amino group from glutamine to oxaloacetate to form aspartate and ±-ketoglutarate. There are a number of transaminase that will transfer an amino group from an amino acid such as aspartate or alanine and transfer it to ±- ketoglutarate to form glutamate. Alanine aminotransferase transfers the amino group of glutamine to pyruvate to form alanine and ±-ketoglutarate. All of these transamination reactions are catalyzed by enzymes that require pyridoxal phosphate as a cofactor. Review the pyridoxal phosphate mechanism.

The enzyme reaction begins with the formation of the external aldimine between PLP and the amino group donor displacing the active site lysine residue. The amino group is transferred to PLP to form the pyridixoamine phosphate intermediate releasing the ±-ketoacid remnant of the donor. The second half of the reaction is the reverse of the first. The amino group acceptor is another ±-ketoacid which forms a Schiff base with the PMP to form a ketimine intermediate. The amino group is transferred to the acceptor to form the amino acid. The important step in the transamination reaction is the protonation of the ±-carbon of the quinonoid intermiate. This protonation determines the stereochemistry at the ±-carbon. In the case of aspartate amino transferase (the postor child of transaminases) the chirality of the amino acid is determined by binding interactions with the substrate. Particularly, the interaction of the conserved arginine residue (Arg-386) with the carboxylate group of the substrate. This interaction orients the substrate so that when Lys-268 protonates the ±"carbon of the quinonoid intermediate it generates an external aldimine with the L-configuration at the ±-carbon.

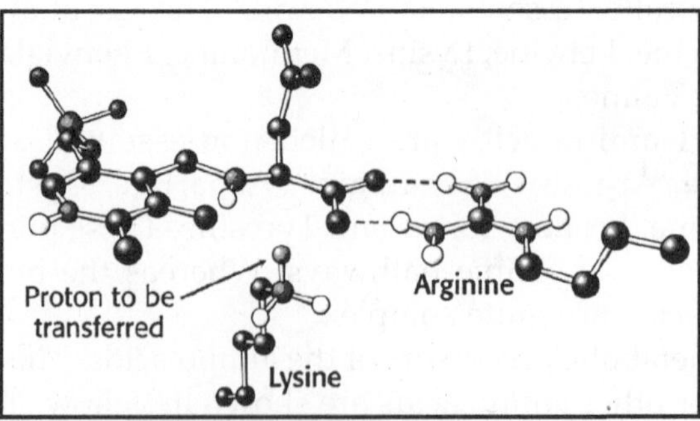

SERINE, CYSTEINE AND GLYCINE BIOSYNTHESIS

Serine is synthesized from 3-phosphoglycerate. Serine is a precursor for both cysteine and glycine. The first step is the oxidation of 3-phosphoglycerate into 3-phosphohydroxy-pyruvate. The enzyme that catalyzes the oxidation is

3-phosphoglycerate dehydrogenase. This oxidation generates an ±-ketoacid which can now be transaminated by transferring an amino group from glutamate to 3-phosphoglycerate to generate 3-phosphoserine and ±-ketoglutarate. The phosphate group is removed by phosphoserine phosphatase to produce serine. Serine is converted into glycine by serine hydroxy-methyltransferase which contains both a pyridoxal phosphate cofactor and a tetrahydrofolate cofactor. It is time to take a closer look as this remarkable cofactor and this remarkable enzyme.

AMINOACETIC ACID

A nonessential amino acid, derived from many proteins, used as a gastric antacid and dietary supplement, and in the treatment of various myopathies and peripheral vascular insufficiencies. A solution is used as an irrigation fluid. Its hydrochloride salt is use as a source of hydrochloric acid in the treatment of achlorhydria.

Alanine

A natural amino acid occurring in two forms, alpha-Alanine, and beta-Alanine.

- Hypoglycemia (Prolongs the stabilization of blood glucose over long periods of time.)
- Alanine may be used for the production of glucose in gluconeogenesis.

Arginine

An amino acid occurring in proteins; it is involved in the urea cycle, which converts ammonia to urea.

- Induces growth hormone release from the pituitary gland.
- Major component of seminal fluid.
- Helpful in burn treatment, elevated ammonia levels, and cirrhosis of the liver.
- Stimulates immune response by enhancing the production of T-Cells.
- Protective effect of toxicity of hydrocarbons and intravenous diuretics.

(Should be used with caution in schizophrenic cases.)

Intake should be kept low in persons with Herpes simplex and Epstein Barr Virus (EBV. Mechanism of interaction is believed because of possible replications of the Herpes and EB viruses.)

Aspartic Acid

A nonessential, natural dibasic amino acid, involved in transamination reactions, the ornithine cycle, and the formation of carnosine, anserine, purines, and pyrimidines.

- Has a protective function over the liver.

- Helps in detoxification of ammonia.
- Promotes mineral uptake in the intestinal tract.
- It is part of the sweetener Aspartain that is a dypeptide of the amino acids; Aspartic Acid and Phenylalanine.

Carnitine

A betaine derivative involved in the transport of fatty acids into mitochondria, where they are metabolized.

- Suggested as useful in mobilization of surface fats such as cellulite.
- Helpful in treatment of fatigue and when muscle weakness is present. Useful in the oxidation of long chain fatty acids. A major source of energy for tissues.
- Reported as useful in myocardial ischemia (low energy metabolism by the heart).
- Useful in clearing triglycerides from the blood.
- Useful in metabolic abnormalities.

Citrulline

An alpha-amino acid involved in urea production; formed from ornithine and is itself converted into Arginine in the urea cycle.

- Produces the amino acids, Arginine and Ornithine.
- Detoxifies nitrogen containing waste products such as ammonia (a cellular toxin) and is therefore part of ammonia detoxification (urea cycle).
- Stimulates growth hormone production.

Cysteine

A sulfur-containing amino acid produced by enzymatic or acid hydrolysis of proteins, readily oxidized to Cystine, sometimes found in urine.

- Assists tissues damaged by alcohol abuse, cigarette smoke and air pollution through the detoxification of Acetalaldahide.
- Helps maintain skin flexibility and texture by slowing abnormal cross linking of collagen (connective tissue protein which holds the skin together).
- Promotes red and white blood cell reproduction and tissue restoration in lung affecting diseases.
- Useful in iron deficient anemia and promotion of iron absorption.
- Helpful in prevention of peroxidized fats and free radicals.
- Converts to Cystinine in the absence of vitamin C.
- One of the amino acids found in tripeptide Glutathione and GTF.
- Hair contains between 10 to 16 per cent Cysteine.
- Heavy metal chelator. (Should be used with caution in diabetics due to possible three dimensional structure changes in the insulin cycle)

Cystine

An essential, Sulfur-containing amino acid, produced by digestion or acid hydrolysis of proteins, sometimes found in the urine and kidneys, and readily reduced to two molecules of Cysteine.

- Reported helpful in dermatological conditions.
- Promotes faster recovery of tissue after surgery.
- Part of the Insulin molecule.
- Found high in hair (sulfur bonds).

GABA - (Gamma Amino Butyric Acid)

An amino acid that is one of the principle inhibitory neurotransmitters in the central nervous system.

- Useful in schizophrenia and epilepsy, depression, high blood pressure, high stress levels, manic Behaviour, and acute agitation due to increased regulation of nerve firings and enhancement of Niacinamide binding receptors.
- May induce calmness and tranquility by inhibiting neurotransmitters, which decrease the activity in neurons.
- May be useful in reducing an enlarged prostate due to the suppression of Prolactia released by the pituitary gland.

Glutamic Acid

Glutamate is a salt of Glutamic Acid; in biochemistry, the term is often used interchangeably with Glutamic Acid. Glutamic Acid is a crystalline dibasic nonessential amino acid, widely distributed in proteins, which is thought to be a neurotransmitter, inhibiting neural excitation in the central nervous system; its hydrochloride salt is used as a gastric acidifier. The monosodium salt of L-Glutamic acid (sodium glutamate) is used in treating encephalopathies associated with hepatic diseases, and to enhance the flavour of foods and tobacco.

- Reported useful in Muscular Dystrophy.
- Does not cross blood brain barriers.
- It is an excitatory neurotransmitter in the CNS; (Precursor of GABA.)

Glutamine

The monoamide of Glutamic Acid, an amino acid occurring in proteins; it is an important carrier of urinary ammonia and is broken down in the kidney by the enzyme glutaminase.

- Is useful in treatment of Alcoholism by reducing the desire to drink.
- May be helpful in the improvement of Autism and mental retardation in children.
- Reported improvement in mental functions such as memory and dexterity.

- Aids peptic ulcer healing due to antacid quality.
- Crosses Blood brain barriers.
- Participates in nucleonic acid synthesis.
- Converts to Glutanic acid.

Glutathione

Reduced Glutathione, a tripeptide of Glutamic Acid, Cysteine, and Glycine, which serves as a reducing agent in many biochemical reactions being converted to oxidized Glutathione (GSSG) in which the Cysteine residues of two Glutathione molecules are connected by a disulfide bridge. Reduced Glutathione is important in protecting erythrocytes (red blood cells) from oxidation and hemolysis; deficiency causes sensitivity to oxidant drugs.

- It is a tripeptide containing amino acid that contains: Cysteine, Glycine, and Glutamate.
- Inhibits peroxide formations, reducing free radical damage.
- Detoxifies aromatic hydrocarbons common in air pollution, such as chlorine. Reported to have protective function against radiation therapy.
- Transports other amino acids into the interior of the cell.

Glycine

A nonessential amino acid, occurring as a constituent of proteins and functioning as an inhibitory neurotransmitter in the central nervous system; used as a gastric antacid and dietary supplement, and in the treatment of various myopathies.

- Is the simplest and sweetest of the amino acids. Can be used as a sweetener in herbal beverages.
- Reported useful in degenerative diseases such as MD (Muscular Dystrophy)
- Detoxifies benzoic acid (a common food additive) and aromatic hydrocarbons in the liver.
- Involved in synthesis of nucleic acids and bio-acids. May be useful in conditions characterized by abnormal nerve firing such as epilepsy, inhibition of tripeptides Glutathione and GTF.
- Is one of the amino acids in the tripeptides Glutathione and GTF; low brain concentrations of Glycine have recently been found in ALS. (Amyotrophic Lateral Sclerosis or"Lue Gehrig's Disease.")

Histidine

Is not listed as essential for adults, but is very essential for infants. An amino acid obtainable from many proteins by the action of sulfuric acid and water; it is ESSENTIAL for optimal growth in INFANTS. Its decarboxylation results in formation of histamine.

- Release of histamines from body stores are required for sexual arousal.

- Reported useful in alleviating pain associated with rheumatoid arthritis. (Should be taken with vitamin C; Should be used with caution with Manic Depressives with elevated histamines; Should be used with caution in woman with severe depression or suicidal tendencies due to Premenstrual Syndrome; Is considered a mild neurotransmitter).

Hydroxyproline

An amino acid produced in the digestion of hydrolytic decomposition of proteins, especially collagens.

Isoleucine

An essential amino acid produced by hydrolysis of fibrin and other proteins; ESSESNTIAL for optimal INFANT growth and for nitrogen equilibrium in adults.

- Needed along with other branch chain amino acids for all rebuilding of muscle tissue.
- Role in the release of energy during muscular work. (Metabolizes along the same pathway as fat.)

Leucine

An essential amino acid, ESSENTIAL for optimal growth in INFANTS, and for nitrogen equilibrium in adults.

- Metabolized along the same pathways as fat.
- Precursors of cholesterol.
- Involved in the role of energy release during any work of the muscles.

Lysine

An essential, naturally occurring amino acid, ESSESNTIAL for optimal growth in human INFANTS, and for maintenance of nitrogen equilibrium in adults.

- Reported to inhibit growth and replication of herpes simplex and Epstein Barr viruses (EBV).
- Promotes bone growth in infants.
- Stimulates secretion of gastric juices.
- Found in abundance in muscle tissue, connective tissue and collagen.
- Low in vegetarian diets.

Methionine

An essential, naturally occurring amino acid, which is an essential component of the diet, furnishing both methyl groups and sulfur necessary for normal metabolism.

- Prevents deposits and cohesion of fats in the liver.

- Gives rise to Taurine (an important inhibitory neuro modulator in the brain).
- Involved in synthesis of Choline (must be given with Vitamin B6 (paradoxial-5-phosphate) to inhibit synthesis of homocysteine which promotes plaque deposition in the arteries)

Autistic Patients

"Taurine is deficient in Autistic patients; Methionine has an altered abnormally normal to high level, with anything distal to the Methionine pathway being low. Since Taurine is so important in the intelligence quotients of most species, this seems to be a significant." Alpha A - Leverton, Protein and Amino Acid Nutrition (1959) B - Rose, Journalism of Biological Chemistry 217:977, 1955.

Ornithine

An amino acid obtained from Arginine by the splitting of urea; it is an intermediate in urea biosynthesis.

- May reduce fat and increase muscle mass by promoting fat metabolism and stimulating growth hormone production.
- Helps in detoxification of ammonia in the urea cycle.
- May be useful in autoimmune disease such as arthritis.

Phenylalanine

An essential, naturally occurring amino acid, ESSESNTIAL for optimal growth in INFANTS, and for nitrogen equilibrium in human adults.

- May be useful in appetite control by stimulating CCK (Chaleceptokinin Enzyme) secretion.
- Shown to be useful in management of certain types of depression.
- Increases blood pressure in hypotension.
- Gives rise to Tyrosine
- It is one of the amino acids in the dypeptide sweetener Aspartain.

Proline

A cyclic amino acid occurring in proteins; it is a major constituent of collagen.

- Is an anti-hypotensive agent in lowering high blood pressure.
- Reported helpful in repairing muscle and tendon damage.
- Reported useful in promoting skin flexibility in relation to aging and sun exposure.
- A major amino acid found in collagen (connective tissue) in the presence of Vitamin C.

Serine

A naturally occurring Amino Acid, present in many proteins.

Taurine

A crystallized acid, ethylamine sulfonic acid, from the bile; found also in small quantities in lung and muscle tissue.

- Low levels seen in newborn infants fed low Taurine diets.
- Associated with retinal degenerations.
- The role of Taurine as a nutrient is to protect the cell membranes by attenuating such toxic compounds as oxidants, secondary bioacids and antibiotics.
- Recommended for children on long-term parenteral nutrition.
- Helpful in balancing calcium and potassium flux in heart muscle.
- Helps patients suffering from congenitive heart failure by alleviating their physical signs and symptoms.
- Increases left ventricular performance without any significant changes in atrial pressure.
- Often times considered a neuro modulator.
- Helpful in treating some types of epilepsy.

Autistic Patients:"Taurine is deficient in autistic patients; Methionine has an altered abnormally normal to high level, with anything distal to the Methionine pathway being low. Since Taurine is so important in the intelligence quotients of most species, this seems to be a significant.

Threonine

An essential, naturally occurring amino acid, essential for human metabolism.

- Rises to three times it's normal value at pregnancy.
- Acts as a lipotropic factor.
- Recently found to increase brain Glycine content, greatly reducing ALS symptoms.

Tryptophan

An essential, naturally occurring amino acid, existing in proteins and essential for human metabolism.

- Reported useful in the management of depression and schizophrenia.
- Produces Serotonin, which induces sleep. Has a Serotonengenic effect.
- Precursor of the vitamin Niacin.
- Vasoconstrictor, which appears to aid in blood clotting mechanism, aids in elevating the threshold of pain.

(Should be taken with Vitamin B6 (paradoxial-5-phosphate) and in the presence of carbohydrates such as fruit or vegetable juice to maximize uptake in the brain.)

Tyrosine

An essential, naturally occurring amino acid present in most proteins. It is a product of Phenylalanine metabolism and a precursor of thyroid hormones, catecholamines, and melanin.

- Along with Phenylalanine, it is a useful anti depressant due to increased production of Tacolamine. (Recommended to be taken with Vitamin B6 (paradoxial-5-phosphate).)
- Reported to stabilize blood pressure by lowering energy in some cases and elevating it in others.
- Involved in tissue pigmentation
- Is important in the formation of thyroid hormone (Should not be used when MAO's are prescribed or when cancer melanoma is present)

Valine

An essential, naturrally occurring amino acid, essential for human metabolism.

- Needed for all muscle building.
- Required in the precursors of cholesterol.

REACTIONS OF AMINO ACIDS

CARBOXYL AND AMINO GROUP REACTIONS

The α–carboxyl and α–amino groups of all amino acids exhibit similar chemical reactivity. The side chains, however, exhibit specific chemical reactivities, depending on the nature of the functional groups. Whereas all of these reactivities are important in the study and analysis of isolated amino acids, it is the characteristic behaviour of the side chain that governs the reactivity of amino acids incorporated into proteins. There are three reasons to consider these reactivities. Proteins can be chemically modified in very specific ways by taking advantage of the chemical reactivity of certain amino acid side chains.

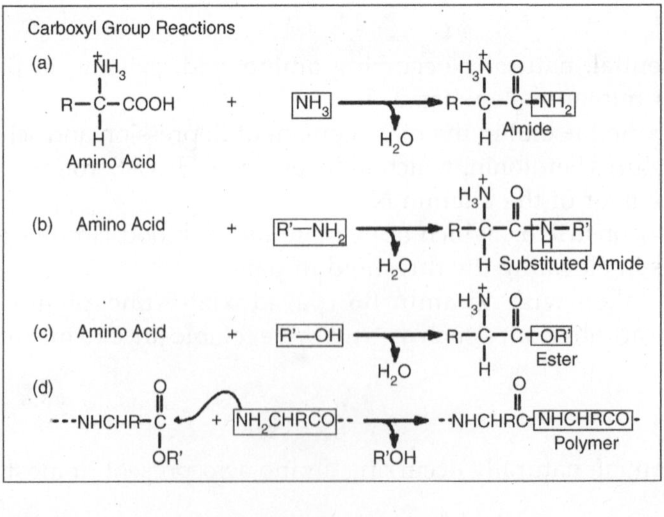

Carboxyl Group Reactions

(a) $R-C(H)(\overset{+}{N}H_3)-COOH$ (Amino Acid) + NH_3 → $R-C(H)(H_3\overset{+}{N})-C(=O)-NH_2$ (Amide) + H_2O

(b) Amino Acid + $R'-NH_2$ → $R-C(H)(H_3\overset{+}{N})-C(=O)-NH-R'$ (Substituted Amide) + H_2O

(c) Amino Acid + $R'-OH$ → $R-C(H)(H_3\overset{+}{N})-C(=O)-OR'$ (Ester) + H_2O

(d) ---NHCHR–C(=O)–OR' + NH_2CHRCO-- → --NHCHRC(=O)–NHCHRCO- (Polymer) + R'OH

Amino Group Reactions

(e) Amino Acid + $R'-CHO$ → ($-H_2O$) Schiff Base + H^+

(f) Amino Acid + $R'-CO-Cl$ → ($-HCl$) Substituted Amide + H^+

Fig. Typical Reactions of the Common Amino Acids

The detection and quantification of amino acids and proteins often depend on reactions that are specific to one or more amino acids and that result in colour, radioactivity, or some other quantity that can be easily measured. Finally and most importantly, the biological functions of proteins depend on the behaviour and reactivity of specific R groups.

The carboxyl groups of amino acids undergo all the simple reactions common to this functional group. Reaction with ammonia and primary amines yields unsubstituted and substituted amides, respectively (Figure). Esters and acid chlorides are also readily formed. Esterification proceeds in the presence of the appropriate alcohol and a strong acid. Polymerization can occur by repetition of the reaction shown in Figure. Free amino groups may react with aldehydes to form Schiff bases (Figure) and can be acylated with acid anhydrides and acid halides.

THE NINHYDRIN REACTION

Amino acids can be readily detected and quantified by reaction with ninhydrin. As shown in Figure, *ninhydrin*, or triketohydrindene hydrate, is a strong oxidizing agent and causes the oxidative deamination of the α–amino function. The products of the reaction are the resulting aldehyde, ammonia, carbon dioxide, and hydrindantin, a reduced derivative of ninhydrin. The ammonia produced in this way can react with the hydrindantin and another molecule of ninhydrin to yield a purple product (Ruhemann's Purple) that can be quantified spectrophotometrically at 570 nm. The appearance of CO_2 can also be monitored. Indeed, CO_2 evolution is diagnostic of the presence of an α–amino acid. α–Imino acids, such as proline and hydroxyproline, give bright yellow ninhydrin products with absorption maxima at 440 nm, allowing these to be distinguished from the α–amino acids. Because amino acids are one of the components of human skin secretions, the ninhydrin reaction was once used extensively by law enforcement and forensic personnel for fingerprint detection. (Fingerprints as old as 15 years can be successfully identified using the ninhydrin reaction.) More sensitive fluorescent reagents are now used routinely for this purpose.

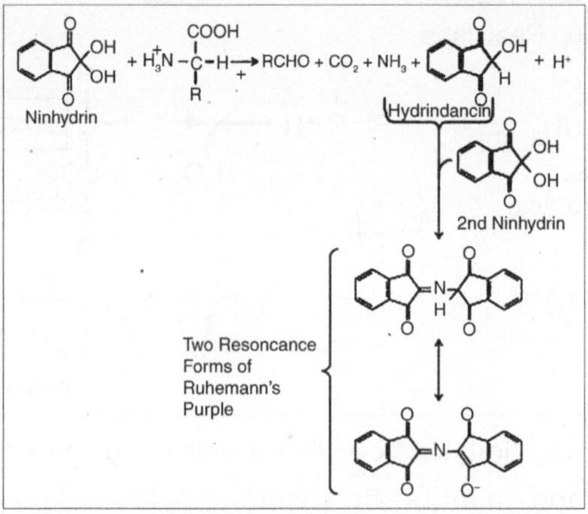

Fig. The Pathway of the Ninhydrin Reaction, which Produces a Coloured Product called "Ruhemann's Purple" that Absorbs Light at 570 nm

Note:

- That the reaction involves and consumes two molecules of ninhydrin.

SPECIFIC REACTIONS OF AMINO ACID SIDE CHAINS

A number of reactions of amino acids have become important in recent years because they are essential to the degradation, sequencing, and chemical synthesis of peptides and proteins. In recent years, biochemists have developed an arsenal of reactions that are relatively specific to the side chains of particular amino acids.

These reactions can be used to identify functional amino acids at the active sites of enzymes or to label proteins with appropriate reagents for further study. Cysteine residues in proteins, for example, react with one another to form disulfide species and also react with a number of reagents, including maleimides (typically *N*–ethylmaleimide), as shown in Figure.

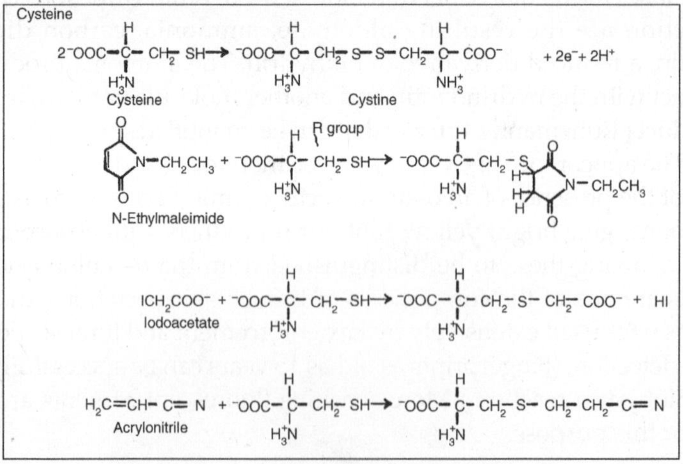

Fig. Reactions of Amino Acid Side–chain Functional Groups

Crysteines also react effectively with iodoacetic acid to yield *S*–carboxymethyl cysteine derivatives. There are numerous other reactions involving specialized reagents specific for particular side chain functional groups.

Figure presents a representative list of these reagents and the products that result. It is important to realise that few if any of these reactions are truly specific for one functional group; consequently, care must be exercised in their use.

AMINO ACIDS DEGRADATED TO OXALOACETATE

Aspartate and asparagines are both degraded into oxaloacetate. Asparagine is hydrolyzed into aspartate and ammonia by asparaginase.

Aspartate is converted into oxaloacetate by aspartate amino transferase which is a PLP enzyme that transfers an amino group from aspartate to ±"ketoglutarate to form glutamate and oxaloacetate.

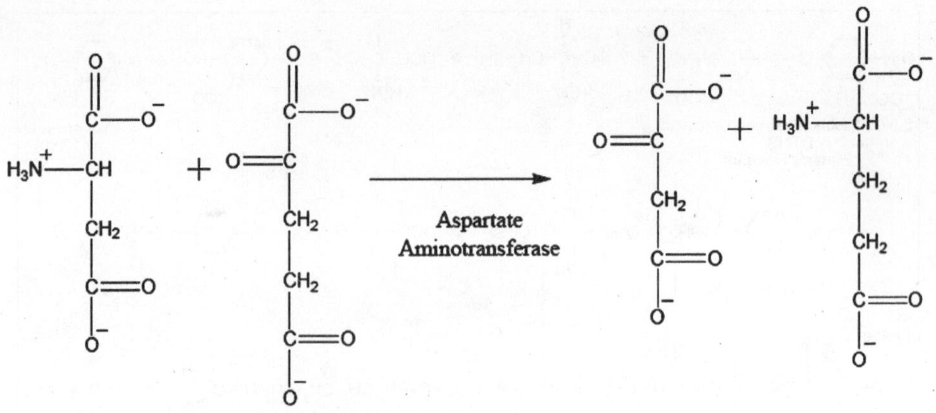

Aspartate Aminotransferase

Amino Acids Degraded to ±-Ketoglutarate.

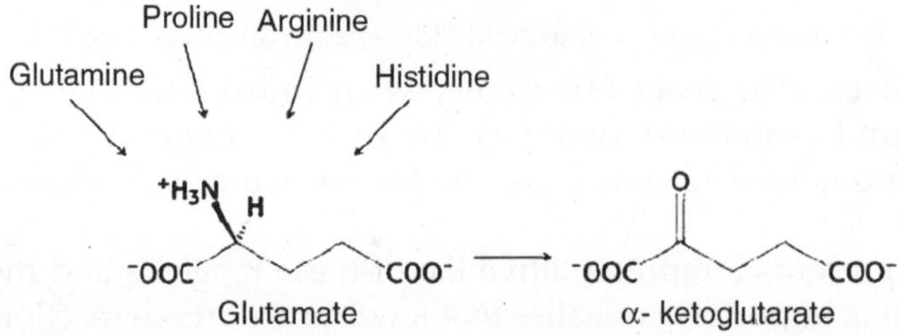

Proline Arginine Glutamine Histidine

Glutamate → α- ketoglutarate

Glutamine, proline, arginine and histidine are converted into glutamate which is then deaminated by a transaminase to form ±-ketoglutarate.

Glutamine is converted into glutamate by glutaminase. Proline is oxidized by proline oxidase to form pyrroline 5-carboxylate which spontaneously hydrolyzes to from glutamate ^{3}-semialdehyde. From the urea cycle we know that arginase converts arginine into ornithine and urea. Ornithine ′-aminotransferase transfers the ′-amino group of ornithine to ±-ketoglutarate to form glutamate ^{3}-semialdehyde and glutamate. Glutamate 3-semialdehyde is oxidized to form glutamate by glutamate 5-semialdehyde dehydrogenase. Histidine is deaminated by histidine ammonia lyase which forms urocanate. Urocanate hydratase adds water to form 4- Imidazolone -5-propionate which is hydrolyzed by imidazalone propionase to form Mformiminoglutamate. Glutamate formiminotransferase transfers the formimino group to tetrahydrofolate to generate glutamate and N^{5-}formimino-THF.

Branched Chain Amino Acids

The degradation of branched chain amino acids uses some of the enzymes we have already encountered in the citric acid cycle or ^{2}-oxidation c.

Most amino acid catabolism occurs in the liver. The branched chain amino acids are not catabolized in the liver. Branched chain amino acids are catabolized mainly in the muscle, adipose, kidney and the brain. The liver does not contain the branched amino acid aminotransferase enzyme which these other tissues contain. Branched chain ±-ketoacid dehydrogenase is a huge multienzyme complex homologous to pyruvate dehydrogenase and ±-ketoglutarate dehydrogenase.

This enzyme contains a thiamine pyrophosphate cofactor, a lipoamide cofactor, a FAD prosthetic group. The chemistry, mechanism and structure of these enzymes is very similar. Branched chain ±-ketoacid dehydrogenase is phosphorylated by a kinase which inactivates the enzyme in a similar manner that pyruvate dehydrogenase is phosphorylated and inactivated. The intake of dietary branched amino acids activates a phosphatase which activates this enzyme.

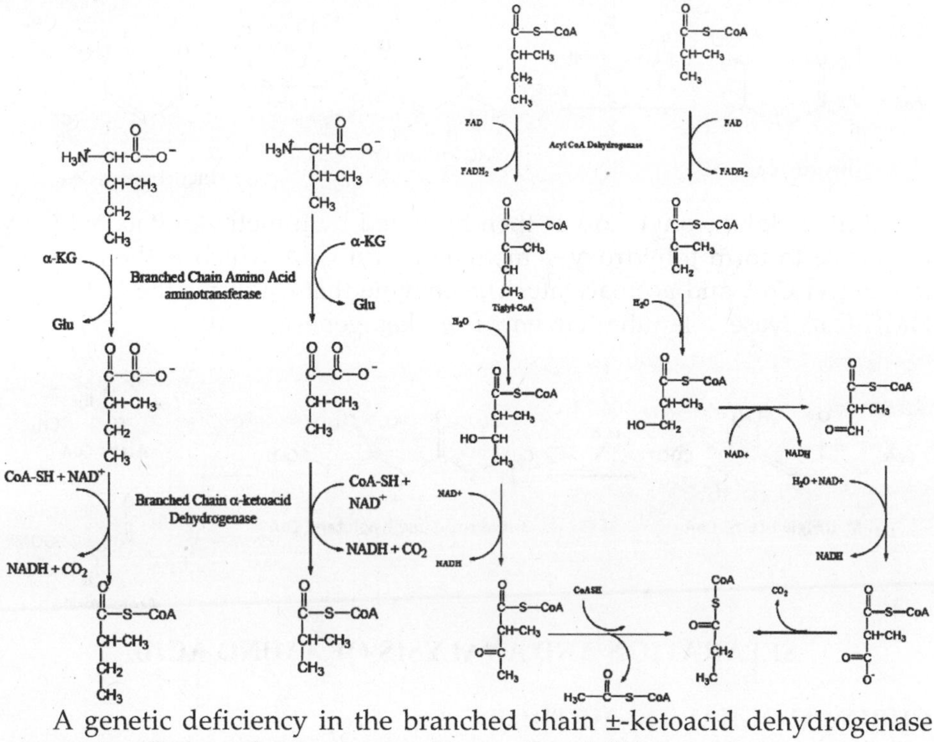

A genetic deficiency in the branched chain ±-ketoacid dehydrogenase enzyme is called maple syrup urine disease. The deficiency causes an excessive buildup of branched ±-ketoacids in the blood and the urine.

The urine of these patients has the odour of maple syrup and hence the name of the disease. Maple syrup disease usually leads to mental retardation unless the patient is placed on diet that is low in valine, isoleucine and leucine early in life.

Amino Acids that are Degraded into Acetyl CoA and Acetoacetate

There are only two amino acids that are purely ketogenic, lysine and leucine. Leucine catabolism is similar to the branched amino acids valine and isoleucine.

First leucine is transaminatedby branched amino acid aminotransferase to form ±-ketoisocaproate which is then oxidatively decarboxylated to form isovaleryl CoA by the branched chain ±-ketoacid dehydrogenase In the next step isovaleryl CoA is dehydrogenated to form β- methylcrotonyl CoA.

The enzyme that catalyzes this dehydrogenation is isovaleryl CoA dehydrogenase. β-methylcrotonyl CoA is then carboxylated by a biotin containing enzyme called methylcrotonyl CoA carboxylase to form β-methylglutaconyl CoA.

ATP + HCO_3^{2-} ; ADP + Pi ; FAD ; $FADH_2$

Isovaleryl CoA → β-Methylcrotonyl CoA → β-Methylglutaconyl CoA

β-methylglutaconyl CoA is then hydrated by β-methylglutaconyl CoA. hydratase to form β-hydroxy-²- methylglutaryl CoA which is then cleaved into acetyl CoA and acetoacetate. The enzyme that catalyzes the last step is HMG-CoA lyase, a familiar enzyme from ketogenesis.

H_2O

β-Methylglutaconyl CoA → 3-Hydroxy-3-methylglutaryl CoA → Acetyl CoA + Acetoacetate

SEPARATION AND ANALYSIS OF AMINO ACID

CHROMATOGRAPHIC METHODS

Chromatography is by far the most useful general group of techniques available for the separation of closely related compounds in a mixture. Here the separation is effected by differences in the equilibrium distribution of the components between two immiscible phases, *viz.*, the stationary and the mobile phases. These differences in the equilibrium distribution are a result of nature and degree of interaction of the components with these two phases. The stationary phase is a porous medium like silica or alumina, through which the sample mixture percolates under the influence of a moving solvent (the mobile phase). There are a number of interactions between the sample and the stationary phase and these have been well exploited to effect the separation of compounds.

The purification and analysis of individual amino acids from complex mixtures was once a very difficult process. Today, however, the biochemist has a wide variety of methods available for the separation and analysis of amino acids, or for that matter, any of the other biological molecules and macromolecules we encounter. All of these methods take advantage of the relative differences in the physical and chemical characteristics of amino acids, particularly ionization behaviour and solubility characteristics. The methods important for amino acids include separations based on partition properties (the tendency to associate with one solvent or phase over another) and separations based on electrical charge.

In all of the partition methods discussed here, the molecules of interest are allowed (or forced) to flow through a medium consisting of two phases—solid–liquid, liquid–liquid, or gas–liquid. In all of these methods, the molecules must show a preference for associating with one or the other phase. In this manner, the molecules partition, or distribute themselves, between the two phases in a manner based on their particular properties. The ratio of the concentrations of the amino acid (or other species) in the two phases is designated the *partition coefficient*.

In 1903, a separation technique based on repeated partitioning between phases was developed by Mikhail Tswett for the separation of plant pigments (carotenes and chlorophylls). Tswett, a Russian botanist, poured solutions of the pigments through columns of finely divided alumina and other solid media, allowing the pigments to partition between the liquid solvent and the solid support. Owing to the colourful nature of the pigments thus separated, Tswett called his technique chromatography. This term is now applied to a wide variety of separation methods, regardless of whether the products are coloured or not. The success of all chromatography techniques depends on the repeated microscopic partitioning of a solute mixture between the available phases.

The more frequently this partitioning can be made to occur within a given time span or over a given volume, the more efficient is the resulting separation. Chromatographic methods have advanced rapidly in recent years, due in part to the development of sophisticated new solid–phase materials. Methods important for amino acid separations include ion exchange chromatography, gas chromatography (GC), and high-performance liquid chromatography (HPLC).

ESSENTIAL AND NON-ESSENTIAL AMINO ACIDS

Salmon, trout and channel catfish fed diets devoid of arginine, histidine, isoleucine, leucine, lysine, methionine, phenylalanine, threonine, tryptophan or valine failed to grow. These same fish fed diets devoid of other L-amino acids grew as well as fish receiving all 18 amino acids tested. The nitrogen component in the test diets was made up of 18 L-amino acids in the pattern found in whole egg protein. All fish on test recovered rapidly when the missing amino acid was replaced in the diet. The slope of the growth curve of the recovery group was identical with that of fish receiving the complete amino acid test diet. Dispensable amino acids tested were alanine, aspartic acid, cystine, glutamic acid, glycine, proline, serine and tyrosine.

These amino acids were found to be not essential for the growth of salmon, trout and channel catfish. Quantitative studies on the requirements of the 10 indispensable amino acids used a casein-gelatin mixture supplemented with crystalline L-amino acids. The test diet had an amino acid pattern of 40 per cent whole egg protein for the nitrogen component. Experiments conducted with carp and eel showed a similar lack of growth when an indispensable amino acid was absent from the diet.

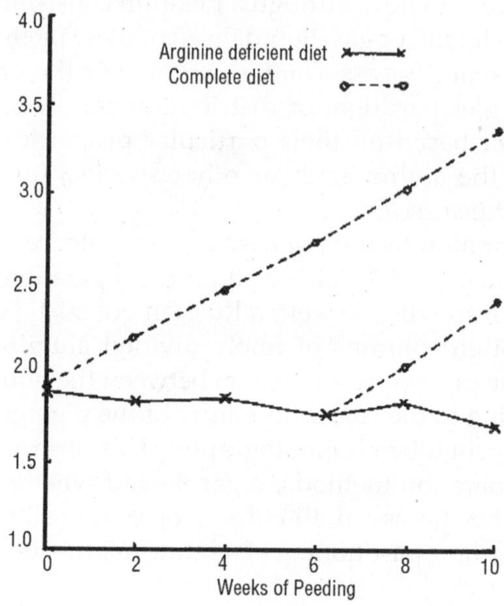

Fig. Growth of Arginine Deficient Fish. The Deficient Group was Divided After Six Weeks on the Deficient Diet and the Missing Amino Acid was Replaced in One of the Two Sub-lots

Essential Amino Acids and Protein Quality

If the essential amino acid requirements of fish are known, it should be possible to meet these needs in culture systems in a number of ways from different food proteins or combinations of food proteins. Phenylalanine is spared by tyrosine. It is not known to be chemically modified nor rendered unavailable by the harsh conditions to which feedstuff proteins are normally subjected during processing.

Measurement of phenylalanine in proteins is uncomplicated so that the provision and evaluation of phenylalanine in proteins in practical diets presents little difficulty. Lysine is a basic amino acid. In addition to the a -

amino acid group normally bound in peptide linkage, it also contains a second, a -amino group. This a -amino group must be free and reactive, otherwise the lysine, although chemically measurable, will not be biologically available. During the processing of feedstuff proteins the a -amino group of lysine may react with non-protein molecules present in the feedstuff to form additional compounds that render the lysine biologically unavailable.

Methionine is spared by cystine. However, measurement of the methionine content of feed proteins is not easy as the amino acid is subject to oxidation during processing. After processing, methionine may be present as such or as the sulphoxide or as the sulphone.

The sulphoxide may be formed from methionine during acid hydrolysis of the feed protein prior to measurement of its any-no acid composition. Acid hydrolysis of proteins before analysis disturbs the original equilibrium between the two compounds so that the composition of the hydrolysate no longer reflects that of the protein. In determining the methionine content of pure proteins, oxidation of the amino acid to methionine sulphone is normally quantitative. In the case of feed proteins, however, this will not reveal how much methionine or methionine sulphoxide was present in the protein prior to performate oxidation and hydrolysis.

Methionine sulphoxide may have some biological value for fish which may have some capability of reconverting it to methionine and thus partially make up for some of the methionine oxidized during processing. Methods have recently been reported for measurement of methionine in proteins using an iodoplatinate reagent before and after reduction with titanium trichloride, to give values for both methionine and the sulphoxide in the original protein. A method for measuring methionine specifically by cyanogen bromide cleavage has also been described. Both methods remain to be independently assessed. Microbiological assay of methionine in feed proteins is a valuable tool although there is the danger that oxides of methionine may differ in their activity for micro-organisms and misrepresent values.

SUPPLEMENTING DIETS WITH AMINO ACIDS

One solution to the use of proteins that are relatively deficient in one or more amino acids is to supplement the protein with appropriate amounts of the amino acid needed in practical diets. Fish appear to utilize free amino acids at various degrees of efficiency. Young carp, Cyprinus carpio, were shown to be unable to grow on diets in which the protein component (casein, gelatin) was replaced by a mixture of amino acids similar in overall composition. A trypsin hydrolyzate of casein was equally ineffective. However, if a diet containing free amino acids as the protein component is carefully neutralized with NaOH to pH 6.5-6.7 then some growth of young carp does occur. This growth was markedly inferior to that occurring on a comparable casein diet under the same conditions. Channel catfish are also unable to utilize free amino acids given as supplements to deficient proteins. When soybean meal was substituted isonitrogenously for menhaden meal, growth and feed efficiency of channel catfish were substantially reduced. Addition of free methionine, cystine or lysine, the most limiting amino acids, to these soy-substituted diets did not enhance weight gain. Raising the arginine level of catfish diets from 11 to 17 g/kg by isonitrogenous substitution of gelatin for casein enhanced weight gain significantly but the addition of free arginine, cystine, tryptophan or methionine to casein had little effect on growth or food conversion. Salmonids are able to utilize free amino acids for growth. A zein-gelatin diet supplemented with lysine and trytophan was shown to be markedly superior to an unsupplemented zein-gelatin diet for rainbow trout when weight gain and protein utilization were used as criteria.

Several investigators have demonstrated the potential of supplementing amino acid deficient proteins with limiting amino acids in diets for salmonids. Casein supplemented with six amino acids produced feed conversion ratios with Atlantic salmon similar to those obtained when an isolated fish protein was used as the dietary protein source. Soybean meal supplemented with five or more amino acids (including methionine and lysine) was a superior protein source to soybean meal alone for rainbow trout. Single additions of methionine and lysine did not, however, improve the value of soybean meal.

These results suggest that the amino acid spectrum of the isolated fish protein they used may possibly approximate the amino acid requirement of rainbow trout. The nutritional value of a soy protein isolate could be enhanced by supplementing it with the first limiting amino acid; *i.e.*, methionine. Diets containing, as protein component, fishmeal, meat and bone meal and yeast and soybean meal could be improved by supplementing with cystine (10 g/kg) and tryptophan (5 g/kg) together. Fishmeal can be entirely replaced without a reduction in food conversion rate in diets for rainbow trout by a mixture of poultry by-product meal and feather meal together with 17 g lysine HCL/kg, 4.8 g DL-methionine/kg and 1.44 g DL-tryptophan/kg.

STRUCTURE OF AMINO ACIDS

Proteins are polymers of amino acids.. Amino acids, with rare exception, contain an α carbon that is connected to an amino (NH_3) group, a carboxyl group (COOH) and a variable side group (R). The side group gives each amino acid its distinctive properties and helps to dictate the folding of the protein.

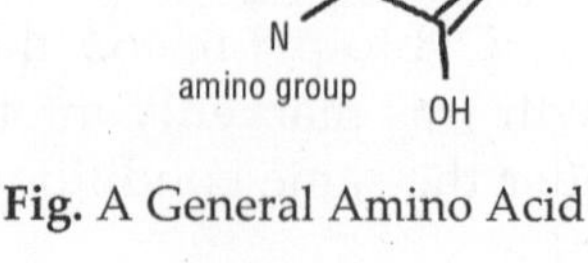

Fig. A General Amino Acid

Each amino acid contains a carboxyl group, an amino group and a variable side group (R). These all connect to a central carbon termed the α-carbon.

SPECTROSCOPIC PROPERTIES OF AMINO ACIDS

One of the most important and exciting advances in modern biochemistry has been the application of spectroscopic methods, which measure the absorption and emission of energy of different frequencies by molecules and atoms. Spectroscopic studies of proteins, nucleic acids, and other biomolecules are providing many new insights into the structure and dynamic processes in these molecules.

ULTRAVIOLET SPECTRA

Many details of the structure and chemistry of the amino acids have been elucidated or at least confirmed by spectroscopic measurements. None of the amino acids absorbs light in the visible region of the electromagnetic spectrum. Several of the amino acids, however, do absorbultraviolet radiation, and all absorb in the infrared region. The absorption of energy by electrons as they rise to higher energy states occurs in the ultraviolet/visible region of the energy spectrum. Only the aromatic amino acids phenylalanine, tyrosine, and tryptophan exhibit significant ultraviolet absorption above 250 nm, as shown in Figure. These strong absorptions can be used for spectroscopic determinations of protein concentration. The aromatic amino acids also exhibit relatively weak fluorescence, and it has recently been shown that tryptophan can exhibit *phosphorescence*—a relatively long-lived emission of light. These fluorescence and phosphorescence properties are especially useful in the study of protein structure and dynamics.

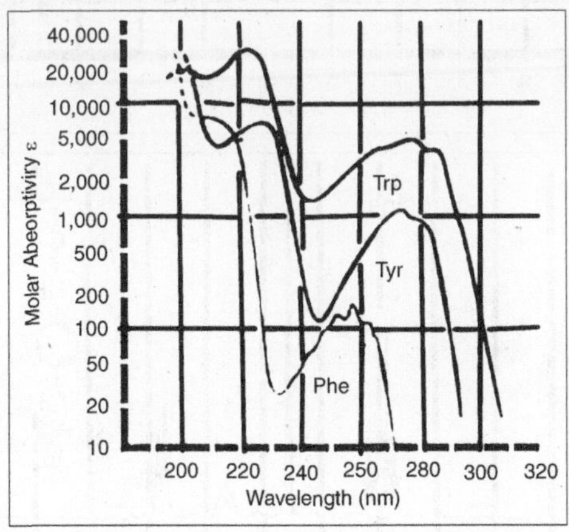

Fig. The Ultraviolet Absorption Spectra of the Aromatic Amino Acids at pH 6

NUCLEAR MAGNETIC RESONANCE SPECTRA

The development in the 1950s of nuclear magnetic resonance (NMR), a spectroscopic technique that involves the absorption of radio frequency energy by certain nuclei in the presence of a magnetic field, played an important part in the chemical characterization of amino acids and proteins. Several important principles rapidly emerged from these studies.

First, the chemical shift of amino acid protons depends on their particular chemical environment and thus on the state of ionization of the amino acid. Second, the change in electron density during a titration is transmitted throughout the carbon chain in the aliphatic amino acids and the aliphatic portions of aromatic amino acids, as evidenced by changes in the chemical shifts of relevant protons. Finally, the magnitude of the coupling constants between protons on adjacent carbons depends in some cases on the ionization state of the amino acid. This apparently reflects differences in the preferred conformations in different ionization states. Proton NMR spectra of two amino acids are shown in Figure. Because they are highly sensitive to their environment, the chemical shifts of individual NMR signals can detect the pH–dependent ionizations of amino acids.

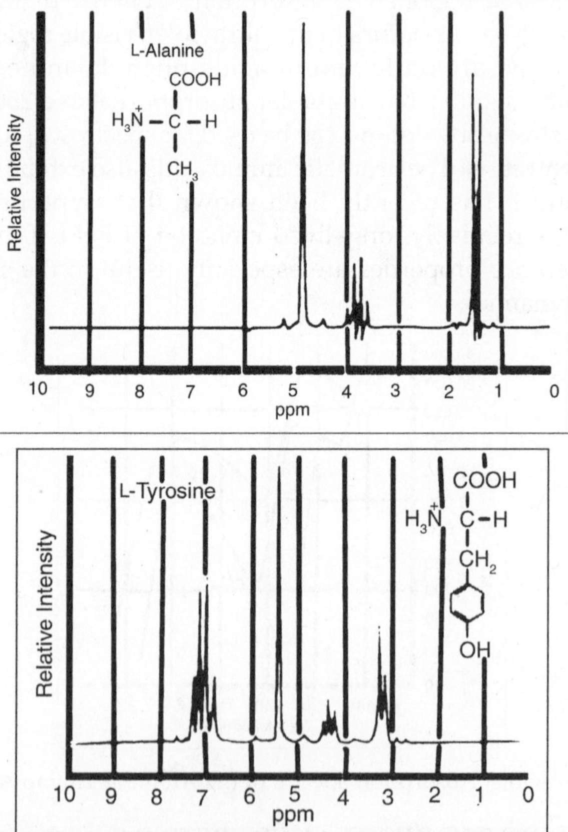

Fig. Proton NMR Spectra of Several Amino Acids. Zero on the Chemical Shift Scale is Defined by the Resonance of Tetramethylsilane (TMS)

Figure shows the ^{13}C chemical shifts occurring in a titration of lysine. Note that the chemical shifts of the carboxyl C, $C_{\alpha'}$, and C_α carbons of lysine are sensitive to dissociation of the nearby α–COOH and α–NH_3^+ protons (with p$K_a$values of about 2 and 9, respectively), whereas the C_δ and C_ε carbons are sensitive to dissociation of the ε–NH_3^+ group.

Such measurements have been very useful for studies of the ionization behaviour of amino acid residues in proteins. More sophisticated NMR measurements at very high magnetic fields are also used to determine the three–dimensional structures of peptides and even small proteins.

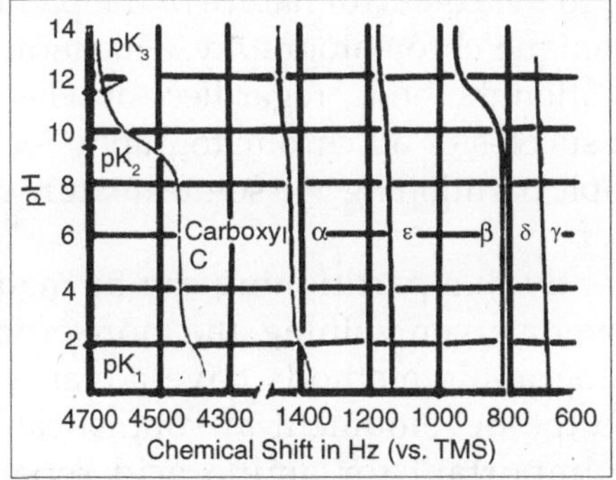

Fig. A Plot of Chemical Shifts Versus pH for the Carbons of Lysine

Changes in chemical shift are most pronounced for atoms near the titrating groups. Note the correspondence between the pK_a values and the particular chemical shift changes. All chemical shifts are defined relative to tetramethylsilane (TMS).

SEPARATION AND ANALYSIS OF AMINO ACID MIXTURES

CHROMATOGRAPHIC METHODS

The purification and analysis of individual amino acids from complex mixtures was once a very difficult process. Today, however, the biochemist has a wide variety of methods available for the separation and analysis of amino acids, or for that matter, any of the other biological molecules and macromolecules we encounter. All of these methods take advantage of the relative differences in the physical and chemical characteristics of amino acids, particularly ionization behaviour and solubility characteristics. The methods important for amino acids include separations based on partition properties (the tendency to associate with one solvent or phase over another) and separations based on electrical charge.

In all of the partition methods discussed here, the molecules of interest are allowed (or forced) to flow through a medium consisting of two phases—solid–liquid, liquid–liquid, or gas–liquid. In all of these methods, the molecules must show a preference for associating with one or the other phase. In this manner, the molecules partition, or distribute themselves, between the two phases in a manner based on their particular properties. The ratio of the concentrations of the amino acid (or other species) in the two phases is designated the *partition coefficient*.

In 1903, a separation technique based on repeated partitioning between phases was developed by Mikhail Tswett for the separation of plant pigments (carotenes and chlorophylls). Tswett, a Russian botanist, poured solutions of the pigments through columns of finely divided alumina and other solid media, allowing the pigments to partition between the liquid solvent and the solid support. Owing to the colourful nature of the pigments thus separated, Tswett called his technique chromatography. This term is now applied to a wide variety of separation methods, regardless of whether the products are coloured or not. The success of all chromatography techniques depends on the repeated microscopic partitioning of a solute mixture between the available phases.

The more frequently this partitioning can be made to occur within a given time span or over a given volume, the more efficient is the resulting separation. Chromatographic methods have advanced rapidly in recent years, due in part to the development of sophisticated new solid–phase materials. Methods important for amino acid separations include ion exchange chromatography, gas chromatography (GC), and high–performance liquid chromatography (HPLC).

6

Lipids

All Lipids are hydrophobic: that's the one property they have in common. This group of molecules includes fats and oils, waxes, phospholipids, steroids (like cholesterol), and some other related compounds.

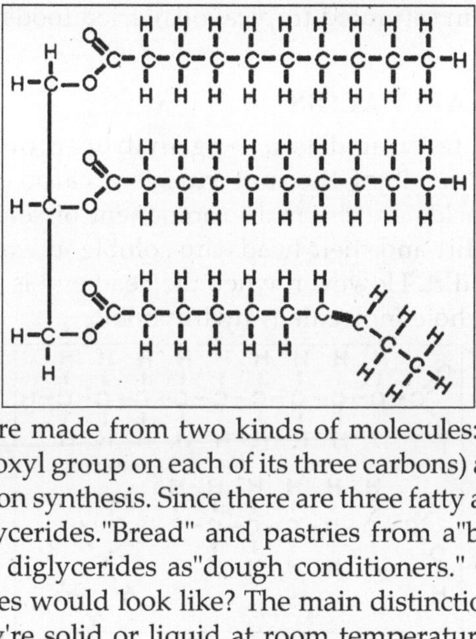

Fats and oils are made from two kinds of molecules: glycerol (a type of alcohol with a hydroxyl group on each of its three carbons) and three fatty acids joined by dehydration synthesis. Since there are three fatty acids attached, these are known as triglycerides."Bread" and pastries from a"bread factory" often contain mono- and diglycerides as"dough conditioners." Can you figure out what these molecules would look like? The main distinction between fats and oils is whether they're solid or liquid at room temperature, and this, as we'll soon see, is based on differences in the structures of the fatty acids they contain.

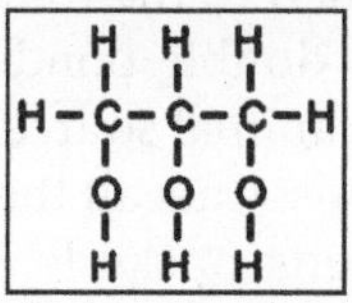

LIPID STRUCTURE

Lipids have different levels of saturation, a term that refers to the number of double bonds in the carbon chain. Fatty acids with no double bonds are saturated, those with one double bond are monounsaturated, and those with more than one double bond are polyunsaturated. Single bonds are stronger and less chemically reactive than double bonds, so the greater the number of double bonds, the greater the opportunity for the fatty acid to react with its chemical environment. It is this differential reactive capacity that makes the number of double bonds an important factor in human nutrition.

Saturated fatty acids are most prevalent in fats of animal origin, palm kernel oil, and coconut oil. Monounsaturated fats are highest in olive oil and canola oil but are also present in fats of animal origin. Polyunsaturated fats are highest in vegetable oils (with the exception of olive oil, which is more than 75 Per cent monounsaturated). In the context of a fat intake that does not exceed 35 Per cent of total calories, mono- and polyunsaturated fats should make up the majority of the fats consumed. Saturated fats are associated with higher cholesterol levels, so they should be minimized when possible. This is most easily achieved by reducing the consumption of animal fats, chocolate candies (often high in saturated tropical oils), fried foods, and high-fat dairy products.

STRUCTURE OF FATTY ACIDS

The"tail" of a fatty acid is a long hydrocarbon chain, making it hydrophobic. The"head" of the molecule is a carboxyl group which is hydrophilic. Fatty acids are the main component of soap, where their tails are soluble in oily dirt and their heads are soluble in water to emulsify and wash away the oily dirt. However, when the head end is attached to glycerol to form a fat, that whole molecule is hydrophobic.

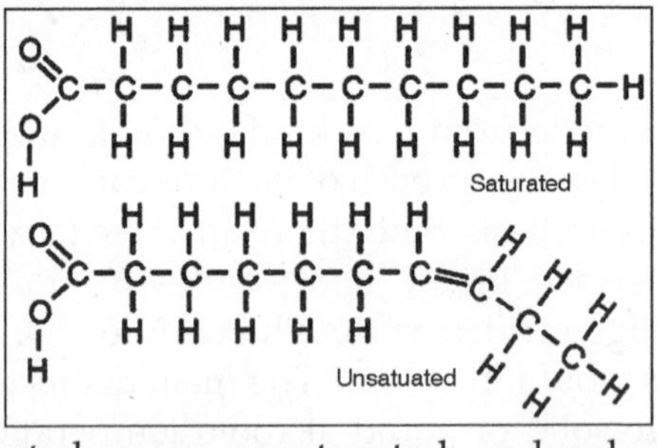

Saturated

Unsatuated

The terms saturated, mono-unsaturated, and poly-unsaturated refer to the number of hydrogens attached to the hydrocarbon tails of the fatty acids as compared to the number of double bonds between carbon atoms in the tail. Fats, which are mostly from animal sources, have all single bonds between the carbons in their fatty acid tails, thus all the carbons are also bonded to the maximum number of hydrogens possible. Since the fatty acids in these triglycerides contain the maximum possible amouunt of hydrogens, these would be calledsaturated fats.

The hydrocarbon chains in these fatty acids are, thus, fairly straight and can pack closely together, making these fats solid at room temperature. Oils, mostly from plant sources, have some double bonds between some of the carbons in the hydrocarbon tail, causing bends or"kinks" in the shape of the molecules. Because some of the carbons share double bonds, they're not bonded to as many hydrogens as they could if they weren't double bonded to each other. Therefore these oils are called unsaturated fats.

Because of the kinks in the hydrocarbon tails, unsaturated fats can't pack as closely together, making them liquid at room temperature. Many people have heard that the unsaturated fats are"healthier" than the saturated ones. Hydrogenated vegetable oil (as in shortening and commercial peanut butters where a solid consistency is sought) started out as"good" unsaturated oil. However, this commercial product has had all the double bonds artificially broken and hydrogens artificially added (in a chemistry lab-type setting) to turn it into saturated fat that bears no resemblance to the original oil from which it came (so it will be solid at room temperature).

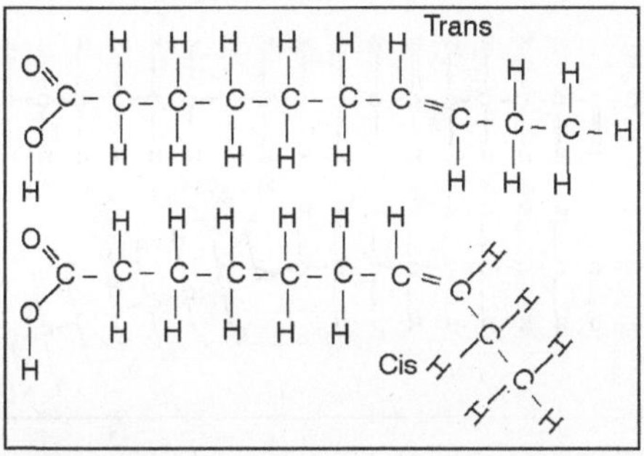

In unsaturated fatty acids, there are two ways the pieces of the hydrocarbon tail can be arranged around a C=C double bond. In cis bonds, the two pieces of the carbon chain on either side of the double bond are either both"up" or both"down," such that both are on the same side of the molecule. In trans bonds, the two pieces of the molecule are on opposite sides of the double bond, that is, one"up" and one"down" across from each other.

Naturally-occurring unsaturated vegetable oils have almost all cis bonds, but using oil for frying causes some of the cis bonds to convert to trans bonds. If oil is used only once like when you fry an egg, only a few of the bonds do this so it's not too bad. However, if oil is constantly reused, like in fast food French fry machines, more and more of the cis bonds are changed to trans until significant numbers of fatty acids with trans bonds build up. The reason this is of concern is that fatty acids with trans bonds are carcinogenic, or cancer-causing.

The levels of trans fatty acids in highly-processed, lipid-containing products such as margarine are quite high, and I have heard that the government is considering requiring that the amounts of trans fatty acids in such products be listed on the labels.

We need fats in our bodies and in our diet. Animals in general use fat for energy storage because fat stores 9 KCal/g of energy. Plants, which don't move around, can afford to store food for energy in a less compact but more easily accessible form, so they use starch (a carbohydrate, NOT A LIPID) for energy storage. Carbohydrates and proteins store only 4 KCal/g of energy, so fat stores over twice as much energy/gram as fat. By the way, this is also related to the idea behind some of the high-carbohydrate weight loss diets. The human body burns carbohydrates and fats for fuel in a given proportion to each other. The theory behind these diets is that if they supply carbohydrates but not fats, then it is hoped that the fat needed to balance with the sugar will be taken from the dieter's body stores. Fat is also is used in our bodies to a) cushion vital organs like the kidneys and b) serve as insulation, especially just beneath the skin.

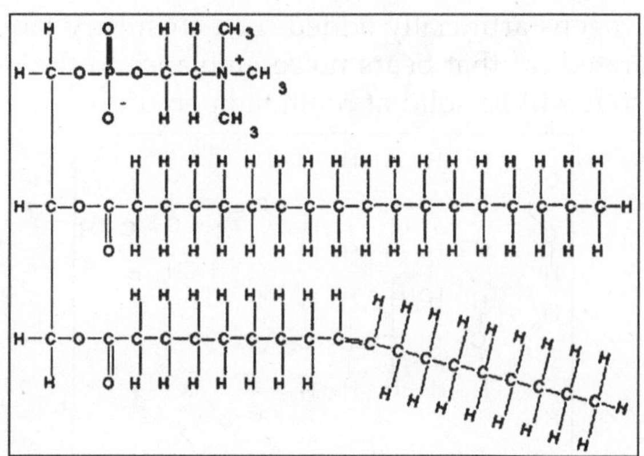

PHOSPHOLIPIDS

Phospholipids are made from glycerol, two fatty acids, and (in place of the third fatty acid) a phosphate group with some other molecule attached to its other end. The hydrocarbon tails of the fatty acids are still hydrophobic, but the phosphate group end of the molecule is hydrophilic because of the oxygens with all of their pairs of unshared electrons. This means that phospholipids are soluble in both water and oil.

An emulsifying agent is a substance which is soluble in both oil and water, thus enabling the two to mix. A"famous" phospholipid is lecithin which is found in egg yolk and soybeans. Egg yolk is mostly water but has a lot of lipids, especially cholesterol, which are needed by the developing chick. Lecithin is used to emulsify the lipids and hold them in the water as an emulsion. Lecithin is the basis of the classic emulsion known as mayonnaise.

Our cell membranes are made mostly of phospholipids arranged in a double layer with the tails from both layers"inside" (facing towards each other) and the heads facing"out" (towards the watery environment) on both surfaces.

STEROIDS

The general structure of cholesterol consists of two six-membered rings side-by-side and sharing one side in common, a third six-membered ring off the top corner of the right ring, and a five-membered ring attached to the right side of that. The central core of this molecule, consisting of four fused rings, is shared by all steroids, including estrogen (estradiol), progesterone, corticosteroids such as cortisol (cortisone), aldosterone, testosterone, and Vitamin D. In the various types of steroids, various other groups/molecules are attached around the edges. Know how to draw the four rings that make up the central structure. Cholesterol is not a"bad guy!" Our bodies make about 2 g of cholesterol per day, and that makes up about 85 per cent of blood cholesterol, while only about 15 per cent comes from dietary sources. Cholesterol is the precursor to our sex hormones and Vitamin D. Vitamin D is formed by the action of UV light in sunlight on cholesterol molecules that have"risen" to near the surface of the skin. At least one source I read suggested that people not shower immediately after being in the sun, but wait at least ½ hour for the new Vitamin D to be absorbed deeper into the skin. Our cell membranes contain a lot of cholesterol (in between the phospholipids) to help keep them"fluid" even when our cells are exposed to cooler temperatures.

Many people have hear the claims that egg yolk contains too much cholesterol, thus should not be eaten. An interesting study was done at Purdue University a number of years ago to test this. Men in one group each ate an egg a day, while men in another group were not allowed to eat eggs. Each of these groups was further subdivided such that half the men got"lots" of exercise while the other half were"couch potatoes." The results of this experiment showed no significant difference in blood cholesterol levels between egg-eaters and non-egg-eaters while there was a very significant difference between the men who got exercise and those who didn't.

Lipoproteins are clusters of proteins and lipids all tangled up together. These act as a means of carrying lipids, including cholesterol, around in our blood. There are two main categories of lipoproteins distinguished by how compact/dense they are. LDL or low density lipoprotein is the"bad guy," being associated with deposition of"cholesterol" on the walls of someone's arteries. HDL or high density lipoprotein is the"good guy," being associated with carrying"cholesterol" out of the blood system, and is more dense/more compact than LDL.

Triglycerides

The majority of consumed lipids are triglycerides, which contain three fatty acids and a glycerol molecule. Fat is stored in the form of triglycerides, which we manufacture when excess energy is consumed. We store triglycerides in adipose tissues (groups of fat cells) and inside muscle cells (intramuscular tri glyceride), both of which are available as an energy source when needed. When fat is burned as a source of energy, the stored triglycerides are taken out of storage, and each molecule is cleaved into its component fatty acids and glycerol molecule. Each fatty acid can then be broken apart (two carbon units at a time) and thrown into the cellular furnaces for the creation of ATP to form heat and provide the energy for muscular work. This process is referred to as the beta-oxidative metabolic pathway because burning fat, besides requiring some carbohydrate for its complete oxidation, also requires oxygen.

Glycerol is a unique lipid that is burned like a carbohydrate rather than a fat and is also an effective humectant (it holds water). Some long-endurance athletes find that adding glycerol to water helps them retain more water (*i.e.,* to superhydrate) than if they consumed water alone. In an extremely hot and humid environment, water loss is likely to be higher than the athlete's fluid-replacement capacity, so beginning a competition in a superhydrated state may provide some advantages.

In studies of tennis and Olympic distance triathlon, athletes consuming glycerol in water prior to competition experienced some protective hyperhydration benefits when exercising in high heat.17,18 However, being superhydrated is likely to impart a degree of discomfort that requires some adaptation. Athletes who consume fluids containing glycerol often describe the feeling of holding extra body water as making them feel"like a water bag," or"heavy," or"stiff." Still, they maintain that how they feel at the end of a race is more important than how they feel at the beginning, so adding glycerol to precompetition fluids has become a standard protocol for some athletes.

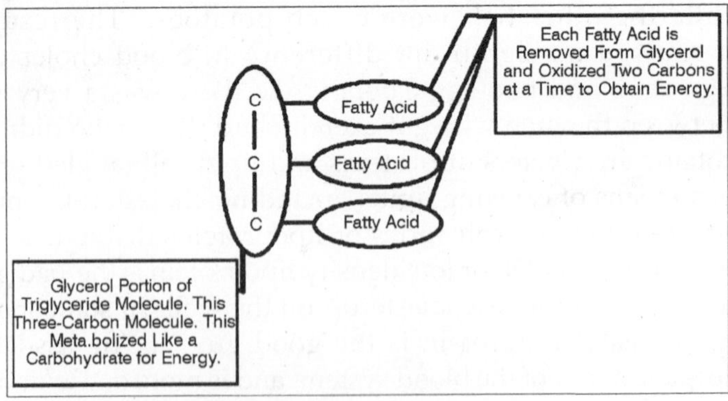

LIPID METABOLISM

Fats (lipids) are an important source of energy for the body. The body's store of fat is constantly broken down and reassembled to balance the body's energy needs with the food available. Groups of specific enzymes help the body break down and process fats. Certain abnormalities in these enzymes can lead to the buildup of specific fatty substances that normally would have been broken down by the enzymes. Over time, accumulations of these substances can be harmful to many organs of the body. Disorders caused by the accumulation of lipids are called lipidoses. Other enzyme abnormalities prevent the body from converting fats into energy normally. These abnormalities are called fatty acid oxidation disorders.

GAUCHER'S DISEASE

Gaucher's disease is caused by a buildup of glucocerebrosides in tissues. Children who have the infantile form usually die within a year, but children and adults who develop the disease later in life may survive for many years.

In Gaucher's disease, glucocerebrosides, which are a product of fat metabolism, accumulate in tissues. Gaucher's disease is the most common lipidosis. The disease is most common among Ashkenazi (Eastern European) Jews. Gaucher's disease leads to an enlarged liver and spleen and a brownish pigmentation of the skin. Accumulations of glucocerebrosides in the eyes cause yellow spots called pingueculae to appear. Accumulations in the bone marrow can cause pain and destroy bone.

RARE HEREDITARY DISORDERS OF LIPID METABOLISM

Wolman's disease results when specific types of cholesterol and glycerides accumulate in tissues. This disease causes enlargement of the spleen and liver. Calcium deposits in the adrenal glands cause them to harden, and fatty diarrhea (steatorrhea) also occurs. Infants with Wolman's disease usually die by 6 months of age.

Cerebrotendinous xanthomatosis occurs when cholestanol, a product of cholesterol metabolism, accumulates in tissues. This disease eventually leads to uncoordinated movements, dementia, cataracts, and fatty growths (xanthomas) on tendons. The disabling symptoms often appear after age 30. If started early, the drug chenodiol helps prevent progression of the disease, but it cannot undo any damage already done.

In sitosterolemia, fats from fruits and vegetables accumulate in blood and tissues. The buildup of fats leads to atherosclerosis, abnormal red blood cells, and xanthomas on tendons. Treatment consists of reducing the intake of foods that are rich in plant fats, such as vegetable oils, and takingcholestyramine resin.

In Refsum's disease, phytanic acid, which is a product of fat metabolism, accumulates in tissues. A buildup of phytanic acid leads to nerve and retinal damage, spastic movements, and changes in the bone and skin. Treatment involves avoiding eating green fruits and vegetables that contain chlorophyll. Plasmapheresis, in which phytanic acid is removed from the blood, may be helpful.

- Type 1, the chronic form of Gaucher's disease, is the most common. It results in an enlarged liver and spleen and bone abnormalities. Most commonly diagnosed during adulthood, type 1 Gaucher's disease may lead to severe liver disease, including increased risk of bleeding from the stomach and esophagus and liver cancer. Nervous system problems can also occur.
- Type 2, the infantile form, usually causes death in the first year of life. Affected infants have an enlarged spleen and severe nervous system problems.
- Type 3, the juvenile form, can begin at any time during childhood. Children with type 3 disease have an enlarged liver and spleen, bone abnormalities, and slowly progressive nervous system problems. Children who survive to adolescence may live for many years.

Many people with Gaucher's disease can be treated with enzyme replacement therapy, in which enzymes are given by vein, usually every 2 weeks. Enzyme replacement therapy is most effective for people who do not have nervous system complications.

TAY-SACHS DISEASE

Tay-Sachs disease is caused by a buildup of gangliosides in the tissues. This disease results in early death.

In Tay-Sachs disease, gangliosides, which are products of fat metabolism, accumulate in tissues. The disease is most common among families of Eastern European Jewish origin. At a very early age, children with this disease become progressively intellectually disabled and appear to have floppy muscle tone. Spasticity develops and is followed by paralysis, dementia, and blindness. These children usually die by age 3 or 4. The disease cannot be treated or cured. Before conception, parents can find out whether they carry the gene that causes the disease. During pregnancy, Tay-Sachs disease can be identified in the fetus by chorionic villus sampling or amniocentesis.

NIEMANN-PICK DISEASE

Niemann-Pick disease is caused by a buildup of sphingomyelin or cholesterol in the tissues. This disease causes many neurologic problems.

In Niemann-Pick disease, the deficiency of a specific enzyme results in the accumulation of sphingomyelin (a product of fat metabolism) or cholesterol. Niemann-Pick disease has several forms, depending on the severity of the enzyme deficiency, which determines how much sphingomyelin or cholesterol accumulates. The most severe forms tend to occur in Jewish people. The milder forms occur in all ethnic groups. In the most severe form (type A), children fail to grow normally and have several neurologic problems. These children usually die by age 3. Children with type B disease develop fatty growths in the skin, areas of dark pigmentation, and an enlarged liver, spleen, and lymph nodes. They may be intellectually disabled. Children with type C disease develop symptoms during childhood, with seizures and neurologic deterioration.

Some forms of Niemann-Pick disease can be diagnosed in the fetus by chorionic villus sampling or amniocentesis. After birth, the diagnosis can be made by a liver biopsy (removal of a tissue specimen for examination under a microscope). None of the types of Niemann-Pick disease can be cured, and children tend to die of infection or progressive dysfunction of the central nervous system. Currently, some therapies that may slow or halt the progression of symptoms in types B and C are being studied.

FABRY'S DISEASE

Fabry's disease is caused by a buildup of glycolipid in tissues. This disease causes skin growths, pain in the extremities, poor vision, recurrent episodes of fever, and kidney or heart failure.

In Fabry's disease, glycolipid, which is a product of fat metabolism, accumulates in tissues. Because the defective gene for this rare disorder is carried on the X chromosome, the full-blown disease occurs only in males (Genetics: X-Linked Inheritance). The accumulation of glycolipid causes noncancerous (benign) skin growths (angiokeratomas) to form on the lower part of the trunk. The corneas become cloudy, resulting in poor vision. A burning pain may develop in the arms and legs, and children may have episodes of fever. Children with Fabry's disease eventually develop kidney failure and heart disease, although most often, they live into adulthood.

Kidney failure may lead to high blood pressure, which may result in stroke. Fabry's disease can be diagnosed in the fetus by chorionic villus sampling or amniocentesis. The disease cannot be cured or even treated directly, but researchers are investigating a treatment in which the deficient enzyme is replaced by transfusion. Treatment consists of taking analgesics to help relieve pain and fever or anticonvulsants. People with kidney failure may need a kidney transplant.

LIPIDS AND PHYSICAL ACTIVITY

Even the leanest, healthiest athletes have a substantial energy pool of stored lipids. The average storage in adipose tissue ranges between 50,000 and 100,000 calories, or enough energy to walk or run 500 to 1,000 miles19 (800 to 1,600 kilometres) without a refueling stop. In addition, athletes store approximately 2,000 to 3,000 calories of lipids inside the muscle tissue.20 These lipids, which are stored in the form of triglycerides, are available as a fuel under the proper conditions of oxygen availability. Maximal fat oxidation occurs at 60 to 65 Per cent VO_2 max, but at higher levels of VO2max there is insufficient oxygen to derive the majority of energy used from fat catabolism.

Triglycerides stored in adipose tissue are cleaved into their component molecules of glycerol and fatty acids and transported to the blood plasma. The glycerol is available to all tissues for energy metabolism, and the free fatty acids are transported to working muscles where they are oxidized for energy.

The triglycerides stored in working muscles are cleaved into glycerol and fatty acids, and the fatty acids can be oxidized for energy where they are resident. The glycerol can also be burned for energy in the working muscle or can be transported to the blood plasma as a source of energy for other tissues.

The lower the exercise intensity, the greater the proportion of fat burned to satisfy energy needs. As exercise intensity increases, the proportion of fat burned decreases and the proportion of carbohydrate burned increases. It is this basic reality that is behind why so many people do low-intensity activity to burn fat and lower body fat levels. However, the proportion of fat burned should not be confused by the total amount of fat burned at different intensities of physical activity.

As exercise intensity increases, the total number of calories burned per unit of time also increases. Although there may be a decrease in the proportion of fat burned to satisfy total energy needs in higher-intensity activity, the total volume of fat burned is greater because the total energy requirement is higher.

The take-away message from this metabolic reality is that athletes interested in lowering body fat should exercise at least as high as 65 Per cent of VO2 max for the duration of their workouts to optimize the total mass of fat that is burned. Exercising at lower intensities burns a greater proportion of fat but less total fat than exercising at higher intensities.

BIOSYNTHESIS OF FATTY ACID

Identification of Genes in Biosynthesis of Fatty Acids and Cholesterol Pathways Using Microarrays Sterol regulatory binding proteins (SREBP) are a family of basic helix-loop-helix leucine zipper transcription factors that regulate biosynthesis of fatty acids and cholesterol. To identify candidate genes in the fatty acid synthesis pathway, transgenic mice over expressing SREBP-1a and SREBP-2 isofroms in liver were used. Labeled cRNA was prepared from liver and hybridized to mouse expression microarrays interrogating ~30,000 genes/ESTs. Some of the known genes, ACC, FAS, and ATP citrate lyase regulated by over expression of SREBP were identified by microarray analysis.

A novel gene, long chain fatty acyl elongase (LCE) regulated by the SREBP1a and 2 was also identified. LCE was over expressed with a fold change of 25 and 9 in SREBP-1a and SREBP-2 mice respectively. The confirmation of LCE was done by northern blot analysis, sequence homology searches and biochemical tests. This enzyme specifically catalyzes the elongation of 12, 14 and 16 carbons but does not elongate the C-18 fatty acids or very long chain fatty acids.

DEGRADATION OF FATTY ACIDS

Multi-million pounds of viscera are generated from the catfish filleting process. Our previous study showed that catfish viscera contained 30-35 per cent crude fat, and catfish visceral oil contained health-promoting fatty acids. Catfish visceral oils show potential for use as nutrient supplement and as cooking oil. Information on thermal stability and degradation of catfish visceral oils is not available.

The objective of this study was to determine weight loss (per cent) of catfish visceral oils and fatty acids after having been subjected to heat treatment. Oil samples included crude, degummed, neutralized, bleached, and deodorized oils.

The oils and individual fatty acids, FAs (C14:0, C16:0, C18:0, C20:0, C16:1, C18:1, C18:2, C18:3, C20:2, C20:4, C22:6) were subjected to the Thermogravimetric Analyser. Samples (0.5-1 mg) was placed in the furnace and heated up to 600°C at the rate of 5°C/min. Weight loss (per cent) of oils and fatty acids were calculated based on the original weight of non-heated samples. Duplicate experiments were conducted.

Over 90 per cent weight loss of all FAs (except C18:1 and C20:2) was observed at 250°C. Less than 2 per cent remaining weight of FAs (except C20:2) was observed at 350°C. Saturated FAs completely decomposed above 400°C. Comparing C18:1 with C18:2 and C18:3, weight loss increased with increased double bonds. Between 200-550°C, weight loss of oils slightly increased with increased temperature, regardless of the purification process. There was, however, no distinct weight loss of all oils below 300°C. Oils with less impurity had greater weight loss especially at 450-500°C. At 550°C, all oil samples completely decomposed.

Knowledge of thermal stability or degradation (weight loss through eVapouration or decomposition) of catfish oils during purification or cooking is essential for the design of processing units, analysis of production cost, and cooking quality if to be used as cooking or seasoning oil.

LIPOPROTEINS

NTESTINAL UPTAKE OF LIPIDS

In order for the body to make use of dietary lipids, they must first be absorbed from the small intestine. Since these molecules are oils, they are essentially insoluble in the aqueous environment of the intestine. The solubilization (or emulsification) of dietary lipids is therefore accomplished by means of bile salts, which are synthesized from cholesterol in the liver and then stored in the gallbladder; they are secreted following the ingestion of fat.

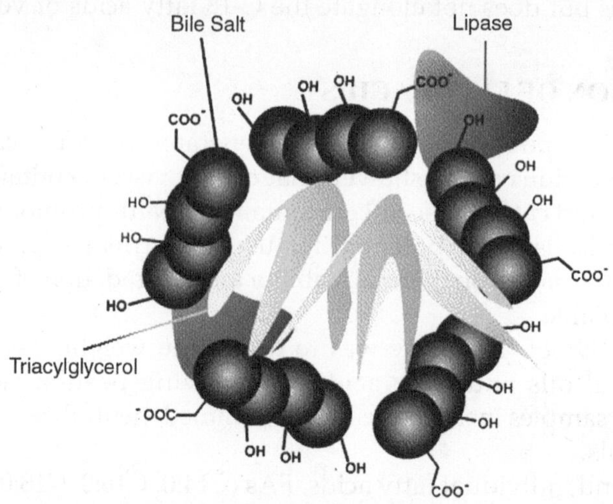

Mixed micelle formed by bile salts, triacylglycerols and pancreatic lipase.

The emulsification of dietary fats renders them accessible to pancreatic lipases (primarily lipase and phospholipase A2). These enzymes, secreted into the intestine from the pancreas, generate free fatty acids and a mixtures of mono- and diacylglycerols from dietary triacylglycerols. Pancreatic lipase degrades triacylglycerols at the 1 and 3 positions sequentially to generate 1,2-diacylglycerols and 2-acylglycerols. Phospholipids are degraded at the 2 position by pancreatic phospholipase A2 releasing a free fatty acid and the lysophospholipid. The products of pancreatic lipases then diffuse into the intestinal epithelial cells, where the re-synthesis of triacyglycerols occurs.

Dietary triacylglycerols and cholesterol, as well as triacylglycerols and cholesterol synthesized by the liver, are solubilized in lipid-protein complexes. These complexes contain triacylglycerol lipid droplets and cholesteryl esters surrounded by the polar phospholipids and proteins identified as apolipoproteins. These lipid-protein complexes vary in their content of lipid and protein.

INTERMEDIATE DENSITY LIPOPROTEINS, IDLS

IDLs are formed as triacylglycerols are removed from VLDLs. The fate of IDLs is either conversion to LDLs or direct uptake by the liver.

Conversion of IDLs to LDLs occurs as more triacylglycerols are removed. The liver takes up IDLs after they have interacted with the LDL receptor to form a complex, which is endocytosed by the cell. For LDL receptors in the liver to recognize IDLs requires the presence of both apo-B-100 and apo-E (the LDL receptor is also called the apo-B-100/apo-E receptor). The importance of apo-E in cholesterol uptake by LDL receptors has been demonstrated in transgenic mice lacking functional apo-E genes. These mice develop severe atherosclerotic lesions at 10 weeks of age.

LOW DENSITY LIPOPROTEINS, LDLS

The cellular requirement for cholesterol as a membrane component is satisfied in one of two way: either it is synthesized de novo within the cell, or it is supplied from extra-cellular sources, namely, chylomicrons and LDLs. As indicated above, the dietary cholesterol that goes into chylomicrons is supplied to the liver by the interaction of chylomicron remnants with the remnant receptor.

In addition, cholesterol synthesized by the liver can be transported to extra-hepatic tissues if packaged in VLDLs. In the circulation VLDLs are converted to LDLs through the action of lipoprotein lipase. LDLs are the primary plasma carriers of cholesterol for delivery to all tissues.

The exclusive apolipoprotein of LDLs is apo-B-100. LDLs are taken up by cells via LDL receptor-mediated endocytosis, as described above for IDL uptake. The uptake of LDLs occurs predominantly in liver (75/ per cent), adrenals and adipose tissue. As with IDLs, the interaction of LDLs with LDL receptors requires the presence of apo-B-100. The endocytosed membrane vesicles (endosomes) fuse with lysosomes, in which the apoproteins are degraded and the cholesterol esters are hydrolyzed to yield free cholesterol. The cholesterol is then incorporated into the plasma membranes as necessary. Excess intracellular cholesterol is re-esterified by acyl-CoA-cholesterol acyltransferase (ACAT), for intracellular storage. The activity of ACAT is enhanced by the presence of intracellular cholesterol.

Insulin and tri-iodothyronine (T3) increase the binding of LDLs to liver cells, whereas glucocorticoids (*e.g.*, dexamethasone) have the opposite effect. The precise mechanism for these effects is unclear but may be mediated through the regulation of apo-B degradation. The effects of insulin and T3 on hepatic LDL binding may explain the hypercholesterolemia and increased risk of athersclerosis that have been shown to be associated with uncontrolled diabetes or hypothyroidism.

An abnormal form of LDL, identified as lipoprotein-X (Lp-X), predominates in the circulation of patients suffering from lecithin-cholesterol acyl transferase (LCAT, see HDL discussion for LCAT function) deficiency or cholestatic liver disease. In both cases there is an elevation in the level of circulating free cholesterol and phospholipids.

HIGH DENSITY LIPOPROTEINS, HDLS

HDLs are synthesized de novo in the liver and small intestine, as primarily protein-rich disc-shaped particles. These newly formed HDLs are nearly devoid of any cholesterol and cholesteryl esters. The primary apoproteins of

HDLs are apo-A-I, apo-C-I, apo-C-II and apo-E. In fact, a major function of HDLs is to act as circulating stores of apo-C-I, apo-C-II and apo-E.

HDLs are converted into spherical lipoprotein particles through the accumulation of cholesteryl esters. This accumulation converts nascent HDLs to HDL2 and HDL3. Any free cholesterol present in chylomicron remnants and VLDL remnants (IDLs) can be esterified through the action of the HDL-associated enzyme, lecithin-cholesterol acyl transferase, LCAT. LCAT is synthesized in the liver and so named because it transfers a fatty acid from the C-2 position of lecithin to the C-3-OH of cholesterol, generating a cholesteryl ester and lysolecithin. The activity of LCAT requires interaction with apo-A-I, which is found on the surface of HDLs.

Cholesterol-rich HDLs return to the liver, where they are endocytosed. Hepatic uptake of HDLs, or reverse cholesterol transport, may be mediated through an HDL-specific apo-A-I receptor or through lipid-lipid interactions. Macrophages also take up HDLs through apo-A-I receptor interaction. HDLs can then acquire cholesterol and apo-E from the macrophages; cholesterol-enriched HDLs are then secreted from the macrophages. The added apo-E in these HDLs leads to an increase in their uptake and catabolism by the liver.

HDLs also acquire cholesterol by extracting it from cell surface membranes. This process has the effect of lowering the level of intracellular cholesterol, since the cholesterol stored within cells as cholesteryl esters will be mobilized to replace the cholesterol removed from the plasma membrane.

The cholesterol esters of HDLs can also be transferred to VLDLs and LDLs through the action of the HDL-associated enzyme, cholesterol ester transfer protein (CETP, also identified as apo-D). This has the added effect of allowing the excess cellular cholesterol to be returned to the liver through the LDL-receptor pathway as well as the HDL-receptor pathway.

LDL RECEPTORS

LDLs are the principal plasma carriers of cholesterol delivering cholesterol from the liver (via hepatic synthesis of VLDLs) to peripheral tissues, primarily the adrenals and adipose tissue. LDLs also return cholesterol to the liver. The cellular uptake of cholesterol from LDLs occurs following the interaction of LDLs with the LDL receptor (also called the apo-B-100/apo-E receptor). The sole apoprotein present in LDLs is apo-B-100, which is required for interaction with the LDL receptor.

The LDL receptor is a polypeptide of 839 amino acids that spans the plasma membrane. An extracellular domain is responsible for apo-B-100/apo-E binding. The intracellular domain is responsible for the clustering of LDL receptors into regions of the plasma membrane termed coated pits. Once LDL binds the receptor, the complexes are rapidly internalized (endocytosed). ATP-dependent proton pumps lower the pH in the endosomes, which results in dissociation of the LDL from the receptor. The portion of the endosomal membranes harbouring the receptor are then recycled to the plasma membrane and the LDL-containing endosomes fuse with lysosomes. Acid hydrolases of the lysosomes degrade the apoproteins and release free fatty acids and cholesterol. As indicated above, the free cholesterol is either incorporated into plasma membranes or esterified (by ACAT) and stored within the cell.

The level of intracellular cholesterol is regulated through cholesterol-induced suppression of LDL receptor synthesis and cholesterol-induced inhibition of cholesterol synthesis. The increased level of intracellular cholesterol that results from LDL uptake has the additional effect of activating ACAT, thereby allowing the storage of excess cholesterol within cells. However, the effect of cholesterol-induced suppression of LDL receptor synthesis is a decrease in the rate at which LDLs and IDLs are removed from the serum. This can lead to excess circulating levels of cholesterol and cholesteryl esters when the dietary intake of fat and cholesterol exceeds the needs of the body. The excess cholesterol tends to be deposited in the skin, tendons and (more gravely) within the arteries, leading to atherosclerosis.

CLINICAL SIGNIFICANCES OF LIPOPROTEIN METABOLISM

Fortunately, few individuals carry the inherited defects in lipoprotein metabolism that lead to hyper- or hypolipoproteinemias (see Tables below for brief descriptions). Persons suffering from diabetes mellitus, hypothyroidism and kidney disease often exhibit abnormal lipoprotein metabolism as a result of secondary effects of their disorders. For example, because lipoprotein lipase (LPL) synthesis is regulated by insulin, LPL deficiencies leading to Type I hyperlipoproteinemia may occur as a secondary outcome of diabetes mellitus. Additionally, insulin and thyroid hormones positively affect hepatic LDL-receptor interactions; therefore, the hypercholesterolemia and increased risk of athersclerosis associated with uncontrolled diabetes or hypothyroidism is likely due to decreased hepatic LDL uptake and metabolism.

Of the many disorders of lipoprotein metabolism, familial hypercholesterolemia (FH) may be the most prevalent in the general population. Heterozygosity at the FH locus occurs in 1:500 individuals, whereas, homozygosity is observed in 1:1,000,000 individuals. FH is an inherited disorder comprising four different classes of mutation in the LDL receptor gene.

The class 1 defect (the most common) results in a complete loss of receptor synthesis. The class 2 defect results in the synthesis of a receptor protein that is not properly processed in the Golgi apparatus and therefore is not transported to the plasma membrane. The class 3 defect results in an LDL receptor that is incapable of binding LDLs. The class 4 defect results in receptors that bind LDLs but do not cluster in coated pits and are, therefore, not internalized.

FH sufferers may be either heterozygous or homologous for a particular mutation in the receptor gene. Homozygotes exhibit grossly elevated serum cholesterol (primarily in LDLs). The elevated levels of LDLs result in their phagocytosis by macrophages. These lipid-laden phagocytic cells tend to deposit within the skin and tendons, leading to xanthomas. A greater complication results from cholesterol deposition within the arteries, leading to atherosclerosis, the major contributing factor of nearly all cardiovascular diseases.

PHARMACOLOGIC INTERVENTION

Drug treatment to lower plasma lipoproteins and/or cholesterol is primarily aimed at reducing the risk of athersclerosis and subsequent coronary artery disease that exists in patients with elevated circulating lipids. Drug therapy usually is considered as an option only if non-pharmacologic interventions (altered diet and exercise) have failed to lower plasma lipids.

- Mevinolin, Mevastatin, Lovastatin: These drugs are fungal HMG-CoA reductase inhibitors. The net result of treatment is an increased cellular uptake of LDLs, since the intracellular synthesis of cholesterol is inhibited and cells are therefore dependent on extracellular sources of cholesterol. However, since mevalonate (the product of the HMG-CoA reductase reaction) is required for the synthesis of other important isoprenoid compounds besides cholesterol, long-term treatments carry some risk of toxicity.
- Nicotinic acid: Nicotinic acid reduces the plasma levels of both VLDLs and LDLs by inhibiting hepatic VLDL secretion, as well as suppressing the flux of FFA release from adipose tissue by inhibiting lipolysis. Because of its ability to cause large reductions in circulating levels of cholesterol, nicotinic acid is used to treat Type II, III, IV and V hyperlipoproteinemias.
- Clofibrate, Gemfibrozil, Fenofibrate: These compounds are derivatives of fibric acid and promote rapid VLDL turnover by activating lipoprotein lipase. They also induce the diversion of hepatic free fatty acids from esterification reactions to those of oxidation, thereby decreasing the liver's secretion of triacylglycerol- and cholesterol-rich VLDLs.
- Probucol: Probucol increases the rate of LDL metabolism and may block the intestinal transport of cholesterol. The net result is a significant reduction in plasma cholesterol levels.
- Cholestyramine or colestipol (resins): These compounds are nonabsorbable resins that bind bile acids which are then not reabsorbed by the liver but excreted. The drop in hepatic reabsorption of bile acids releases a feedback inhibitory mechanism that had been inhibiting bile acid synthesis. As a result, a greater

amount of cholesterol is converted to bile acids to maintain a steady level in circulation. Additionally, the synthesis of LDL receptors increases to allow increased cholesterol uptake for bile acid synthesis, and the overall effect is a reduction in plasma cholesterol. (This treatment is ineffective in homozygous FH patients, since they are completely deficient in LDL receptor

MEMBRANE LIPIDS AND BILAYER MEMBRANES

The simpler lipids are used primarily for energy storage by living organisms, but more complex lipids are used to form biological membranes. The most important and abundant class of biological membrane lipids is the class of phospholipids. Most phospholipids are derived from glycerol and so they are called phosphoglycerides. The amino alcohol sphingosine gives rise to sphingomyelin, the only significant membrane phospholipid which is not a phosphoglyceride.

In phosphoglycerides, the hydroxy groups at C-1 and C-2 of glycerol are esterified to fatty acids while the hydroxy group at C-3 of glycerol is esterified to phosphoric acid; this forms diacylglycerol-3-phosphates. The two fatty acids found in a diacylglycerol-3-phosphate molecule need not be the same. Diacylglycerol-3-phosphates are only a minor constituent of membranes. The major constituents of membranes are derivatives of diacylglycerol-3-phosphates in which the esterified phosphate also forms an ester linkage with an alcohol. The alcohols commonly found in phosphoglycerides include glycerol, ethanolamine, and choline.

The glycolipids are lipids which contain sugars. Glycolipids are derived from sphingosine, as is the phospholipid sphingomyelin. Glycolipids contain no esterified phosphate.

Both phospholipids and glycolipids are amphipathic molecules, that is, molecules in which one end is hydrophobic and one end is hydrophilic. The fatty acid chains form the hydrophobic end of the molecule while the polar glycerol-phosphate-alcohol or sphingosine-sugar portion forms the hydrophilic end of the molecule.

BILAYER MEMBRANES

When amphipathic molecules are placed in aqueous solution, their hydrophobic tails attempt to orient themselves towards each other. The two ways in which they do so are to form small spherical micelles or to form a planar lipid bilayer. A micelle is a small structure, generally less than two micrometer in diameter, while a lipid bilayer typically has a thickness of about 0.5 micrometer and can have an area of several square millimeters. Sonication or other treatment of lipids can produce spherical liposomes, or lipid vesicles, which contain aqueous solution within an enclosing lipid bilayer. Liposomes are considerably larger than micelles.

Both liposomes and planar bilayer membranes can be prepared from simple solutions of phospholipids such as the diacylglycerol-3-phosphatidyl cholines. Studies of such artificially prepared membranes have given us much insight into the functions and operation of membranes in living cells. Simple bilayer membranes are highly permeable to water molecules, while ions such as sodium ion and potassium ion can traverse them only more slowly (by nine orders of magnitude!). For small molecules, the rates of permeation through simple bilayer membranes increase with their solubility in nonpolar solvents relative to their solubility in water. These observations strongly suggest that transfer of substances through a membrane requires desolvation, transfer through the anhydrous membrane interior, and then resolvation on the other side of the membrane. The differences in transport behaviour between biological membranes and simple bilayer membranes are now believed to be due to the incorporation of protein molecules or other specifically transporting molecules directly into or onto the surface of a bilayer cell membrane. Considerable research in biology and in chemistry is being directed towards the transport properties of membranes.

FATTY ACID OXIDATION

Utilization of dietary lipids requires that they first be absorbed through the intestine. As these molecules are oils they would be essentially insoluble in the aqueous intestinal environment. Solubilization (emulsification) of dietary lipid is accomplished via bile salts that are synthesized in the liver and secreted from the gallbladder.

The emulsified fats can then be degraded by pancreatic lipases (lipase and phospholipase A2). These enzymes, secreted into the intestine from the pancreas, generate free fatty acids and a mixtures of mono- and diacylglycerols from dietary triacylglycerols. Pancreatic lipase degrades triacylglycerols at the 1 and 3 positions sequentially to generate 1,2-diacylglycerols and 2-acylglycerols. Phospholipids are degraded at the 2 position by pancreatic phospholipase A2 releasing a free fatty acid and the lysophospholipid. Following absorption of the products of pancreatic lipase by the intestinal mucosal cells, the resynthesis of triacylglycerols occurs. The triacylglycerols are then solubilized in lipoprotein complexes (complexes of lipid and protein) called chylomicrons. A chylomicron contains lipid droplets surrounded by the more polar lipids and finally a layer of proteins. Triacylglycerols synthesized in the liver are packaged into VLDLs and released into the blood directly. Chylomicrons from the intestine are then released into the blood via the lymph system for delivery to the various tissues for storage or production of energy through oxidation. The triacylglycerol components of VLDLs and chylomicrons are hydrolyzed to free fatty acids and glycerol in the capillaries of adipose tissue and skeletal muscle by the action of lipoprotein lipase. The free fatty acids are then absorbed by the cells and the glycerol is returned via the blood to the liver (and kidneys). The glycerol is then converted to the glycolytic intermediate DHAP. The classification of blood lipids is distinguished based upon the density of the different lipoproteins. As lipid is less dense than protein, the lower the density of lipoprotein the less protein there is.

MOBILIZATION OF FAT STORES

The primary sources of fatty acids for oxidation are dietary and mobilization from cellular stores. Fatty acids from the diet can are delivered from the gut to cells via transport in the blood. Fatty acids are stored in the form of triacylglycerols primarily within adipocytes of adipose tissue. In response to energy demands, the fatty acids of stored triacylglycerols can be mobilized for use by peripheral tissues. The release of metabolic energy, in the form of fatty acids, is controlled by a complex series of interrelated cascades that result in the activation of hormone-sensitive lipase. The stimulus to activate this cascade, in adipocytes, can be glucagon, epinephrine or b-corticotropin. These hormones bind cell-surface receptors that are coupled to the activation of adenylate cyclase upon ligand binding. The resultant increase in cAMP leads to activation of PKA, which in turn phosphorylates and activates hormone-sensitive lipase. This enzyme hydrolyzes fatty acids from carbon atoms 1 or 3 of triacylglycerols. The resulting diacylglycerols are substrates for either hormone-sensitive lipase or for the non-inducible enzyme diacylglycerol lipase. Finally the monoacylglycerols are substrates for monoacylglycerol lipase. The net result of the action of these enzymes is three moles of free fatty acid and one mole of glycerol. The free fatty acids diffuse from adipose cells, combine with albumin in the blood, and are thereby transported to other tissues, where they passively diffuse into cells.

Hormone Induced Fatty Acid Mobilization in Adipocytes

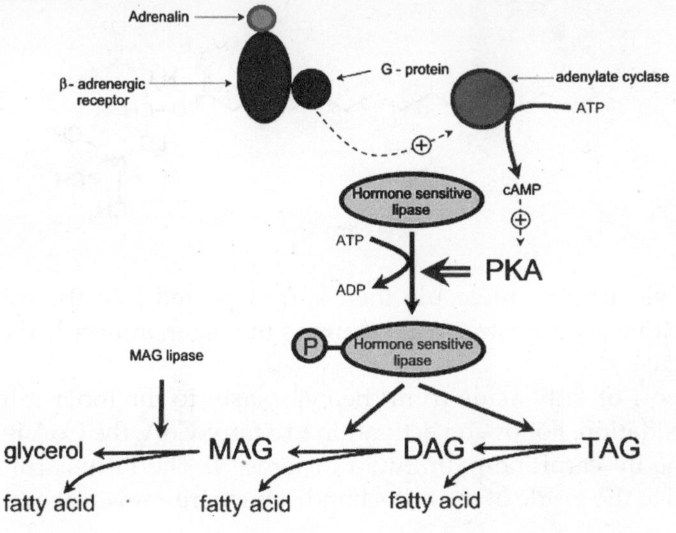

Model for the activation of hormone-sensitive lipase by epinephrine. Epinephrine binds its' receptor and leads to the activation of adenylate cyclase.

The resultant increase in cAMP activates PKA which then phosphorylates and activates hormone-sensitive lipase. Hormone-sensitive lipase hydrolyzes fatty acids from triacylglycerols and diacylglycerols. The final fatty acid is released from monoacylglycerols through the action of monoacylglycerol lipase, an enzyme active in the absence of hormonal stimulation.

In contrast to the hormonal activation of adenylate cyclase and (subsequently) hormone-sensitive lipase in adipocytes, the mobilization of fat from adipose tissue is inhibited by numerous stimuli. The most significant inhibition is that exerted upon adenylate cyclase by insulin. When an individual is well fed state, insulin released from the pancreas prevents the inappropriate mobilization of stored fat. Instead, any excess fat and carbohydrate are incorporated into the triacylglycerol pool within adipose tissue.

ACTIVATION AND TRANSFER OF FATTY ACID

Fatty acids must be activated in the cytoplasm before being oxidized in the mitochondria. Activation is catalyzed by fatty acyl-CoA ligase (also called acyl-CoA synthetase or thiokinase). The net result of this activation process is the consumption of 2 molar equivalents of ATP.

$$\text{Fatty acid} + \text{ATP} + \text{CoA} \rightarrow \text{Acyl-CoA} + \text{PPi} + \text{AMP}$$

Oxidation of fatty acids occurs in the mitochondria. The transport of fatty acyl-CoA into the mitochondria is accomplished via an acyl-carnitine intermediate, which itself is generated by the action of carnitine acyltransferase I, an enzyme that resides in the outer mitochondrial membrane, from acyl-CoA and carnitine.

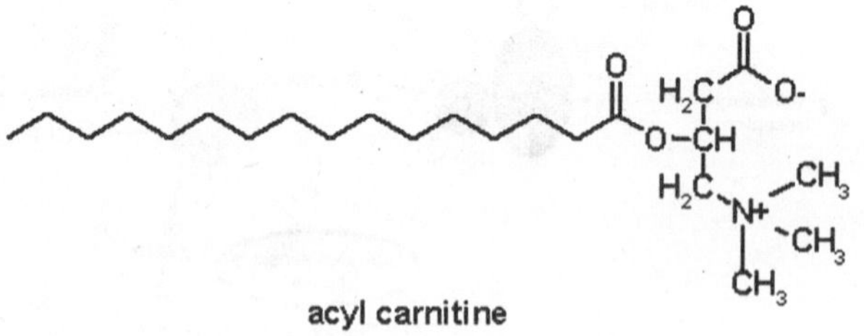

acyl carnitine

The acyl-carnitine molecule then is transported into the mitochondria where carnitine acyltransferase II catalyzes the regeneration of the fatty acyl-CoA molecule.

Transport of fatty acids from the cytoplasm to the inner mitochondrial space for oxidation. Following activation to a fatty-CoA, the CoA is exchanged for carnitine by carnitine-palmitoyltransferase I. The fatty-carnitine is then transported to the inside of the mitochondrion where a reversal exchange takes place through the action of carnitine-palmitoyltransferase II. Once inside the mitochondrion the fatty-CoA is a substrate for the b-oxidation machinery.

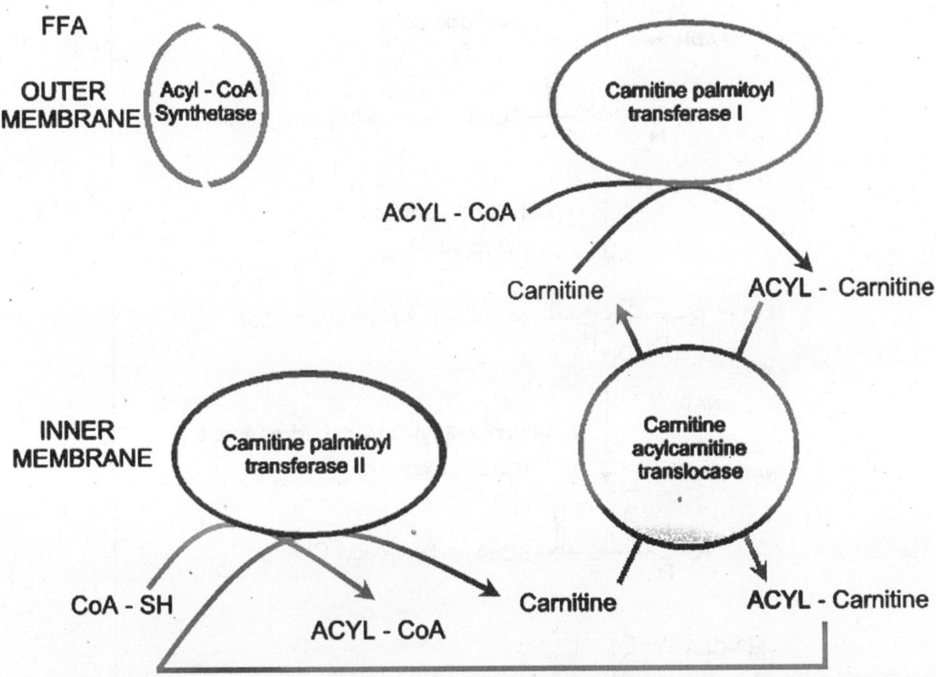

The human body produces carnitine through an enzymatically catalyzed biosynthetic pathway out of lysine and methionine.

REACTION SEQUENCE IN THE -OXIDATION

Each round of -oxidation produces one mole of NADH, one mole of FADH2 and one mole of acetyl-CoA. The acetyl-CoA--- the end product of each round of -oxidation--- then enters the TCA cycle, where it is further oxidized to CO_2 with the concomitant generation of three moles of NADH, one mole of $FADH_2$ and one mole of ATP. The NADH and FADH2 generated during the fat oxidation and acetyl-CoA oxidation in the TCA cycle then can enter the respiratory pathway for the production of ATP.

The process of fatty acid oxidation is termed -oxidation since it occurs through the sequential removal of 2-carbon units by oxidation at the b-carbon position of the fatty acyl-CoA molecule. The oxidation of fatty acids yields significantly more energy per carbon atom than does the oxidation of carbohydrates. The net result of the oxidation of one mole of oleic acid (an 18-carbon fatty acid) will be 146 moles of ATP (2 mole equivalents are used during the activation of the fatty acid), as compared with 114 moles from an equivalent number of glucose carbon atoms.

ALTERNATIVE OXIDATION PATHWAYS

The majority of natural lipids contain an even number of carbon atoms. A small proportion that contain odd numbers; upon complete -oxidation, these yield acetyl-CoA units plus a single mole of propionyl-CoA. The propionyl-CoA is converted, in an ATP-dependent pathway, to succinyl-CoA. The succinyl-CoA can then enter the TCA cycle for further oxidation.

The oxidation of unsaturated fatty acids is essentially the same process as for saturated fats, except when a double bond is encountered. In such a case, the bond is isomerized by a specific enoyl-CoA isomerase and oxidation continues. In the case of linoleate, the presence of the -12 unsaturation results in the formation of a dienoyl-CoA during oxidation. This molecule is the substrate for an additional oxidizing enzyme, the NADPH requiring 2,4-dienoyl-CoA reductase.

Phytanic acid is a fatty acid present in the tissues of ruminants and in dairy products and is, therefore, an important dietary component of fatty acid intake.

Because phytanic acid is methylated, it cannot act as a substrate for the first enzyme of the -oxidation pathway (acyl-CoA dehydrogenase). An additional mitochondrial enzyme, -hydroxylase, adds a hydroxyl group to the -carbon of phytanic acid, which then serves as a substrate for the remainder of the normal oxidative enzymes. This process is termed -oxidation.

REGULATION OF FATTY ACID METABOLISM

In order to understand how the synthesis and degradation of fats needs to be exquisitely regulated, one must consider the energy requirements of the organism as a whole. The blood is the carrier of triacylglycerols in the form of VLDLs and chylomicrons, fatty acids bound to albumin, amino acids, lactate, ketone bodies and glucose. The pancreas is the primary organ involved in sensing the organism's dietary and energetic states by monitoring glucose concentrations in the blood. Low blood glucose stimulates the secretion of glucagon, whereas, elevated blood glucose calls for the secretion of insulin.

The metabolism of fat is regulated by two distinct mechanisms. One is short-term regulation, which can come about through events such as substrate availability, allosteric effectors and/or enzyme modification. The other mechanism, long-term regulation, is achieved by alteration of the rate of enzyme synthesis and turn-over.

ACC (acetyl-CoA carboxylase) is the rate-limiting (committed) step in fatty acid synthesis. This enzyme is activated by citrate and inhibited by palmitoyl-CoA and other long-chain fatty acyl-CoAs. ACC activity can also be affected by phosphorylation. For instance, glucagon-stimulated increases in PKA activity result in the phosphorylation of certain serine residues in ACC leading to decreased activity of the enzyme. By contrast, insulin leads to PKA-independent phosphorylation of ACC at sites distinct from glucagon, which bring about increased ACC activity. Both of these reaction chains are examples of short-term regulation.

Insulin, a product of the well-fed state, stimulates ACC and FAS synthesis, whereas starvation leads to a decrease in the synthesis of these enzymes. Adipose tissue levels of lipoprotein lipase also are increased by insulin and decreased by starvation.

However, the effects of insulin and starvation on lipoprotein lipase in the heart are just the inverse of those in adipose tissue. This sensitivity allows the heart to absorb any available fatty acids in the blood in order to oxidize them for energy production. Starvation also leads to increases in the levels of cardiac enzymes of fatty acid oxidation, and to decreases in FAS and related enzymes of synthesis.

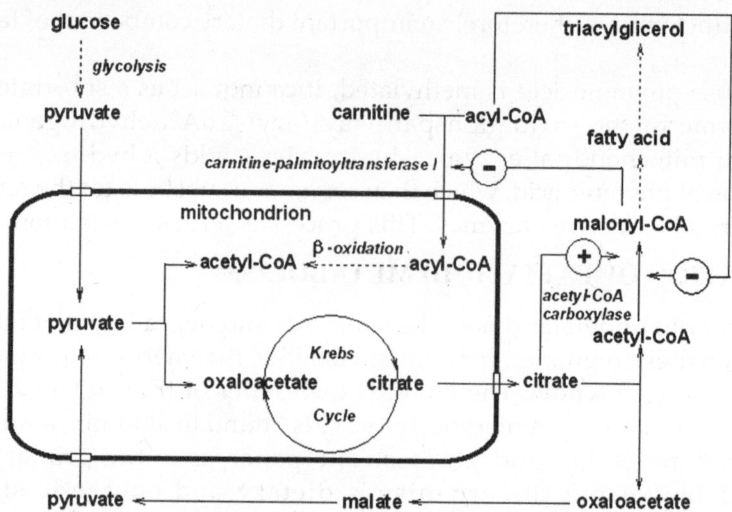

Adipose tissue contains hormone-sensitive lipase, which is activated by PKA-dependent phosphorylation; this activation increases the release of fatty acids into the blood. This in turn leads to the increased oxidation of fatty acids in other tissues such as muscle and liver. In the liver, the net result (due to increased acetyl-CoA levels) is the production of ketone bodies (see below). This would occur under conditions in which the carbohydrate stores and gluconeogenic precursors available in the liver are not sufficient to allow increased glucose production. The increased levels of fatty acid that become available in response to glucagon or epinephrine are assured of being completely oxidized, because PKA also phosphorylates ACC; the synthesis of fatty acid is thereby inhibited.

Insulin has the opposite effect to glucagon and epinephrine: it increases the synthesis of triacylglycerols (and glycogen). One of the many effects of insulin is to lower cAMP levels, which leads to increased dephosphorylation through the enhanced activity of protein phosphatases such as PP-1. With respect to fatty acid metabolism, this yields dephosphorylated and inactive hormone-sensitive lipase. Insulin also stimulates certain phosphorylation events. This occurs through activation of several cAMP-independent kinases, one of which phosphorylates and thereby stimulates the activity of ACC.

Fat metabolism can also be regulated by malonyl-CoA-mediated inhibition of carnitine acyltransferase I. Such regulation serves to prevent de novo synthesized fatty acids from entering the mitochondria and being oxidized.

CLINICAL SIGNIFICANCE OF FATTY ACIDS

The majority of clinical problems related to fatty acid metabolism are associated with processes of oxidation.

These disorders fall into four main groups:

- Deficiencies in Carnitine: Deficiencies in carnitine lead to an inability to transport fatty acids into the mitochondria for oxidation. This can occur in newborns and particularly in pre-term infants. Carnitine deficiencies also are found in patients undergoing hemodialysis or exhibiting organic aciduria. Carnitine deficiencies may manifest systemic symptomology or may be limited to only muscles. Symptoms can range from mild occasional muscle cramping to severe weakness or even death. Treatment is by oral carnitine administration.
- Carnitine Palmitoyltransferase I (CPT I) Deficiency: Deficiencies in this enzyme affect primarily the liver and lead to reduced fatty acid oxidation and ketogenesis. Carnitine Palmitoylransferase II (CPT II) deficiency results in recurrent muscle pain and fatigue and myoglobinuria following strenuous exercise. Carnitine acyltransferases may also be inhibited by sulfonylurea drugs such as tolbutamide and glyburide.
- Deficiencies in Acyl-CoA Dehydrogenases: A group of inherited diseases that impair -oxidation result from deficiencies in acyl-CoA dehydrogenases. The enzymes affected may belong to one of four categories:
  - Very long-chain acyl-CoA dehydrogenase (VLCAD)
  - Long-chain acyl-CoA dehydrogenase (LCAD)
  - Medium-chain acyl-CoA dehydrogenase (MCAD)
  - Short-chain acyl-CoA dehydrogenase (SCAD)

  MCAD deficiency is the most common form of this disease. In the first years of life this deficiency will become apparent following a prolonged fasting period. Symptoms include vomiting, lethargy and frequently coma. Excessive urinary excretion of medium-chain dicarboxylic acids as well as their glycine and carnitine esters is diagnostic of this condition. In the case of this enzyme deficiency. taking care to avoid prolonged fasting is sufficient to prevent clinical problems.
- Refsum's Disease: Refsum's disease is a rare inherited disorder in which patients lack the mitochondrial -oxidizing enzyme. As a consequence, they accumulate large quantities of phytanic acid in their tissues and serum. This leads to severe symptoms, including cerebellar ataxia, retinitis pigmentosa, nerve deafness and peripheral neuropathy. As expected, the restriction of dairy products and ruminant meat from the diet can ameliorate the symptoms of this disease.

KETOGENESIS

During high rates of fatty acid oxidation, primarily in the liver, large amounts of acetyl-CoA are generated. These exceed the capacity of the TCA cycle, and one result is the synthesis of ketone bodies, or ketogenesis. The ketone bodies are acetoacetate, -hydroxybutyrate, and acetone. The formation of acetoacetyl-CoA occurs by condensation of two moles of acetyl-CoA through a reversal of the thiolase catalyzed reaction of fat oxidation. Acetoacetyl-CoA and an additional acetyl-CoA are converted to -hydroxy- -methylglutaryl-CoA (HMG-CoA) by HMG-CoA synthase, an enzyme found in large amounts only in the liver.

Some of the HMG-CoA leaves the mitochondria, where it is converted to mevalonate (the precursor for cholesterol synthesis) by HMG-CoA reductase. HMG-CoA in the mitochondria is converted to acetoacetate by the action of HMG-CoA lyase. Acetoacetate can undergo spontaneous decarboxylation to acetone, or be enzymatically converted to -hydroxybutyrate through the action of -hydroxybutyrate dehydrogenase.

SCoA SCoA

thiolase

HSCoA

SCoA

Acetoacetyl-CoA

H_2O

HMG-CoA-synthetase

HSCoA

SCoA

HMG-CoA

HMG-CoA-lyase

SCoA

CO_2

non-enzymatic reaction

Acetoacetate

Acetone

NADH + H^+

beta-hydroxybutyrate dehydrogenase

NAD^+

beta-hydroxybutyrate

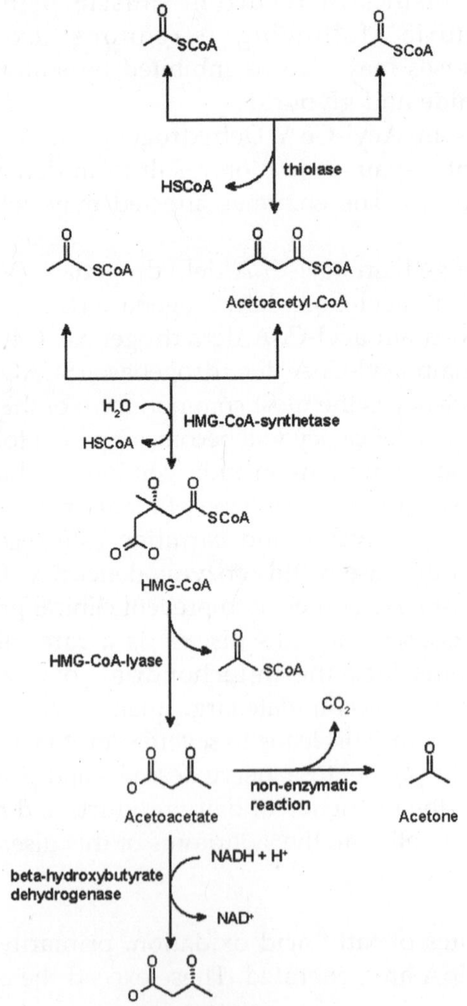

When the level of glycogen in the liver is high the production of -hydroxybutyrate increases. When carbohydrate utilization is low or deficient, the level of oxaloacetate will also be low, resulting in a reduced flux through the TCA cycle. This in turn leads to increased release of ketone bodies from the liver for use as fuel by other tissues. In early stages of starvation, when the last remnants of fat are oxidized, heart and skeletal muscle will consume primarily ketone bodies to preserve glucose for use by the brain. Acetoacetate and -hydroxybutyrate, in particular, also serve as major substrates for the biosynthesis of neonatal cerebral lipids.

Ketone bodies are utilized by extrahepatic tissues through the conversion of -hydroxybutyrate to acetoacetate and of acetoacetate to acetoacetyl-CoA. The first step involves the reversal of the -hydroxybutyrate dehydrogenase reaction, and the second involves the action (shown below) of acetoacetate:succinyl-CoA transferase, also called ketoacyl-CoA-transferase.

Acetoacetate + Succinyl-CoA ↔ Acetoacetyl-CoA + succinate

The latter enzyme is present in all tissues except the liver. Importantly, its absence allows the liver to produce ketone bodies but not to utilize them. This ensures that extrahepatic tissues have access to ketone bodies as a fuel source during prolonged starvation. Acetoacetyl-CoA is converted into two molecules of acetyl-CoA by a thiolase: Acetoacetyl-CoA + HS-CoA ↔ 2 Acetyl-CoA

REGULATION OF KETOGENESIS

The fate of the products of fatty acid metabolism is determined by an individual's physiological status. Ketogenesis takes place primarily in the liver and may by affected by several factors:

- Control in the release of free fatty acids from adipose tissue directly affects the level of ketogenesis in the liver. This is, of course, substrate-level regulation.
- Once fats enter the liver, they have two distinct fates. They may be activated to acyl-CoAs and oxidized, or esterified to glycerol in the production of triacylglycerols. If the liver has sufficient supplies of glycerol-3-phosphate, most of the fats will be turned to the production of triacylglycerols.
- The generation of acetyl-CoA by oxidation of fats can be completely oxidized in the TCA cycle. Therefore, if the demand for ATP is high the fate of acetyl-CoA is likely to be further oxidation to CO_2.
- The level of fat oxidation is regulated hormonally through phosphorylation of ACC, which may activate it (in response to glucagon) or inhibit it (in the case of insulin).

FATTY ACID SYNTHESIS

One might predict that the pathway for the synthesis of fatty acids would be the reversal of the oxidation pathway. However, this would not allow distinct regulation of the two pathways to occur even given the fact that the pathways are separated within different cellular compartments. The pathway for fatty acid synthesis occurs in the cytoplasm, whereas, oxidation occurs in the mitochondria. The other major difference is the use of nucleotide co-factors. Oxidation of fats involves the reduction of $FADH^+$ and NAD^+. Synthesis of fats involves the oxidation of NADPH. However, the essential chemistry of the two processes are reversals of each other. Both oxidation and synthesis of fats utilize an activated two carbon intermediate, acetyl-CoA. However, the acetyl-CoA in fat synthesis exists temporarily bound to the enzyme complex as malonyl-CoA. The synthesis of malonyl-CoA is the first committed step of fatty acid synthesis and the enzyme that catalyzes this reaction, acetyl-CoA carboxylase (ACC), is the major site of regulation of fatty acid synthesis. Like other enzymes that transfer CO_2 to substrates, ACC requires a biotin co-factor.

$$\underset{\text{bicarbonate}}{HCO_3^{\ominus}} + \underset{\text{acetyl-CoA}}{H_3C-\overset{O}{\overset{\|}{C}}-SCoA} \xrightarrow[\text{ATP} \quad \text{ADP + Pi}]{\text{ACC}} \underset{\text{malonyl-CoA}}{{}^{\ominus}OOC-CH_2-\overset{O}{\overset{\|}{C}}-SCoA}$$

The rate of fatty acid synthesis is controlled by the equilibrium between monomeric ACC and polymeric ACC. The activity of ACC requires polymerization. This conformational change is enhanced by citrate and inhibited by long-chain fatty acids. ACC is also controlled through hormone mediated phosphorylation (see below). The acetyl groups that are the products of fatty acid oxidation are linked to CoASH. As you should recall, CoA contains a phosphopantetheine group coupled to AMP. The carrier of acetyl groups (and elongating acyl groups) during fatty acid synthesis is also a phosphopantetheine prosthetic group, however, it is attached a serine hydroxyl in the synthetic enzyme complex.

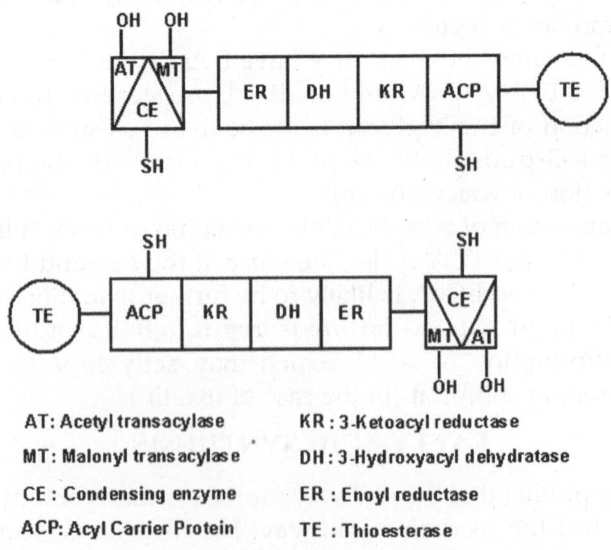

The carrier portion of the synthetic complex is called acyl carrier protein, ACP. This is somewhat of a misnomer in eukaryotic fatty acid synthesis since the ACP portion of the synthetic complex is simply one of many domains of a single polypeptide. The acetyl-CoA and malonyl-CoA are transferred to ACP by the action of acetyl-CoA transacylase and malonyl-CoA transacylase, respectively.

The synthesis of fatty acids from acetyl-CoA and malonyl-CoA is carried out by fatty acid synthase, FAS.

The active enzyme is a dimer of identical subunits.

All of the reactions of fatty acid synthesis are carried out by the multiple enzymatic activities of FAS. Like fat oxidation, fat synthesis involves 4 enzymatic activities. These are, b-keto-ACP synthase (condensing enzyme), b-keto-ACP reductase, 3-OH acyl-ACP dehydratase and enoyl-ACP reductase(give a look to the FAS reaction mechanism). The two reduction reactions require NADPH oxidation to $NADP^+$.

The primary fatty acid synthesized by FAS is palmitate. Palmitate is then released from the enzyme by ***thioesterase*** reaction and can then undergo separate elongation and/or unsaturation to yield other fatty acid molecules.

ORIGIN OF CYTOPLASMIC ACETYL

Acetyl-CoA is generated in the mitochondria primarily from two sources, the pyruvate dehydrogenase (PDH) reaction and fatty acid oxidation. In order for these acetyl units to be utilized for fatty acid synthesis they must be present in the cytoplasm. The shift from fatty acid oxidation and glycolytic oxidation occurs when the need for energy diminishes. This results in reduced oxidation of acetyl-CoA in the TCA cycle and the oxidative phosphorylation pathway. Under these conditions the mitochondrial acetyl units can be stored as fat for future energy demands.

Acetyl-CoA enters the cytoplasm in the form of citrate via the tricarboxylate transport system as diagrammed. In the cytoplasm, citrate is converted to oxaloacetate and acetyl-CoA by the ATP driven *ATP-citrate lyase* reaction. This reaction is essentially the reverse of that catalyzed by the TCA enzyme *citrate synthase* except it requires the energy of ATP hydrolysis to drive it forward. The resultant oxaloacetate is converted to malate by *malate dehydrogenase* (MDH).The malate produced by this pathway can undergo oxidative decarboxylation by *malic enzyme*. The co-enzyme for this reaction is $NADP^+$ generating NADPH. The advantage of this series of reactions for converting mitochondrial acetyl-CoA into cytoplasmic acetyl-CoA is that the NADPH produced by the *malic enzyme* reaction can be a major source of reducing co-factor for the *fatty acid synthase* activities.

REGULATION OF FATTY ACID METABOLISM

One must consider the global organismal energy requirements in order to effectively understand how the synthesis and degradation of fats (and also carbohydrates) needs to be exquisitely regulated. The blood is the carrier of triacylglycerols in the form of VLDLs and chylomicrons, fatty acids bound to albumin, amino acids, lactate, ketone bodies and glucose.

The pancreas is the primary organ involved in sensing the organisms dietary and energetic states via glucose concentrations in the blood. In response to low blood glucose, glucagon is secreted, whereas, in response to elevated blood glucose insulin is secreted.

The regulation of fat metabolism occurs via two distinct mechanisms. One is short term regulation which is regulation effected by events such as substrate availability, allosteric effectors and/or enzyme modification. ACC is the rate limiting (committed) step in fatty acid synthesis. This enzyme is activated by citrate and inhibited by palmitoyl-CoA and other long chain fatty acyl-CoAs. ACC activity also is affected by phosphorylation. Glucagon stimulated increases in PKA activity result in phosphorylation of certain serine residues in ACC leading to decreased activity of the enzyme. On the other hand insulin leads to phosphorylation of ACC at sites distinct from glucagon that result in increased ACC activity. These forms of regulation are all defined as short term regulation.

Control of a given pathways' regulatory enzymes can also occur by alteration of enzyme synthesis and turn-over rates. These changes are long term regulatory effects. Insulin stimulates ACC and FAS synthesis, whereas, starvation leads to decreased synthesis of these enzymes. Adipose tissue lipoprotein lipase levels also are increased by insulin and decreased by starvation. However, in contrast to the effects of insulin and starvation on adipose tissue, their effects on heart lipoprotein lipase are just the inverse. This allows the heart to absorb any available fatty acids in the blood in order to oxidize them for energy production. Starvation also leads to increases in the levels of fatty acid oxidation enzymes in the heart as well as a decrease in FAS and related enzymes of synthesis.

Adipose tissue contains hormone-sensitive lipase, that is activated by PKA-dependent phosphorylation leading to increased fatty acid release to the blood. This leads to increased fatty acid oxidation in other tissues such as muscle and liver. In the liver the net result (due to increased acetyl-CoA levels) is the production of ketone bodies. This would occur under conditions where insufficient carbohydrate stores and gluconeogenic precursors were available in liver for increased glucose production. The increased fatty acid availability in response to glucagon or epinephrine is assured of being completely oxidized since PKA also phosphorylates (and as a result inhibits) ACC, thus inhibiting fatty acid synthesis.

Insulin, on the other hand, has the opposite effect to glucagon and epinephrine leading to increased glycogen and triacylglyceride synthesis. One of the many effects of insulin is to lower cAMP levels which leads to increased dephosphorylation through the enhanced activity of protein phosphatases such as PP-1. With respect to fatty acid metabolism this yields dephosphorylated and inactive hormone sensitive lipase. Insulin also stimulates certain phosphorylation events. This occurs through activation of several cAMP-independent kinases. Insulin stimulated phosphorylation of ACC activates this enzyme.

Regulation of fat metabolism also occurs through malonyl-CoA induced inhibition of carnitine acyltransferase I. This functions to prevent the newly synthesized fatty acids from entering the mitochondria and being oxidized.

ELONGATION AND DESATURATION

The fatty acid product released from FAS is palmitate (via the action of palmitoyl thioesterase) which is a 16:0 fatty acid, *i.e.* 16 carbons and no sites of unsaturation. Elongation and unsaturation of fatty acids occurs in both the mitochondria and endoplasmic reticulum (microsomal membranes). The predominant site of these processes is in the ER membranes. Elongation involves condensation of acyl-CoA groups with malonyl-CoA. The resultant product is two carbons longer (CO2 is released from malonyl-CoA as in the FAS reaction) which undergoes reduction, dehydration and reduction yielding a saturated fatty acid. The reduction reactions of elongation require NADPH as co-factor just as for the similar reactions catalyzed by FAS. Mitochondrial elongation involves acetyl-CoA units and is a reversal of oxidation except that the final reduction utilizes NADPH instead of FADH2 as co-factor.

Desaturation occurs in the ER membranes as well and in mammalian cells involves 4 broad specificity fatty acyl-CoA desaturases (non-heme iron containing enzymes). These enzymes introduce unsaturation at C4, C5, C6 or C9.

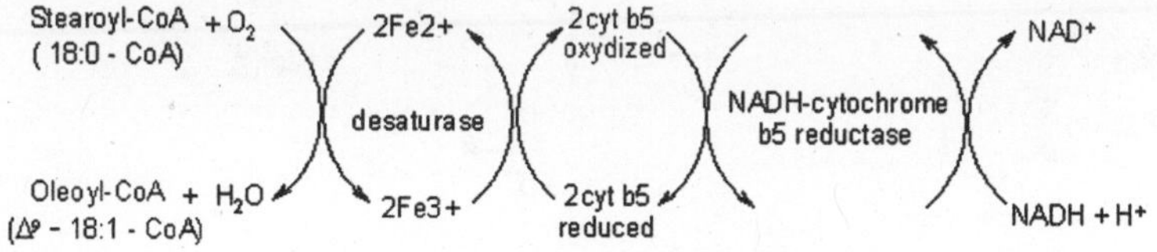

The electrons from the reduced fatty acid (saturated) and from the desaturase are transferred to molecular oxygen to form water and oxydized fatty acid (unsaturated). NADH is the donor of two electrons which keep the desaturase in the reduced form through the action of the enzyme NADH-cytochrome b5 reductase. These electrons are un-coupled from mitochondrial oxidative-phosphorylation and, therefore, do not yield ATP.

Since these enzymes cannot introduce sites of unsaturation beyond C9 they cannot synthesize either linoleate (18:2 9, 12) or linolenate (18:3 9, 12, 15). These fatty acids must be acquired from the diet and are, therefore, referred to as essential fatty acids. Linoleic is especially important in that it required for the synthesis of arachidonic acid. As we shall encounter later, arachindonate is a precursor for the eicosanoids (the prostaglandins and thromboxanes). It is this role of fatty acids in eicosanoid synthesis that leads to poor growth, wound healing and dermatitis in persons on fat free diets. Also, linoleic acid is a constituent of epidermal cell sphingolipids that function as the skins water permeability barrier.

SYNTHESIS OF TRIGLYCERIDES

Fatty acids are stored for future use as triacylglycerols in all cells, but primarily in adipocytes of adipose tissue. Triacylglycerols constitute molecules of glycerol to which three fatty acids have been esterified. The fatty acids present in triacylglycerols are predominantly saturated. The major building block for the synthesis of triacylglycerols, in tissues other than adipose tissue,is glycerol-3-phosphate. Adipocytes lack glycerol kinase, therefore, dihydroxyacetone phosphate (DHAP), produced during glycolysis, is the precursor for triacylglycerol synthesis in adipose tissue. This means that adipoctes must have glucose to oxidize in order to store fatty acids in the form of triacylglycerols. DHAP can also serve as a backbone precursor for triacylglycerol synthesis in tissues other than adipose, but does so to a much lesser extent than glycerol.

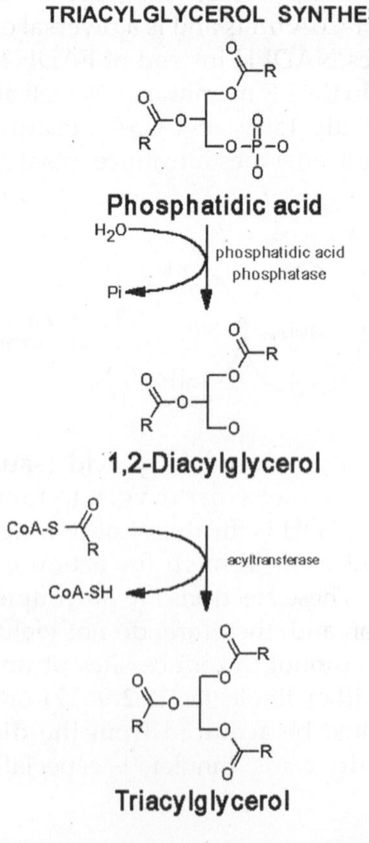

TRIACYLGLYCEROL SYNTHESIS

Phosphatidic acid

H_2O → Pi (phosphatidic acid phosphatase)

1,2-Diacylglycerol

CoA-S-CO-R → CoA-SH (acyltransferase)

Triacylglycerol

Phoshatidic Acid Synthesis

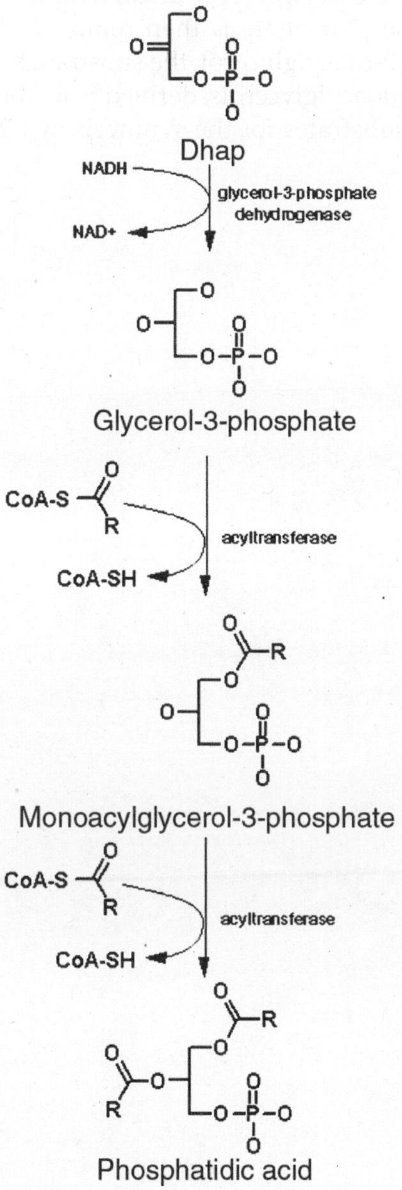

Dhap

NADH → NAD+ glycerol-3-phosphate dehydrogenase

Glycerol-3-phosphate

CoA-S(C=O)R → CoA-SH acyltransferase

Monoacylglycerol-3-phosphate

CoA-S(C=O)R → CoA-SH acyltransferase

Phosphatidic acid

The glycerol backbone of triacylglycerols is activated by phosphorylation at the C-3 position by glycerol kinase.

The utilization of DHAP for the backbone is carried out through the action of glycerol-3-phosphate dehydrogenase, a reaction that requires NADH (the same reaction as that used in the glycerol-phosphate shuttle). The fatty acids incorporated into triacylglycerols are activated to acyl-CoAs through the action of acyl-CoA synthetases. Two molecules of acyl-CoA are esterified to glycerol-3-phosphate to yield 1,2-diacylglycerol phosphate (commonly identified as phosphatidic acid). The phosphate is then removed, by phosphatidic acid phosphatase, to yield 1,2-diacylglycerol, the substrate for addition of the third fatty acid. Intestinal monoacylglycerols, derived from the hydrolysis of dietary fats, can also serve as substrates for the synthesis of 1,2-diacylglycerols.

7

Vitamins: The Fat Soluble

Vitamin, any of the organic (carbon-containing) compounds that the body requires in small amounts to maintain health and function properly. Children additionally need vitamins to grow. The body gets most of its vitamins from the foods we eat. A healthy diet with plenty of fruits and vegetables should provide nearly all of the vitamins a person needs. Scientists have classified 13 compounds as vitamins. They have given most of these vitamins letter or letter plus number names, such as A, B12, and D. Most vitamins are produced by plants. Some, such as vitamin D, are produced only by animals. A few vitamins are made by the body itself. For example, bacteria in the digestive tract help produce vitamin K, and the skin uses sunlight to produce vitamin D.

Vitamins are also manufactured for sale as supplements for people who need additional vitamins to meet their body's requirements. For example, doctors often prescribe vitamin supplements for pregnant and nursing women to provide the additional nutrients needed by a rapidly growing fetus or infant. Older adults may not meet their vitamin requirements through food because the body's ability to absorb vitamins is impaired with age.

Expressed in milligrams (mg) or international units (IU) for adults and children of normal health, these recommendations provide useful guidelines for daily intake. Such guidelines are useful not only for professionals in nutrition but also for the growing number of families and individuals who eat irregular meals and rely on prepared foods, most of which are now required to carry nutritional labeling.

VITAMINS ARE VITAL

The word vitamine, later shortened to vitamin, was coined by Polish-American chemist Casimir Funk in the early 20th century. Funk was searching for the then-unknown substance in foods that prevents such diseases as beriberi, rickets, and scurvy. After experiments on pigeons, Funk guessed that an amine-a compound containing nitrogen and hydrogen-in foods was responsible. He called this substance a vital (necessary for life) amine, or vitamine. Scientists later discovered that not all vitamins were amines, however.

Vitamins help the body carry out essential biochemical processes. Vitamins generally combine with proteins to create enzymes that promote chemical reactions. These enzymes play an important role in metabolism-the reactions that break down the fats, carbohydrates, and proteins in food so that the body can use them for energy and cell repair.

The enzymes also promote reactions involved in the formation of bone, hormones, blood cells, nervous-system chemicals, and genetic material. Without vitamins, many of these reactions would slow down or cease. The different vitamins are not chemically related, and most differ in their actions in the body. Deficiency of particular vitamins can lead to various diseases. Too little vitamin C, for example, can cause rickets, a disease in which the bones fail to develop properly. Bowlegs or"knock knees" are signs of rickets. Rickets and the other vitamin-deficiency diseases that interested Casimir Funk have largely disappeared in the developed world as the result of fortified foods and improved nutrition. But these diseases still occur in the developing world, especially where malnutrition is common.

Interest in vitamins grew during the 1990s, especially in those vitamins that act as antioxidants-vitamins A, C, and E. Antioxidants neutralize molecules known free radicals that cause cell damage. Free radicals have been linked to a number of disorders, including Alzheimer's disease, arteriosclerosis, cancer, diabetes, and Parkinson disease. Researchers theorized that increased doses of antioxidants might also prevent and even cure such diseases. These claims have so far not panned out, however, as scientific studies have failed to demonstrate the preventive abilities of antioxidants. Research on antioxidants and other vitamins continues.

FAT-SOLUBLE VITAMINS

Fat-soluble vitamins A, D, E, and K differ from the water-soluble vitamins in their accumulative action. Little evidence has been recorded for hypervitaminosis with the water-soluble vitamins since these compounds are rapidly metabolized and excreted when intake exceeds liver or tissue storage capacity, but hypervitaminosis is a common occurrence in fish and other animals when large quantities of any one of the fat-soluble vitamins are ingested.

Toxicity symptoms involving vitamins A and D are indistinguishable from deficiency symptoms for the same vitamins. On the other hand, symptoms of excess vitamins E and K intake are more discrete. Fish rations may often be enriched with fish oils to increase caloric density of the ration resulting in excessive intake of the fat-soluble vitamins.

VITAMIN A

A fat-soluble factor that promoted growth in rats by Hopkins and by Osborne and Mendel at the beginning of this century. McCollum and Simmonds cured xerophthalmia, a disease of the eye, with this material. The chemical structure of vitamin A and its relationship to -carotene was shown by Von Euler in 1928. Active vitamin A was synthesized in the mid-thirties. The relationship of the vitamin A alcohols to naturally occurring -carotene Retinene, the aldehyde form of vitamin A, has been isolated from the retina of dark adapted eyes and is involved in vision in dim light. Retinoic acid, which is the oxidized form of vitamin A alcohol, has been shown to have some vitamin A activity. Vitamin A1 is found in saltwater fish, whereas vitamin A is more abundant in freshwater fish. Interconversion of the two forms in living fish tissue has been demonstrated.

Fish oils contain vitamin A as free alcohols or esters. Vitamin A alcohol occurs as a light coloured viscous oil which is heat labile and subject to air oxidation. Beta-carotene occurs as an orange, crystalline compound which is more stable to heat and oxidation. Vitamin A is water insoluble but is soluble in fat and organic solvent.

POSITIVE FUNCTIONS

Vitamin A is essential in maintaining epithelial cells, preventing atrophy and keratinization of epithelial tissues. It is combined with a protein in visual purple and is important in night vision. Vitamin A also prevents xerophthalmia in rats and young children. Vitamin A promotes growth of new cells and aids in maintaining resistance to infection.

THE SYNDROME OF DEFICIENCY OR EXCESS

Hypovitaminosis A is characterized by poor growth, poor vision, keratinization of epithelial tissue, xerophthalmia, night blindness, haemorrhage in the anterior chamber of the eye, haemorrhage of the base of the fins, and abnormal bone formation. Nerve degeneration has been reported in pigs, chickens, rats, rabbits, and ducks, but only occasionally observed in fish after long periods of deficiency.

Hypervitaminosis A fish and in other animals and involved enlargement of liver and spleen, abnormal growth, skin lesions, epithelial keratinization, hyperplasia. of head cartilage, and abnormal bone formation resulting in ankylosis and fusion of vertebrae. Hypervitaminosis A is reflected in very high liver oil vitamin A content and elevated serum alkaline phosphatase. Removal of excess vitamin A from the diet promotes rapid recovery.

SOURCES AND PROTECTION

Cod liver oil is the best known source of vitamin A although black sea bass, swordfish, or ling cod oils contain much more of the vitamin. Whale liver oil contains kitol which has little or no biological activity until heated above 200° C when one molecule of biologically active vitamin A is generated per molecule of whale kitol.

This biologically inactive kitol may be deposited in the whale to avoid hypervitaminosis A during excessive vitamin A intake. The possibility of hypervitaminosis A occurs when tuna, shark, or ling cod viscera are used in preparation of moist fish diets. Synthetic vitamin A preparations, such as vitamin A palmitate, are available and are often used to supplement rations low in fish meal, fish viscera, or carotenes. Some fish species seem able to utilize b-carotene as a vitamin A source, whereas others are unable to split the g-carotene molecule and vitamin A must be added to the diet.

CLINICAL ASSESSMENT

Vitamin A status can best be assessed by the absence of deficiency signs and by assay of liver oil for vitamin A content. Assay for the vitamin in blood or plasma has not been found to be useful.

VITAMIN D

Rickets was induced with test diets by Hopkins, and Mellanby cured the disease in dogs by adding cod liver oil to the ration. Ultraviolet light exposure was shown to have an anti-rachitic effect, and provitamin D activity had been ascribed to ergosterol. Crystalline vitamin D was isolated by Angus and activated 7-dehydrocholesterol was isolated by Windaus.

Vitamin D_2 or ergocalciferol, is one of several biologically active forms of vitamin D. Vitamin D_3 or activated 7-dehydrocholesterol has the chemical formula $C_{27}H_{44}O$ and contains a more simplified, unsaturated 8-carbon side chain. Vitamin D_3, also known as cholecalciferol, is formed in most animal tissue by the rupture of one of the ring bonds of 7-dehydrocholesterol by ultraviolet radiation. Cholecalciferol is a white, crystalline compound soluble in fat and organic solvents and is stable to heat and oxidation in mild alkali or acid solution.

POSITIVE FUNCTIONS

Vitamin D is essential for maintaining calcium and inorganic phosphate homeostasis. It is involved in alkaline phosphatase activity, promotes intestinal absorption of calcium, and influences the action of parathyroid hormone on bone. Fish may sequester calcium from water through the gill membrane; thus the major function of vitamin D for other animals may not be necessary to satisfy calcium requirements for fish.

SYNDROME OF DEFICIENCY OR EXCESS

Rickets and abnormal bone formation for animals has not been observed in fish fed low vitamin D diets. However little work has been done under carefully controlled conditions with young growing fish and only alteration in alkaline phosphatase activity has been reported on different vitamin D diet intake.

Hypervitaminosis D, however, has been reported. Brook trout fed large doses of vitamin D showed impaired growth, lethargy, and dark colouration. High intake of vitamin D mobilizes phosphorus and calcium from the bone and tissues and may result in fragile bones, poor growth, and poor appetite related to the nausea in man afflicted with hyper-vitaminosis D. This area needs to be explored because of the potential for hypervitaminosis D in fish fed diets containing various fish viscera which might contain large amounts of the vitamin D. Tuna liver oil may contain, for example, 100 to 1 000 times as much active vitamin D as cod liver oil.

SOURCES AND PROTECTION

Vitamin D requirements of many animals can be met by ultraviolet irradiation of the skin where cholesterol derivatives are converted to calciferol. Since the vitamin is fat soluble and accumulates in lipid stores, fish liver oil is a rich source of the material. Content varies tremendously in liver oil, however, with values of about 25 I.U./g present in shark liver oil and over 200 000 I.U./g in albacore tuna liver oil. Cod liver oil contains from 100-500 I.U./g and animal liver contains some vitamin D. One international unit (I.U.) is equal to 0.025 mg of crystalline vitamin D.

CLINICAL ASSESSMENT

Absorption maxima in the ultraviolet region can be used to detect provitamins D in the non-saponifiable fraction of oils. Concentrated preparations of vitamin D can be assayed by the Carr-Price antimony trichloride reaction when assay is necessary to determine biologically active materials in liver oil of fish on different treatments.

The chick assay may not apply to fish liver storage levels because the biologically active form for fish has not been determined. Therefore, clinical assessment for hyper-vitaminosis A must rely on crude methods for determination of vitamin D in liver oil samples by the Carr-Price reaction or by measuring absorption in the UV spectrum.

VITAMIN E

The existence of an antisterility vitamin was postulated by Evans and Bishop. This factor was named'vitamin E' by Sure. The active tocopherol was isolated, characterized, and synthesized by Karrer. Vitamin E is composed of a class of compounds known as tocopherols. One of the most important tocopherols is -tocopherol.

Eight naturally occurring tocopherol derivatives have been isolated and all belong to the D series. Synthetic -tocopherol is a racemic DL- -tocoperhol mixture. The derivatives of tocol or of tocotrienol are named, alpha, beta, gamma, delta, epsilon, eta, zeta1 and zeta2-tocopherol.

The pure tocopherols are fat-soluble oils which are capable of esterification to form crystalline compounds. The tocopherols are stable to heat and acids in the absence of oxygen, but are rapidly oxidized in the presence of nascent oxygen, peroxides, or other oxidizing agents.

The tocopherols are sensitive to ultraviolet light and are excellent antioxidants in the free form. The esters are more stable and are commonly used as dietary supplements-anticipating hydrolysis in the gut and absorption of the free alcohol to act as an active intra- and intercellular antioxidant. Ethyl derivatives on the aromatic ring are also active. Oxidation products of a-tocopherol can be reduced with hydrosulphite to -tocopherylhydroquinone or, in the presence of ascorbic acid, to -tocopherol.

Table: The Tocopherols

Tocopherol (alpha)	5,7,8-trimethyltocol
Tocopherol (beta)	5,8-dimethyltocol
Tocopherol (gamma)	7,8-dimethyltocol
Tocopherol (zeta$_2$)	5,7-dimethyltocol
Tocopherol (eta)	7-methyltocol
Tocopherol (delta)	8-methyltocol
Tocopherol (epsilon)	5,8-dimethyltocotrienol
Tocopherol (zeta$_1$)	5,7,8-trimethyltocotrienol

POSITIVE FUNCTIONS

The tocopherols act as extra and intracellular antioxidants to maintain homeostasis of labile metabolites in the cell and tissue plasma. As physiological antioxidants, these usually protect oxidizable vitamins and labile unsaturated fatty acids. Vitamin E prevents encephalomalacia in chicks, erythrocyte haemolysis in several animals and steatitis in mink, pigs, and farm animals.

The tocopherols also prevent exudative diathesis in chicks, white muscle disease in lambs and calves, and dietary liver necrosis in rats. They act with selenium and with vitamin C for normal reproductive activity and are involved in prevention of nutritional muscular dystrophy in the chick, the yellowtail and carp.

The tocopherols act as free radical traps to stop the chain reaction during peroxide formation and stabilize unsaturated carbon bonds of polyunsaturated fatty acids and other long-chain labile compounds.

Vitamin E is involved in the maintenance of normal blood capillary permeability and the integrity of heart muscle. It was first shown to be involved in prevention of sterility and foetal resorption in rats. It may likewise affect embryo membrane permeability and hatch-ability of fish eggs.

SYNDROME OF DEFICIENCY OR EXCESS

One of the first signs is erythrocyte fragility closely followed by anaemia, ascites, xerophthalmia, poor growth, poor food conversion, epicarditis, and ceroid deposits in spleen and liver. Muscle dystrophy and xerophthalmia have been in yellowtail and carp. Non-specific forms of degenerative conditions have been described in several species of fish fed large quantities of polyunsaturated fatty acids with inadequate tocopherol in the ration. Hypervitaminosis E results in poor growth, toxic liver reaction, and death.

SOURCES AND PROTECTION

Vegetable oils are rich sources of tocopherols. Synthetic -tocopherol in esterfied acetate or phosphate form is commonly used as a diet supplement. These esters are much more stable than the free form which is rapidly lost by air oxidation or in the presence of labile compounds like the polyunsaturated fish oils. Addition of antioxidants such as BHT (butylated hydroxyanisole) and BHT (butylated hydroxytoluene) protects fats and other labile compounds in the ration from oxidation, but these antioxidants have no vitamin E activity.

CLINICAL ASSESSMENT

The erythrocyte fragility test indicates the physiological state of fish. Absence of histologically detectable ceroid in liver and spleen from representative samples in the population is a good clue on the presence of adequate amounts of physiological antioxidants in the fish. A barbituric acid test for oxidation of components has not been applied to fish tissues as a clinical tool for assessment of nutrition state except when liver oils become saturated with very labile polyunsaturated fatty acids such as when feeding squid or saury oils. These assays are good indicators of the state of oxidation of the finished ration, but the absence of deficiency signs and normal erythrocyte fragility are better clinical tests. Analysis for tocopherol is difficult and time-consuming and only applicable under critical research experiment situation.

VITAMIN K

The name vitamin K (for'koagulation') was proposed by Dam. Dam isolated the vitamin from alfalfa and from fish meal in 1939. It was synthesized later that year. The K vitamins K_2 contain 6,7, or 9 isoprene units in the side chain which varies from 30-45 carbon atoms. Many isosteres of vitamin K have been identified in animal tissues, plant tissues, and micro-organisms. Although fairly stable compounds, they are destroyed upon oxidation and exposure to ultraviolet radition. Menadione is very reactive and is subject in aqueous media to chemical interaction.

POSITIVE FUNCTIONS

Vitamin K is involved in the hepatic synthesis of blood clotting proteins - prothrombin and proconvertin. Substituted forms of vitamin K are strongly bacteriostatic and may serve as an alternate defence mechanism for bacterial infections. Vitamin K is structurally related to flavo compounds which function as electron carriers in the electron transport system. The primary role of vitamin K is to maintain a fast normal blood clotting rate which is so important to fish living in a water environment.

SYNDROME OF DEFICIENCY OR EXCESS

Prothrombin time in salmon fed diets devoid of vitamin K was increased three to five times and, during prolonged deficiency states, anaemia and haemorrhagic areas appeared in the gills, eyes, and vascular tissues.

Increased blood clotting time has also been reported for other fish reared on diets with low vitamin K content. Interrelationships with other vitamins have not been documented in fish where the primary deficiency signs are slow blood clotting and haemorrhage, severe anaemia, or death in wounded fish. Haemorrhagic areas often appear in fragile tissues such as the gills.

SOURCES AND PROTECTION

Vitamin K sources are green, leafy vegetables. Alfalfa leaves are among the best sources of vitamin K. Low levels are found in soybeans and animal liver. Synthetic menadione is also available. Although vitamin K in ground alfalfa is fairly stable, synthetic material should be protected from exposure to ultraviolet light and to excessive oxidizing or reducing conditions. The use of rapidly cured dehydrated alfalfa is essential to minimize formation of the physiological antagonist, dicumarol. The diet should be kept dry, prepared with minimum exposure to air oxidation, and fed as soon as practicable after manufacture to minimize vitamin K loss.

ANTIMETABOLITES AND INACTIVATION

Dicumarol was isolated from spoiled sweet clover hay and shown to prevent the normal function of vitamin K in maintenance of normal blood clotting times. Early work was with cattle and hogs but was soon extended to rats, other experimental animals, and man.

Dicumarol is an anticoagulant and has been used to prevent thrombosis in animals and man. It is not an antimetabolite competing for vitamin K sites but plays another role in preventing normal blood clotting. Vitamin K counteracts the dicumarol effect quantitatively. Another anticoa-gulant, warfarin, is a common rat poison which has five to ten times the anticoagulant activity of dicumarol. The effects of warfarin can also be reversed by administration of vitamin K.

CLINICAL ASSESSMENT

Vitamin K status of fish is determined by measuring blood clotting time. Chemical methods for determining menadione content of active metabolite after storage are available but have not been applied to measure vitamin K activity in the liver or tissues of young, growing fish.

CERTAIN FOODS AND BIOTIN

Certain foods, notably liver and kidney, were found to be protective against a form of dermatitis resulting from the consumption of egg white by rats, in the early thirties. It was soon learned that a specific component of egg albumin, avidin, rendered dietary biotin unavailable, hence producing the symptoms. Biotin was variously called coenzyme R and'vitamin H'. It was isolated by Du Vigneaud in 1941 and synthesized by workers of Merck and Company in 1943. Biotin was once thought to be part of factor H for fish. Blue slime patch disease due to biotin deficiency was reported in trout.

Biotin

Biotin is a monocarboxylic acid slightly soluble in water and alcohol and insoluble in organic solvents. Salts of the acid are soluble in water. Aqueous solutions or the dry material are stable at 100° C and to light.

The vitamin is destroyed by acids and alkalis and by oxidizing agents such as peroxides or permanganate. Biocytin is a bound form of biotin isolated from yeast, plant, and animal tissues. Other bound forms of the vitamin can generally be liberated by peptic digestion. Oxybiotin has partial vitamin activity but oxybiotin sulphonic acid and other analogues are antimetabolites inhibiting the growth of bacteria.

Avidin, a protein found in raw egg white, binds biotin and makes it unavailable to fish and other animals. Heating to denature the protein makes the bound biotin available again to the fish. Biocytin or -biotinyl lysine (the epsilon amino group of lysine and the carboxyl of biotin being combined in a peptide bond) is hydrolyzed by the enzyme biotinase making the protein-bound biotin available.

POSITIVE FUNCTIONS

Biotin is required in several specific carboxylation and decarboxylation reactions, including the carboxylation of pyruvic acid to form oxaloacetic acid. It is part of the coenzyme of several carboxylating enzymes fixing CO_2 such as propionyl coenzyme A involved in the conversion of propionic acid to succinic acid in methylmalonyl coenzyme A.

Biotin is also involved in the conversion of acetyl CO_2 to malonyl coenzyme A in the formation of long chain fatty acids. It has possible involvement in citrulline synthesis and may have effects on purine and pyrimidine synthesis. It is involved in the conversion of unsaturated fatty acids to the stable cis form in the synthesis of biologically active fatty acids.

DEFICIENCY SYNDROME

Some signs of biotin deficiency in salmonids are skin disorders, muscle atrophy, lesions in the colon, loss of appetite, and spastic convulsions. Haematology discloses fragmentation of erythrocytes. Poor growth is a common symptom and has been reported for salmonids, common carp, goldfish (Carassius auratus) and eel.

Blue slime patch disease in brook trout deficient in biotin appears typical for this species. Fish reared in 10 -15° C water exhaust biotin stores in 8-12 weeks and the first signs are anorexia, poor food conversion, and general listlessness before the more acute deficiency symptoms become detectable.

Sources and Protection

Rich sources of biotin are liver, kidney, yeast, milk products, and egg yolks. Nut meats contain good supplies of biotin. The diet should be protected from strong oxidizing agents or conditions which promote oxidation of ingredients. Raw egg white should not be incorporated into moist fish diets. Cooking will inactivate the avidin.

Antimetabolites and Inactivation

Many biotin homologues with different side chain lengths inhibit the growth of bacteria. Oxybiotin, in which sulphur is replaced by oxygen, has about the same biological activity as the natural biotin. On the other hand, oxybiotin sulphonic acid inhibits biotin activity.

Clinical Assessment

Measurement or urinary excretion of biotin in animals does not prove a good clinical method since biotin is synthesized by several organisms in the gut. In fish, this technique needs to be explored and may well be valuable as a clinical tool since the gastrointestinal tract of many freshwater fish contains only small quantities of bacteria. Biotin is one of the most expensive vitamins to add to fish rations. Salmon feeding actively in the oceans had liver biotin concentrations of 10-12 mg of wet liver tissue. The concentrations in the liver of young salmon fingerlings in fresh water fed test diets containing an excess of biotin were between 6-8 mg of the vitamin/g of tissue. Fish with these levels of biotin in the liver should probably be in sound biotin nutritional status.

FOLIC ACID (FOLACIN)

A factor effective in curing metaloblastic anaemia in monkeys was found in yeast and liver concentrates and was designated vitamin M in 1935. In 1939 an anti-anaemic factor was found in liver and called vitamin BC. These were later shown to be the same substance with folic acid as the active ingredient. Folic acid was synthesized in 1946 and was soon used in fish diets for preventing purified diet anaemia.

Folic Acid

Folic acid crystallizes into yellow spear-shaped leaflets which are soluble in water and dilute alcohol. It can be precipitated with heavy metal salts. It is stable to heat in neutral or alkaline solution, but unstable in acid solution. It deteriorates when exposed to sunlight, or during prolonged storage. Several analogues have biological activity including pteroic acid, rhizopterin, folinic acid, xanthopterin and several formyltetrahydropteroyl-glutamic acid derivatives. These have closely allied ring structures and many have been isolated as derivatives in various animals or microbiological preparations. One simple form, xanthopterin, present in the pigments of insects, is shown below and is of special interest because of early work with this compound as the anti-anaemic factor H for fish.

Positive Functions

Folic acid is required for normal blood cell formation and is involved as a coenzyme in one-carbon transfer mechanisms. In the presence of ascorbic acid, folic acid is transformed into the active (5-formyl-5, 6,7,8) tetrahydrofolic acid. Folic acid is involved in many one-carbon metabolism systems such as serine and glycine interconversion, methionine-homocysteine synthesis, histidine synthesis, and pyrimidine synthesis.

Several coenzyme forms of the active vitamin have been isolated. Folic acid is involved in the conversion of megaloblastic bone marrow to normoblastic type. It has a role in blood glucose regulation and improves cell membrane function and hatchability of eggs.

Deficiency Syndrome

Macrocytic normochromic anaemia occurs in several experimental animals, including fish fed diets devoid of the folic acid. Increasing numbers of senile cells are observed as the deficiency progresses until only a few old and degenerating cells are found in the blood of deficient fish.

Anterior kidney imprints disclose only adult cells and no preforms present. Other signs observed have been poor growth, anorexia, general anaemia, lethargy, fragile fins, dark skin pigmentation, and infarction of spleen.

SOURCES AND PROTECTION

Yeast, green vegetables, liver, kidney, glandular tissue, fish tissue, and fish viscera are good sources of folic acid. Insects contain xanthopterin which has folic acid activity. At one time the yellow pigment of xanthopterin was identified as the fish anti-anaemic factor H, but subsequent experiments showed only partial activity and that folic acid itself was a much more potent antimacrocytic anaemia factor.

Insects may contribute significantly to the folic acid requirements of wild fish, but in scientific fish husbandry artificial diets are more reliable sources.

Activity is lost during extended storage and when material is exposed to sunlight. Therefore, dry feeds should be carefully protected during manufacture and moist diet rations should be carefully preserved. Both types of fish diets should be fed soon after manufacture to assure minimal loss of folic acid activity.

ANTIMETABOLITES AND INACTIVATION

One antagonist of folacin is 4-aminopteroylglutamic acid or aminopterin.

This material, when incorporated in the diet of guinea pigs and rats, induces anaemia and leucopenia and has been used to treat leukemia in man. Amethopterin (4-amino-N10-methylpteroylglutamic acid) can also be used to induce deficiency by inhibiting purine, pyrimidine, and nucleic acid protection; *viz.*, inhibition of nucleic acid synthesis results in macroytic anaemia.

CLINICAL ASSESSMENT

Haematology is used as a simple clinical tool to assess haemopoiesis in fish. Anterior kidney imprints easily disclose normal distribution of immature cells and preforms undergoing reticulosis.

Salmon actively feeding in the ocean and young salmon fingerlings raised in fresh water and fed diets rich in folacin show liver storage of 3-4ωg of folic acid/g of wet tissue. Microbiological assay is preferred for assessment of total folic acid in dietary raw materials because the total biological activity it measures includes all the various coenzyme forms and folic acid analogues.

Assessment of the dietary intake of folic acid is important for intensive cold water fish husbandry. In pond culture, aquatic and terrestrial insects, algae, etc., may also be available and folic acid in the supplementary feed may not be as critical. Since folic acid is labile in storage, excess amounts are generally added to manufactured feed in anticipation of storage losses.

However, prudent fish husbandry dictates rapid use of manufactured rations with minimum storage. Routine periodic haematology of fish assures proper nutritional status for maximum production and sound health.

The author has noted in several series of experiments that when fish diseases occur through inadvertent contamination of the water supply, those groups of fish partially or completely deficient in folic acid were among the first to show acute disease symptoms. Therefore, folic acid must also play an important role in resistance to disease.

VITAMIN B12

The antipernicious anaemia factor found to be contained in liver was isolated and crystallized in the mid forties. This substance, named vitamin B12 by its discoverers, was later to be recognized as essential for growth of chicken fed diets entirely of plant origin and was designated animal protein factor (APF).

When anaemic salmon were injected with crystalline B12 in combination with folic acid and xanthopterin positive haemopoiesis occurred within a few days, and the salmon showed rapid recovery from the anaemia. Vitamin B12 or cynacobalamin has the following approximate chemical structure.

Cyanocobalamin

The molecule has a planar group and a nucleotide group lying nearly at right angles to one another. This cobalt-containing vitamin has a net charge of one at the central cobalt atom to which is attached a replaceable cyano group. Vitamin B12 is stable to mild heat in neutral solution, but is rapidly destroyed by heating in dilute acid or alkali. Crude concentrates are more unstable and rapidly lose activity.

The compound is similar to the porphyrins in its spatial configuration with a central cobalt atom linked to four reduced pyrrole rings in the haeme series. Replacing the cyanide ion with a variety of anions produce derivatives which have comparable biological activities; *viz.*, hydroxocobalamin, nitritocobalamin, chlorocobalamin and sulphatocobalamin.

Positive Functions

Cyanocobalamin is involved with folic acid in haemopoiesis. It is required by many micro-organisms and is a growth factor for many animals. The animal protein factor present in fish and animal by-products was not recognized until crystalline vitamin B12 was injected into anaemic chinook salmon fingerlings in 1949 and positive haemopoiesis was observed. A coenzyme incorporating vitamin B12 is involved in the reversible isomerization of methyl-malonyl coenzyme A to succinyl coenzyme A and in the isomerization of methylaspartate to glutomate. Cyanocobalamin is involved in the coenzyme for the methylation of homocystine to form methionine.

It is also involved in several other one-carbon reactions and in the synthesis of labile methyl compounds. One vitamin B12 containing coenzyme acts in methylation of the purine ring during thymine synthesis. Vitamin B12 is also involved in cholesterol metabolism, in purine and pyrimidine biosynthesis, and in the metabolism of glycols.

Deficiency Syndrome

Deficiency signs in young pigs, chicks, and rats show abnormal blood elements, poor growth, and pernicious anaemia. An intrinsic factor is necessary for good absorption of the vitamin from the gut.

This factor is a low molecular weight mucoprotein which normally occurs in gastric juice, and especially in hog gut mucosa. Pernicious anaemia in chinook and coho salmon is characterized by fragmented, erythrocytes with many aberrant forms present.

Haemoglobin levels are inconsistent and erythrocyte counts have a range extending from frank anaemia to a near normal blood pattern. Cyanocobalamin stores in fish tissues are slowly exhausted and only after 12-16 weeks on test do the symptoms appear in deficient salmon populations. Poor appetite, poor growth, poor food conversion, and some dark pigmentation can be observed before frank anaemia is detected.

Sources and Protection

Rich sources of vitamin B12 are found in fish meal, fish viscera, liver, kidney, glandular tissues, and slaughter house wastes. Since vitamin B12 is labile on storage, and in mild acid solution is easily destroyed by heating, care must be exercised in diet preparation containing flesh or meat scraps.

Antimetabolites and Inactivation.

The vitamin B12 coenzymes are very unstable under light, which rapidly decomposes the coenzymes. Photo-sensitivity is increased in dilute acid solutions.

Clinical Assessment

Generalized anaemia with fragmentation of erythrocytes and extremely variable haemoglobin levels and erythrocyte counts indicate possible B12 deficiency. Prompt response in individual fish is obtained by injecting B12 alone or in combination with folic acid in the ratio of 1 part vitamin B12 to 100 parts folic acid. Careful interpretation of haemotological data will enable one to distinguish one form of anaemia from the other.

ASCORBIC ACID

The isolation of ascorbic acid was due to Szent-Györgyi of Hungary and C.G. King of the U.S.A. Vitamin C synthesis was accomplished in 1933 after the chemical structure of ascorbic acid was established by British and Swiss workers. McCay and Tunison reported scoliosis in brook trout fed formalin-preserved meat in 1934 and McLaren observe haemorrhages in trout fed rations low in ascorbic acid. It was not until the sixties that a critical need for L-ascorbic acid by trout and salmon was demonstrated. L-ascorbic acid is a white, odourless, crystalline compound, soluble in water but insoluble in fat solvents.

It is readily oxidized to dehydroascorbic acid, the less biologically potent form. Ascorbic acid is very stable in acid solution because of the preservation of the lactone ring, but in alkaline solution hydrolysis occurs rapidly and vitamin activity is lost. It is very heat labile and prone to atmospheric oxidation, especially in the presence of copper, iron, or several other metallic catalysts. The reduced form is the most biologically active but several derivatives or salts are obtainable which have varying degrees of ascorbate activity.

Positive Functions

L-ascorbic acid acts as a biological reducing agent in hydrogen transport. It is involved in many enzyme systems for hydroxylation; *i.e.*, hydroxylation of tryptophan, tyrosine, or proline. It is involved in the detoxification of aromatic drugs and also acts in the production of adrenal steroids.

Ascorbic acid is necessary for the formation of hydroxy proline which is a constituent of collagen, a component of intercellular material in bones and soft tissues. Ascorbic acid plays a synergistic role with vitamin E as intracellular antioxidants and free radical traps. The conversion of folic acid to folinic acid requires vitamin C. Ascorbic acid is involved in the formation of chondroitin sulphate and intercellular ground substance.

Labelled ascorbic acid fed to fish previously deficient in the vitamin was shown to be rapidly mobilized and fixed in areas of rapid collagen synthesis and became concentrated in the thick collagen of the skin and in cartilagenous bones, as well as in the glands of the anterior kidney. Ascorbic acid is also involved in erythrocyte maturation.

Deficiency Syndrome

Scurvy with impaired collagen formation, perifollicular haemorrhages, loose teeth and poor bone formation, anaemia, and oedema have been reported in other animals. Deficiency signs in fish are generally related to impaired collagen formation. Fish soon show hyperplasia of jaw and snout. The same symptoms have been observed in trout, salmon, yellowtail, carp, guppies and char.

Histologically, hypertrophy of the adrenal tissues and haemorrhage at the bases of fins have been observed in coho salmon. Deficiency signs cease to develop and new growth becomes normal upon replacement of ascorbic acid in the ration. Anaemia eventually develops in extremely deficient fish and scoliosis and lordosis do not repair but are walled off by new growth around the afflicted areas of the spine when ascorbic acid is once again added to the ration.

Requirements

In the rainbow trout, reasonable blood and anterior kidney storage levels were obtained with an intake of about 100 mg of vitamin C/kg of dry ration in 10°, 12°, or 15° C water systems. When wound repair experiments were initiated, however, or when fish were exposed to other stress then the requirements doubled or tripled.

When severe abdominal or intramuscular wounds were inflicted, young fish needed at least 500 mg of active ascorbate for tissue repair comparable with control fish receiving 1 g or more of ascorbate in the diet/kg of dry diet. Coho salmon appear to need about half of these requirements for adequate tissue levels and for maximum severe wound repair rates.

The requirement for ascorbic acid is related to stress, growth rate and size of the animal, as well as to the other nutrients present in the diet. A compromise value of about 200 mg of ascorbic acid/kg diet for trout and salmon raised in freshwater systems between 10-15°C would ensure reasonable tissue storage levels and furnish some excess for mild stress conditions and for ascorbic acid loss from the diet through oxidation.

Large common carp can synthesize some ascorbate and the requirement for this species may be dependent on fish size and the environment in which they are reared.

Table. Growth and Tissue Ascorbate

Vitamin C diet treatment	Trout			Salmon		
	Average weight at 24 weeks	*Ascorbate concentrate* [1/]		*Average weight at 24 weeks*	*Ascorbate concentrate* [1/]	
		blood	*kidney*		*blood*	*kidney*
mg/100 g	*g*	*g/g*	*g/g*	*g*	*g/g*	s/s
0	2.4	[2/]	[2/]	5.0	22.3±2.2	89
5	9.6	34.4±2.9	125	6.0	30.5±1.2	132
10	10.6	34.6±1.3	137	5.7	35.8±1.6	265
20	10.1	38.8±3.3	132	6.1	34.2±2.3	183
40	10.2	46.8±6.2	162	6.3	33.7±2.0	225
100	10.8	51.0±4.6	247	6.0	37.8±2.3	321

1/ Average of five samples for blood (± S.D.) and two for head kidney tissue

2/ No fish available for assay

Sources and Protection

Ascorbic acid is widely distributed in nature with citrus fruits, cabbage, liver, and kidney tissue good sources for the vitamin. Fresh insects and fish tissues contain reasonable amounts of the vitamin. Synthetic ascorbic acid is also readily available. Fish food should be protected from oxidizing agents and kept sealed or frozen until used to prevent loss of the vitamin.

Antimetabolites and Inactivation

D-ascorbic acid, the optical isomer of the active form, has no vitamin activity and competes for sites of several enzyme reactions mediated by L-ascorbic acid. 6-deoxy-L-ascorbic acid, dehydroascorbic acid and L-glucoascorbic acid have very little activity.

Clinical Assessment

Ascorbic acid status for experimental animals is normally attempted by tissue ascorbate analysis. Most of the assays previously used measure total ascorbate and not biologically active L-ascorbic acid. Consequently these are fraught with errors and misconception of true vitamin C status.

In fish tissues, blood and liver do not adequately reflect the ascorbic acid intake and status, whereas assay of the anterior kidney which contains adrenal tissue provides a fairly representative picture of tissue storage of the vitamin. Stress rapidly reduces the ascorbic acid content of this tissue with concurrent production of adrenal steroids. Conversely, dietary repletion is reflected by up to four or five-fold storage levels from the deficient state. Examination of fragile support cartilage in the gill filaments under low magnification will detect early hypovitaminosis before clinically acute symptoms become noticeable.

However, the best tissue for routine clinical analysis to assess vitamin C status in trout and salmon appears to be the anterior kidney with samples selected from the junction of the two forward wings. Tissues are blotted free of blood with filter paper, and then assayed for total ascorbate by one of the improved quick methods to determine total ascorbate.

Inositol

Muscle'sugar' was discovered by Scherer in 1850 and was characterized as inositol in 1887. Inositol was shown to be an alopecia (a form of hairlessness) preventing factor for mice in 1940. Poor growth and poor food passage in inositol-deficient fish was observed in salmon and carp.

Inositol

Seven optically inactive and two optically active isomers of hexahydroxycyclohexane can exist. One of the optically active forms, myo-inositol, is a white crystalline powder soluble in water and insoluble in alcohol and ether. The material can be synthesized, but is easily isolated from biological material in free or combined forms.

The mixed calcium-magnesium salt of the hexophosphate is phytin. Isomers have little biological activity but will compete in chemical reactions. Inositol is a highly stable compound.

POSITIVE FUNCTIONS

Myo-inositol is a structural component in living tissues. It has lipotropic action by preventing accumulation of cholesterol in one type of fatty liver disease and is involved with choline in maintaining normal lipid/metabolism.

It is a growth-promoting substance for micro-organisms and prevents an alopecia in mice. It is a reserve carbohydrate in muscle as well as a major component of phosphoglycerides in animal tissues.

DEFICIENCY SYNDROME

Poor growth, increased gastric emptying time, oedema, dark colour, and distended stomachs are symptoms observed in salmon, trout, carp and catfish held for long periods on inositol-deficient test rations.

A'spectacle-eye' condition for rats has not been observed under the experimental conditions used in fish studies. The major deficiency sign is inefficiency in digestion and food utilization and concomitant poor growth leading to a population of fish with distended abdomens.

SOURCES AND PROTECTION

Myo-inositol occurs ubiquitously in large amounts wherever biological tissue is found. Wheat germ, dried peas, and beans are rich sources. Brain, heart, and glandular tissues are very good sources of biologically active inositol. Citrus fruit pulp and dried yeast also contain inositol. The compound is stable.

ANTIMETABOLITES AND INACTIVATION

Seven optically inactive and one optically active but biologically inactive stereo isomers occur. Since inositol is synthesized in the biologically active form by many microorganisms in the gut, only large quantities of chemically synthesized inactive isomers added to diets would interfere with the metabolism of inositol for growth and normal physiological function.

Biologically inactive stereoisomers of myo-inositol do not compete for critical sites in metabolism. Methyl derivatives and mono-, di- and triphosphoric acid esters occur naturally. Salts of the hexaphosphate or phytin make the bound inositol practically unavailable to the animal.

CLINICAL ASSESSMENT

Assessment in fish has been based on lack of deficiency signs coupled with the most efficient food conversion. Salmon feeding actively in oceans show 1-1.5 mg of inositol/g of fresh liver tissue and young fingerlings raised in freshwater at 10-15°C had 600-700ωg/g of liver tissue. An alternate, better assessment may be based on a standard muscle section or whole carcass analysis for free or for bound inositol. Projection of inositol intake from normal fish diet ingredients should indicate an excess of this particular vitamin.

CHOLINE

Methylation as a basic metabolic process was postulated by Hoffmeister. Methyl transfer was shown in vivo by Thompson and the interrelationships between choline, methionine, and homo-cystine were shown by du Vigneaud in 1939-42. Trout fed low choline rations developed haemorrhagic kidneys and salmon showed an aversion to food in choline-deficient diet.

Choline

Choline is a very strong organic base and forms many derivatives widely distributed in animal and vegetable tissue. One derivative, acetylcholine, is involved in the transmission of nerve impulses across synapses. Choline is very hygro-scopic, very soluble in water, and is stable to heat in acid, but decomposes in alkaline solutions.

Positive Functions

Choline acts as a methyl donor in trans-methylation reactions. It is a lipotropic and antihaemorrhagic factor preventing the development of fatty livers. It is involved in the synthesis of phospholipids and in fat transport. Acetylcholine transmits the excitory state across the ganglionic synapses and neuromuscular junctions. Choline is essential for growth and good food conversion in fish.

Deficiency Syndrome

Deficiency signs include poor growth and poor food conversion accompanied by impaired fat metabolism. Haemorrhagic kidneys and intestines have been reported in trout and increased gastric emptying time has been observed in salmon.

Sources and Protection

Rich sources of choline are wheat germ, beans, brain, and heart tissue. Choline hydrochloride, the commercially available form, may inactivate -tocopherol and vitamin K when in direct contact with these vitamins and care should be exercised in selecting properly protected (gelatin coated) fat soluble vitamins in fish diet preparation.

Clinical Assessment

Choline status of fish can be estimated from the assay of choline content of the dietary ingredients with the absence of deficiency signs. Maximum liver storage may not be the best criteria to determine choline nutritional status but has been used to assess the tentative requirement listed for the two species of salmon.

p-AMINOBENZOIC ACID

Para-aminobenzoic acid is a white crystalline powder which is water soluble and heat and light stable in aqueous and mild alkaline solution. It is a growth promoting vitamin for micro-organisms which require the vitamin for folic acid synthesis. Large intake has been shown to counteract the antimetabolite effect of sulphonamides in bacterial culture. No positive function or deficiency signs have been observed in fish.

LIPOIC ACID

Lipoic acid is both fat soluble and water soluble. Its active form is the amide derivative known as lipoamide. Lipoic acid functions as a coenzyme in a -ketoacid decarboxylation. It was discovered independently in several laboratories during the period 1945-50 and shown to be an essential component of multienzyme, the pyruvate dehydrogenase complex. Pyruvate is converted to'active acetaldehyde' which in turn is picked up by lipoamide.

Subsequent oxidation of the aldehyde results in the reduction of lipoamide to a disulphydryl form. The multi-enzyme unit also includes thiamine pyrophosphate, coenzyme A, and flavin adenine dinucleotide. Glandular tissues are good sources of lipoic acid. No requirements have been determined for fish.

8

Nucleic Acids

The first isolation of what we now refer to as DNA was accomplished by Johann Friedrich Miescher circa 1870. He reported finding a weakly acidic substance of unknown function in the nuclei of human white blood cells, and named this material "nuclein".

A few years later, Miescher separated nuclein into protein and nucleic acid components. In the 1920's nucleic acids were found to be major components of chromosomes, small gene-carrying bodies in the nuclei of complex cells. Elemental analysis of nucleic acids showed the presence of phosphorus, in addition to the usual C, H, N & O.

Unlike proteins, nucleic acids contained no sulfur. Complete hydrolysis of chromosomal nucleic acids gave inorganic phosphate, 2-deoxyribose (a previously unknown sugar) and four different heterocyclic bases. To reflect the unusual sugar component, chromosomal nucleic acids are called deoxyribonucleic acids, abbreviated DNA. Analogous nucleic acids in which the sugar component is ribose are termed ribonucleic acids, abbreviated RNA. The acidic character of the nucleic acids was attributed to the phosphoric acid moiety.

The two monocyclic bases shown here are classified as pyrimidines, and the two bicyclic bases are purines. Each has at least one N-H site at which an organic substituent may be attached. They are all polyfunctional bases, and may exist in tautomeric forms.

Base-catalyzed hydrolysis of DNA gave four nucleoside products, which proved to be N-glycosides of 2'-deoxyribose combined with the heterocyclic amines.The base components are coloured green, and the sugar is black. As noted in the 2'-deoxycytidine structure on the left, the numbering of the sugar carbons makes use of primed numbers to distinguish them from the heterocyclic base sites.

The corresponding N-glycosides of the common sugar ribose are the building blocks of RNA, and are named adenosine, cytidine, guanosine and uridine (a thymidine analog missing the methyl group).

From this evidence, nucleic acids may be formulated as alternating copolymers of phosphoric acid (P) and nucleosides (N),

$$\sim P - N - P - N' - P - N'' - P - N''' - P - N \sim$$

At first the four nucleosides, distinguished by prime marks in this crude formula, were assumed to be present in equal amounts, resulting in a uniform structure, such as that of starch. However, a compound of this kind, presumably common to all organisms, was considered too simple to hold the hereditary information known to reside in the chromosomes. This view was challenged in 1944, when Oswald Avery and colleagues demonstrated that bacterial DNA was likely the genetic agent that carried information from one organism to another in a process called "transformation". He concluded that "nucleic acids must be regarded as possessing biological specificity, the chemical basis of which is as yet undetermined." Despite this finding, many scientists continued to believe that chromosomal proteins, which differ across species, between individuals, and even within a given organism, were the locus of an organism's genetic information.

It should be noted that single celled organisms like bacteria do not have a well-defined nucleus. Instead, their single chromosome is associated with specific proteins in a region called a "nucleoid". Nevertheless, the DNA from bacteria has the same composition and general structure as that from multicellular organisms, including human beings.

Views about the role of DNA in inheritance changed in the late 1940's and early 1950's. By conducting a careful analysis of DNA from many sources, Erwin Chargaff found its composition to be species specific. In addition, he found that the amount of adenine (A) always equaled the amount of thymine (T), and the amount of guanine (G) always equaled the amount of cytosine (C), regardless of the DNA source. As set forth in the following table, the ratio of (A+T) to (C+G) varied from 2.70 to 0.35. The last two organisms are bacteria.

DNA AND THE CHEMISTRY OF INHERITANCE

Living organisms are unique, among the known forms of matter, because they are capable of creating their own specific, highly organized structure out of substances taken from far more disorganized surroundings, and can transmit this capability to their offspring.

Perhaps the oldest and most profound theoretical problem in biology is the effort to explain the curious paradox that, despite its unique capability for self-duplication and inheritance, a living organism is nevertheless a mixture of substances which are separately no more possessed of these properties than are the more prosaic molecules that never occur in living cells.

This question has been at the root of a long train of experiments, debates, and speculations that begins in classical times and continues unbroken through the development of present-day "molecular biology."

The basic issues are simply stated. If the component parts of a cell are not themselves alive, whence come the life-properties exhibited by the whole? Apart from the untenable notion of a mystic non-material "vital force" which supposedly animates the otherwise dead substance of the cell, the debate has elicited two main positions:

- There is, in fact, some special cellular component which possesses the fundamental attribute of self-duplication, and which is therefore a "living molecule" and the basic source of the life-properties of the cell.
- The unique properties of life are inherently connected with the very considerable complexity of living substance and arise from interactions among its separable constituents which are not exhibited unless these components occur together in the complex whole. In this view, only the entire living cell is capable of selfduplication.

There is at this time a widespread impression that this issue has now been resolved and that the cell does indeed contain a component—DNA—which, according to the theory of the "DNA code," possesses the basic attribute of life, self-duplication, and which guides the activities of all the inheritable processes of the cell.

The importance of this conclusion is self-evident, for it would answer, at last, the basic question of the origin of the unique properties of life, and, if correct, should lead to unprecedented technological control over these properties. It is appropriate, therefore, to ask what criteria are required to establish that a molecule, such as DNA, is capable of self-duplication, to examine the degree to which the available evidence meets such criteria, and to determine whether the undoubted importance of DNA in the biology of inheritance may be due to some properties other than those attributed to it by current theory.

SOME IMPLICATIONS

The chief conclusion to be derived from the foregoing considerations is that the unique capability of living organisms for self-duplication and inheritance arises from complex multi-molecular interactions among at least several classes of cellular components.

Neither DNA nor any other cellular component is a "self-duplicating molecule" or the "master chemical of the cell"—terms which sometimes appear in current generalizations. There is no evidence from recent investigations of the bio-chemical aspects of genetics which requires abandonment of the conclusion, long established by biological data, that the least complex agent capable of selfduplication is the intact living cell.

The point of view developed here also suggests alternatives to a number of current views of phenomena related to inheritance:

- Various hypothetical schemes for regulating the "activity" of template genes are invoked in order to explain how the DNA code might operate in cellular differentiation and development. Alternatively, it may be suggested that differentiation is associated with the nucleotide sequestration system, tissue-specific changes in DNA snythesis (*e.g.*, at chromosome "puffs") regulating nucleotide levels and thereby governing the size and metabolic character of the cell.

- The relationship between DNA and a species' characteristic sensitivity to ionizing radiation is often interpreted in terms of radiation-induced mutations in genetic templates. Alternatively, radiation-induced inhibition of DNA synthesis, which results in the accumulation of toxic concentrations of free nucleotides, may be regarded as the mediating process. This view is compatible with Sparrow's observation that in certain non-polyploid plants, radiation sensitivity is proportional to cellular DNA content, and is particularly affected by the relative proportions of heterochromatin.
- Sahasrabudhe has suggested that tumor cells may be characterized by marked changes in free nucleotide level resulting from nucleotide sequestration due to DNA synthesis. In this connection, it is also of interest that tumor and other rapidly growing cells exhibit distinctive pathways of oxidative metabolism, which are in turn regulated by free nucleotide levels.
- The theory of the DNA code suggests a seemingly simple explanation for the emergence of the first forms of life from the prebiotic organic environment: that life began with the fortuitous appearance of a "self-duplicating nucleic acid molecule" which then organized the complex chemistry of life around itself. In contrast, the viewpoint developed here suggests that the primitive function of nucleic acid was to sequester free nucleotides, and thereby regulate over-all metabolic activity. Thus, the nucleic acid template is to be regarded as a late development which improved the precision of the earlier modes of regulation of cellular activity, and which is, in any case, incapable of self-duplication.

The theory of the DNA code, if correct, also leads to important expectations regarding the feasibility of technological control over inheritance. If the nucleotide sequence of DNA were indeed a self-sufficient source of the inherited specificity of living things, then it might be possible, in the not too distant future, to synthesize artificial DNA molecules, which, on being introduced into living organisms, would artificially establish new inheritable characteristics quite outside the range of those normally observed.

However, if, as we have concluded, biological specificity is only partly due to DNA, then it is likely that the specificity represented by proteins may impose severe limits on the acceptability, to the cell, of abnormal DNA nucleotide sequences. This view suggests that technological control over biological inheritance may be far less plausible than indicated by inferences drawn from the code theory. Finally, the viewpoint developed here bears on some fundamental questions regarding the relationship between physical theory and biology. The theory of the DNA code is often regarded as an example of the success with which "modern physics" can solve basic problems about living systems which have eluded the supposedly less critical analyses of classical biologists.

This view leads to the generalization that, despite its obvious complexity, the living cell must be governed by the "laws of physics"—as revealed by analysis of non-living systems—and that the most effective strategy for elucidating its unique properties is to study isolated parts which are sufficiently simple to permit their analyses in physico-chemical terms.

However, several developments in theoretical physics suggest that the fundamental properties of matter are in better harmony with the view that the unique properties of the cell are derived from interactions of its molecular parts and are inherently incapable of being elucidated by a simple summation of the observed properties of the isolated parts.

This viewpoint is a direct outgrowth of Bohr's theory of complementarity. It may also be closely related to the more recent theory of the "Bootstrap Universe." This theory suggests that the unique properties of the atomic nucleus are inherently associated with its complexity, and are not accountable by the observed properties of fragments isolated from disrupted nuclei.

It may be suggested then, that, if biology is to be guided by the insights into the properties of matter that are afforded by modern physical theory, the role of DNA in the living cell must be viewed as subsumed under the complex properties of the system—the living cell—of which functional DNA is a part. The theory of the DNA code is sometimes epitomized by the statement "DNA is the secret of life," an aphorism which appears increasingly to guide the course of current biological investigations. The viewpoint developed here suggests that biology might be more wisely guided by the aphorism, "Life is the secret of DNA."

THE CHEMICAL NATURE OF DNA

The polymeric structure of DNA may be described in terms of monomeric units of increasing complexity. In the top shaded box of the following illustration, the three relatively simple components mentioned earlier are shown. Below that on the left, formulas for phosphoric acid and a nucleoside are drawn. Condensation polymerization of these leads to the DNA formulation outlined above. Finally, a 5'- monophosphate ester, called a nucleotide may be drawn as a single monomer unit, shown in the shaded box to the right. Since a monophosphate ester of this kind is a strong acid (pKa of 1.0), it will be fully ionized at the usual physiological pH. Isomeric 3'-monophospate nucleotides are also known, and both isomers are found in cells. They may be obtained by selective hydrolysis of DNA through the action of nuclease enzymes. Anhydride-like di- and tri-phosphate nucleotides have been identified as important energy carriers in biochemical reactions, the most common being ATP (adenosine 5'-triphosphate).

Table: Names of DNA Base Derivatives

Base	Nucleoside	5'-Nucleotide
Adenine	2'-Deoxyadenosine	2'-Deoxyadenosine-
5'-monophos-phate		
Cytosine	2'-Deoxycytidine	2'-Deoxycytidine-5'-monophos-phate
Guanine	2'-Deoxyguanosine	2'-Deoxyguanosine-
5'-monophos-phate		
Thymine	2'-Deoxythymidine	2'-Deoxythymidine-
5'-monophos-phate		

A complete structural representation of a segment of the DNA polymer formed from 5'-nucleotides. Several important characteristics of this formula should be noted.

- First, the remaining P-OH function is quite acidic and is completely ionized in biological systems.
- Second, the polymer chain is structurally directed.
- Third, although this appears to be a relatively simple polymer, the possible permutations of the four nucleosides in the chain become very large as the chain lengthens.
- Fourth, the DNA polymer is much larger than originally believed. Molecular weights for the DNA from multicellular organisms are commonly 109 or greater.

Information is stored or encoded in the DNA polymer by the pattern in which the four nucleotides are arranged. To access this information the pattern must be "read" in a linear fashion, just as a bar code is read at a supermarket checkout. Because living organisms are extremely complex, a correspondingly large amount of information related to this complexity must be stored in the DNA.

Consequently, the DNA itself must be very large, as noted above. Even the single DNA molecule from an E. coli bacterium is found to have roughly a million nucleotide units in a polymer strand, and would reach a millimeter in length if stretched out.

The nuclei of multicellular organisms incorporate chromosomes; which are composed of DNA combined with nuclear proteins called histones. The fruit fly has 8 chromosomes, humans have 46 and dogs 78 (note that the amount of DNA in a cell's nucleus does not correlate with the number of chromosomes). The DNA from the smallest human chromosome is over ten times larger than E. coli DNA, and it has been estimated that the total DNA in a human cell would extend to 2 meters in length if unraveled. Since the nucleus is only about 5?m in diameter, the chromosomal DNA must be packed tightly to fit in that small volume.

In addition to its role as a stable informational library, chromosomal DNA must be structured or organized in such a way that the chemical machinery of the cell will have easy access to that information, in order to make important molecules such as polypeptides. Furthermore, accurate copies of the DNA code must be created as cells divide, with the replicated DNA molecules passed on to subsequent cell generations, as well as to progeny of the organism. The nature of this DNA organization, or secondary structure, will be discussed in a later section.

RNA, A DIFFERENT NUCLEIC ACID

The high molecular weight nucleic acid, DNA, is found chiefly in the nuclei of complex cells, known as eucaryotic cells, or in the nucleoid regions of procaryotic cells, such as bacteria. It is often associated with proteins that help to pack it in a usable fashion.

In contrast, a lower molecular weight, but much more abundant nucleic acid, RNA, is distributed throughout the cell, most commonly in small numerous organelles called ribosomes. Three kinds of RNA are identified, the largest subgroup being ribosomal RNA, rRNA, the major component of ribosomes, together with proteins. The size of rRNA molecules varies, but is generally less than a thousandth the size of DNA. The other forms of RNA are messenger RNA, mRNA, and transfer RNA, tRNA. Both have a more transient existence and are smaller than rRNA.

All these RNA's have similar constitutions, and differ from DNA in two important respects.The sugar component of RNA is ribose, and the pyrimidine base uracil replaces the thymine base of DNA. The RNA's play a vital role in the transfer of information (transcription) from the DNA library to the protein factories called ribosomes, and in the interpretation of that information (translation) for the synthesis of specific polypeptides.

THE SECONDARY STRUCTURE OF DNA

In the early 1950's the primary structure of DNA was well established, but a firm understanding of its secondary structure was lacking. Indeed, the situation was similar to that occupied by the proteins a decade earlier, before the alpha helix and pleated sheet structures were proposed by Linus Pauling. Many researchers grappled with this problem, and it was generally conceded that the molar equivalences of base pairs (A & T and C & G) discovered by Chargaff would be an important factor.

Rosalind Franklin, working at King's College, London, obtained X-ray diffraction evidence that suggested a long helical structure of uniform thickness. Francis Crick and James Watson, at Cambridge University, considered hydrogen bonded base pairing interactions, and arrived at a double stranded helical model that satisfied most of the known facts, and has been confirmed by subsequent findings.

BASE PAIRING

Careful examination of the purine and pyrimidine base components of the nucleotides reveals that three of them could exist as hydroxy pyrimidine or purine tautomers, having an aromatic heterocyclic ring. Despite the added stabilization of an aromatic ring, these compounds prefer to adopt amide-like structures. These options are shown in the following diagram, with the more stable tautomer drawn in blue.

A simple model for this tautomerism is provided by 2-hydroxypyridine. A compound having this structure might be expected to have phenol-like characteristics, such as an acidic hydroxyl group. However, the boiling point of the actual substance is 100º C greater than phenol and its acidity is 100 times less than expected (pKa = 11.7). These differences agree with the 2-pyridone tautomer, the stable form of the zwitterionic internal salt.

Note that this tautomerism reverses the hydrogen bonding behaviour of the nitrogen and oxygen functions,the N-H group of the pyridone becomes a hydrogen bond donor and the carbonyl oxygen an acceptor.

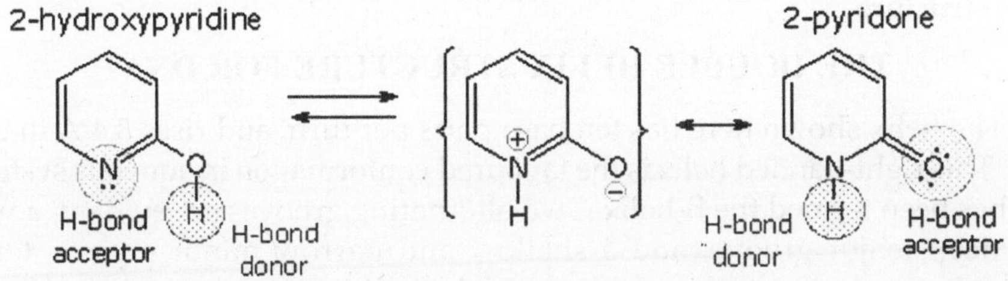

2-hydroxypyridine

2-pyridone

H-bond acceptor

H-bond donor

H-bond donor

H-bond acceptor

The additional evidence for the pyridone tautomer,consists of infrared and carbon nmr absorptions associated with and characteristic of the amide group. Once they had identified the favoured base tautomers in the nucleosides, Watson and Crick were able to propose a complementary pairing, via hydrogen bonding, of guanosine (G) with cytidine (C) and adenosine (A) with thymidine (T). This pairing, which is shown in the following diagram, explained Chargaff's findings beautifully, and led them to suggest a double helix structure for DNA. Before viewing this double helix structure itself, it is instructive to examine the base pairing interactions in greater detail. The G#C association involves three hydrogen bonds (coloured pink), and is therefore stronger than the two-hydrogen bond association of A#T.

These base pairings might appear to be arbitrary, but other possibilities suffer destabilizing steric or electronic interactions. The C#T pairing on the left suffers from carbonyl dipole repulsion, as well as steric crowding of the oxygens. The G#A pairing on the right is also destabilized by steric crowding (circled hydrogens).

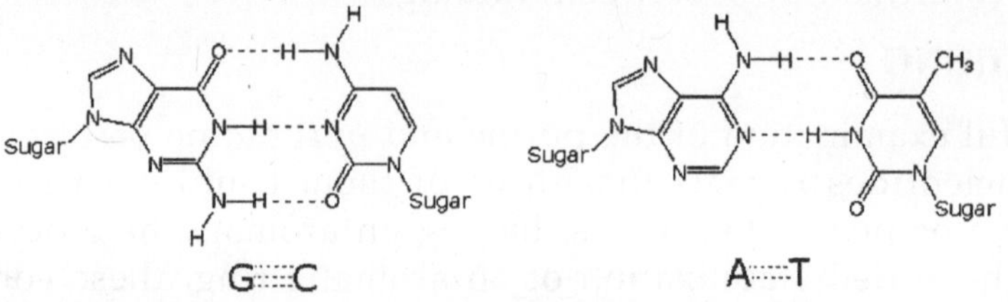

G:::C A:::T

A simple mnemonic device for remembering which bases are paired comes from the line construction of the capital letters used to identify the bases. A and T are made up of intersecting straight lines. In contrast, C and G are largely composed of curved lines. The RNA base uracil corresponds to thymine, since U follows T in the alphabet.

The Double Helix

After many trials and modifications, Watson and Crick conceived an ingenious double helix model for the secondary structure of DNA. Two strands of DNA were aligned anti-parallel to each other.Complementary primary nucleotide structures for each strand allowed intra-strand hydrogen bonding between each pair of bases.Coiling these coupled strands then leads to a double helix structure.

THE DOUBLE HELIX STRUCTURE FOR DNA

The helix shown here has ten base pairs per turn, and rises 3.4 Å in each turn. This right-handed helix is the favoured conformation in aqueous systems, and has been termed the B-helix. Two alternating grooves are evident, a wide and deep major groove, and a shallow and narrow minor groove. Other molecules, including polypeptides, may insert into these grooves, and in so doing perturb the chemistry of DNA. Other helical structures of DNA have also been observed, and are designated by letters.

DNA REPLICATION

A double helix structure for DNA, Watson and Crick stated, "It has not escaped our notice that the specific pairing we have postulated immediately suggests a possible copying mechanism for the genetic material.". The essence of this suggestion is that, if separated, each strand of the molecule might act as a template on which a new complementary strand might be assembled, leading finally to two identical DNA molecules. Indeed, replication does take place in this fashion when cells divide, but the events leading up to the actual synthesis of complementary DNA strands are sufficiently complex that they will not be described in any detail. As depicted in the following drawing, the DNA of a cell is tightly packed into chromosomes. First, the DNA is wrapped around small proteins called histones (coloured pink below). These bead-like structures are then further organized and folded into chromatin aggregates that make up the chromosomes. An overall packing efficiency of 7,000 or more is thus achieved. Clearly a sequence of unfolding events must take place before the information encoded in the DNA can be used or replicated.

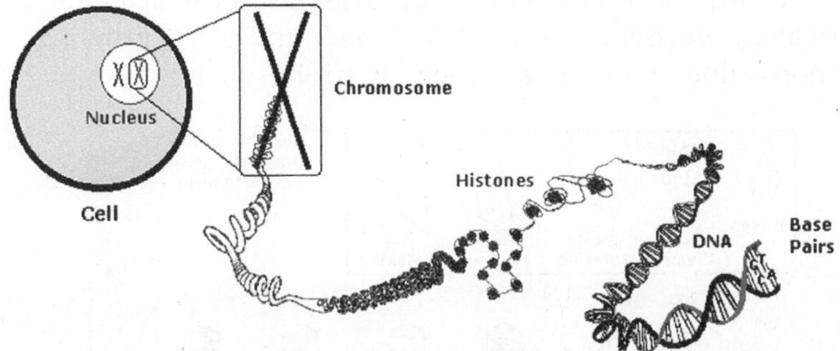

Once the double stranded DNA is exposed, a group of enzymes act to accomplish its replication. These are described briefly here:

Topoisomerase: This enzyme initiates unwinding of the double helix by cutting one of the strands.

Helicase: This enzyme assists the unwinding. Note that many hydrogen bonds must be broken if the strands are to be separated..

SSB: A single-strand binding-protein stabilizes the separated strands, and prevents them from recombining, so that the polymerization chemistry can function on the individual strands.

DNA Polymerase: This family of enzymes link together nucleotide triphosphate monomers as they hydrogen bond to complementary bases. These enzymes also check for errors (roughly ten per billion), and make corrections.

Ligase: Small unattached DNA segments on a strand are united by this enzyme.

Polymerization of nucleotides takes place by the phosphorylation reaction described by the following equation.

Di- and triphosphate esters have anhydride-like structures and are consequently reactive phosphorylating reagents, just as carboxylic anhydrides are acylating reagents. Since the pyrophosphate anion is a better leaving group than phosphate, triphosphates are more powerful phosphorylating agents than are diphosphates.The DNA polymerization process that builds the complementary strands in replication, could in principle take place in two ways. Referring to the general equation above, R1 could represent the next nucleotide unit to be attached to the growing DNA strand, with R2 being this strand. Alternatively, these assignments could be reversed.

In practice, the former proves to be the best arrangement. Since triphosphates are very reactive, the lifetime of such derivatives in an aqueous environment is relatively short. However, such derivatives of the individual nucleosides are repeatedly synthesized by the cell for a variety of purposes, providing a steady supply of these reagents. In contrast, the growing DNA segment must maintain its functionality over the entire replication process, and can not afford to be changed by a spontaneous hydrolysis event.

As a result, these chemical properties are best accomodated by a polymerization process that proceeds at the 3'-end of the growing strand by 5'-phosphorylation involving a nucleotide triphosphate.

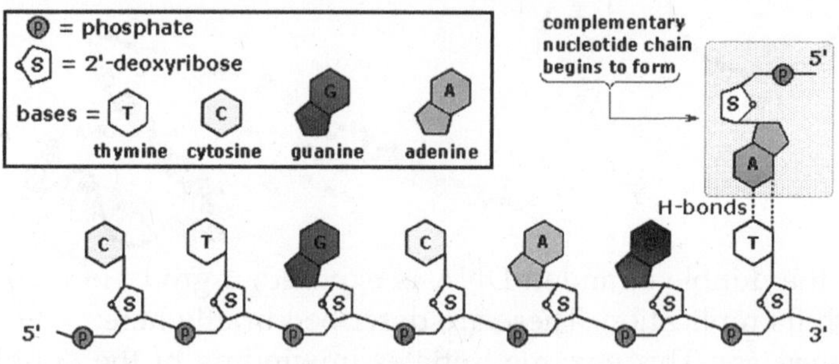

The polymerization mechanism described here is constant. It always extends the developing DNA segment towards the 3'-end. when a nucleotide triphosphate attaches to the free 3'-hydroxyl group of the strand, a new 3'-hydroxyl is generated. There is sometimes confusion on this point, because the original DNA strand that serves as a template is read from the 3'-end towards the 5'-end, and authors may not be completely clear as to which terminology is used.

Because of the directional demand of the polymerization, one of the DNA strands is easily replicated in a continuous fashion, whereas the other strand can only be replicated in short segmental pieces.

This is illustrated in the following diagram. Separation of a portion of the double helix takes place at a site called the replication fork. As replication of the separate strands occurs, the replication fork moves away, unwinding additional lengths of DNA. Since the fork in the diagram is moving towards the 5'-end of the red-coloured strand, replication of this strand may take place in a continuous fashion (building the new green strand in a 5' to 3' direction). This continuously formed new strand is called the leading strand. In contrast, the replication fork moves towards the 3'-end of the original green strand, preventing continuous polymerization of a complementary new red strand. Short segments of complementary DNA, called Okazaki fragments, are produced, and these are linked together later by the enzyme ligase. This new DNA strand is called the lagging strand.

When you consider that a human cell has roughly 109 base pairs in its DNA, and may divide into identical daughter cells in 14 to 24 hours, the efficiency of DNA replication must be extraordinary. The procedure described above will replicate about 50 nucleotides per second, so there must be many thousand such replication sites in action during cell division. A given length of double stranded DNA may undergo strand unwinding at numerous sites in response to promoter actions. The unraveled "bubble" of single stranded DNA has two replication forks, so assembly of new complementary strands may proceed in two directions. The polymerizations associated with several such bubbles fuse together to achieve full replication of the entire DNA double helix

RNA AND PROTEIN SYNTHESIS

The genetic information stored in DNA molecules is used as a blueprint for making proteins. Why proteins? Because these macromolecules have diverse primary, secondary and tertiary structures that equip them to carry out the numerous functions necessary to maintain a living organism. As noted in the protein, these functions include:

- Structural integrity (hair, horn, eye lenses etc.).
- Molecular recognition and signaling (antibodies and hormones).
- Catalysis of reactions (enzymes)..
- Molecular transport (hemoglobin transports oxygen).
- Movement (pumps and motors).

The critical importance of proteins in life processes is demonstrated by numerous genetic diseases, in which small modifications in primary structure produce debilitating and often disasterous consequences. Such genetic diseases incluse Tay-Sachs, phenylketonuria (PKU), sickel cell anemia, achondroplasia, and Parkinson disease.

The unavoidable conclusion is that proteins are of central importance in living cells, and that proteins must therefore be continuously prepared with high structural fidelity by appropriate cellular chemistry.

Early geneticists identifed genes as hereditary units that determined the appearance and/ or function of an organism (*i.e.* its phenotype).

We now define genes as sequences of DNA that occupy specific locations on a chromosome. The original proposal that each gene controlled the formation of a single enzyme has since been modified as: one gene = one polypeptide. The intriguing question of how the information encoded in DNA is converted to the actual construction of a specific polypeptide has been the subject of numerous studies, which have created the modern field of Molecular Biology.

THE CENTRAL DOGMA AND TRANSCRIPTION

Francis Crick proposed that information flows from DNA to RNA in a process called transcription, and is then used to synthesize polypeptides by a process called translation. Transcription takes place in a manner similar to

DNA replication. A characteristic sequence of nucleotides marks the beginning of a gene on the DNA strand, and this region binds to a promoter protein that initiates RNA synthesis. The double stranded structure unwinds at the promoter site., and one of the strands serves as a template for RNA formation, as depicted in the following diagram. The RNA molecule thus formed is single stranded, and serves to carry information from DNA to the protein synthesis machinery called ribosomes. These RNA molecules are therefore called messenger-RNA (mRNA).

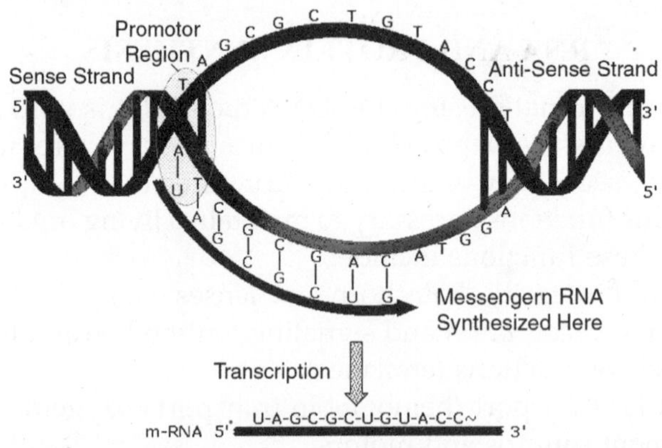

An important distinction must be made here. One of the DNA strands in the double helix holds the genetic information used for protein synthesis. This is called the sense strand, or information strand (coloured red above). The complementary strand that binds to the sense strand is called the anti-sense strand (coloured green), and it serves as a template for generating a mRNA molecule that delivers a copy of the sense strand information to a ribosome.

The promoter protein binds to a specific nucleotide sequence that identifies the sense strand, relative to the anti-sense strand. RNA synthesis is then initiated in the 3' direction, as nucleotide triphosphates bind to complementary bases on the template strand, and are joined by phosphate diester linkages. A characteristic "stop sequence" of nucleotides terminates the RNA synthesis. The messenger molecule (coloured orange above) is released into the cytoplasm to find a ribosome, and the DNA then rewinds to its double helix structure.

In eucaryotic cells the initially transcribed m-RNA molecule is usually modified and shortened by an "editing" process that removes irrelevent material. The DNA of such organisms is often thousands of times larger and more complex than that composing the single chromosome of a procaryotic bacterial cell. This difference is due in part to repetitive nucleotide sequences (ca. 25 per cent iu the human genome). Furthermore, over 95 per cent of human DNA is found in intervening sequences that separate genes and parts of genes. The informational DNA segments that make up genes are called exons, and the noncoding segments are called introns. Before the mRNA molecule leaves the nucleus, the nonsense bases that make up the introns are cut out, and the informationally useful exons are joined together in a step known as RNA splicing. In this fashion shorter mRNA molecules carrying the blueprint for a specific protein are sent on their way to the ribosome factories.

The Central Dogma of molecular biology, which at first was formulated as a simple linear progression of information from DNA to RNA to Protein, is summarized in the following illustration. The replication process on the left consists of passing information from a parent DNA molecule to daughter molecules. The middle transcription process copies this information to a mRNA molecule. Finally, this information is used by the chemical machinery of the ribosome to make polypeptides.

As more has been learned about these relationships, the central dogma has been refined to the representation displayed on the right. The dark blue arrows show the general, well demonstrated, information transfers noted above. The relatively simple RNA-protein structures of a virus may induce RNA replication or a reverse transcription from RNA to DNA (magenta arrow), but these are exceptional, albeit significant transfers.

The AIDS virus, for example, functions by the action of reverse transcriptase enzymes accompanying its RNA. Direct translation of DNA information into protein synthesis (orange arrow) has not yet been observed in a living organism. Finally, proteins appear to be an informational dead end, and do not provide a structural blueprint for either RNA or DNA.

TRANSLATION

Translation is a more complex process than transcription. This would, of course, be expected. After all, the coded messages produced by the German Enigma machine could be copied easily, but required a considerable decoding effort before they could be read with understanding. In a similar sense, DNA replication is simply a complementary base pairing exercise, but the translation of the four letter (bases) alphabet code of RNA to the twenty letter (amino acids) alphabet of protein literature is far from trivial. Clearly, there could not be a direct one-to-one correlation of bases to amino acids, so the nucleotide letters must form short words or codons that define specific amino acids. Many questions pertaining to this genetic code were posed in the late 1950's:

- How many RNA nucleotide bases designate a specific amino acid?
  If separate groups of nucleotides, called codons, serve this purpose, at least three are needed. There are 43 = 64 different nucleotide triplets, compared with 42 = 16 possible pairs.
- Are the codons linked separately or do they overlap?
  Sequentially joined triplet codons will result in a nucleotide chain three times longer than the protein it describes. If overlapping codons are used then fewer total nucleotides would be required.

- If triplet segments of mRNA designate specific amino acids in the protein, how are the codons identified?
  For the sequence ~CUAGGU~ are the codons CUA & GGU or ~C, UAG & GU~ or ~CU, AGG & U~?
- Are all the codon words the same size?

In Morse code the most widely used letters are shorter than less common letters. Perhaps nature employs a similar scheme.

Physicists and mathematicians, as well as chemists and microbiologists all contributed to unravelling the genetic code. Although earlier proposals assumed efficient relationships that correlated the nucleotide codons uniquely with the twenty fundamental amino acids, it is now apparent that there is considerable redundancy in the code as it now operates. Furthermore, the code consists exclusively of non-overlapping triplet codons.

Clever experiments provided some of the earliest breaks in deciphering the genetic code. Marshall Nirenberg found that RNA from many different organisms could initiate specific protein synthesis when combined with broken E.coli cells (the enzymes remain active). A synthetic polyuridine RNA induced synthesis of poly-phenylalanine, so the UUU codon designated phenylalanine. Likewise an alternating ~CACA~ RNA led to synthesis of a ~His-Thr-His-Thr~ polypeptide. The following table presents the present day interpretation of the genetic code. Note that this is the RNA alphabet, and an equivalent DNA codon table would have all the U nucleotides replaced by T. Methionine and tryptophan are uniquely represented by a single codon. At the other extreme, leucine is represented by eight codons. The average redundancy for the twenty amino acids is about three. Also, there are three stop codons that terminate polypeptide synthesis. The translation process is fundamentally straightforward. The mRNA strand bearing the transcribed code for synthesis of a protein interacts with relatively small RNA molecules (about 70-nucleotides) to which individual amino acids have been attached by an ester bond at the 3'-end.

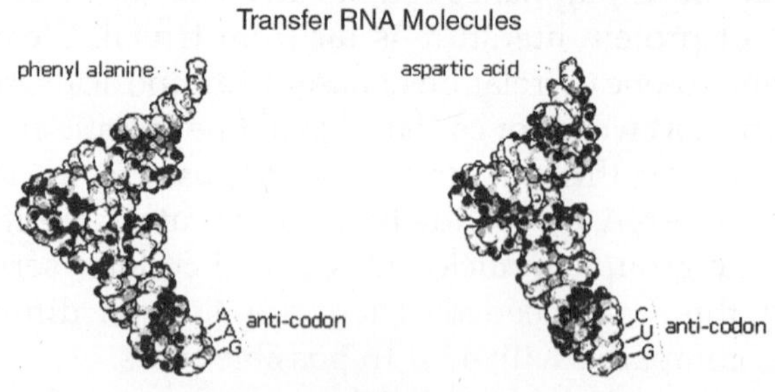

Fig. tRNA Molecules

These transfer RNA's (tRNA) have distinctive three-dimensional structures cosisting of loops of single-stranded RNA connected by double stranded segments. This cloverleaf secondary structure is further wrapped into an "L-shaped" assembly, having the amino acid at the end of one arm, and a characteristic anti-codon region at the other end. The anti-codon consists of a nucleotide triplet that is the complement of the amino acid's codon(s). Models of two such tRNA molecules are shown to the right. When read from the top to the bottom, the anti-codons depicted here should complement a codon in the previous table.

A cell's protein synthesis takes place in organelles called ribosomes. Ribosomes are complex structures made up of two distinct and separable subunits (one about twice the size of the other).

Each subunit is composed of one or two RNA molecules associated with 20 to 40 small proteins. The ribosome accepts a mRNA molecule, binding initially to a characteristic nucleotide sequence at the 5'-end.

This unique binding assures that polypeptide synthesis starts at the e codon. A tRNA molecule with the appropriate ant-codon then attaches at the starting point and this is followed by a series of adjacent tRNA attachments, peptide bond formation and shifts of the ribosome along the mRNA chain to expose new codons to the ribosomal chemistry.

9

Alcohol and Drug

ALCOHOLS

Alcohols exhibit rapid broad-spectrum antimicrobial activity against vegetative bacteria (including mycobacteria), viruses, and fungi but are not sporicidal. They are, however, known to inhibit sporulation and spore germination, but this effect is reversible. Because of the lack of sporicidal activity, alcohols are not recommended for sterilization but are widely used for both hard-surface disinfection and skin antisepsis.

Lower concentrations may also be used as preservatives and to potentiate the activity of other biocides. Many alcohol products include low levels of other biocides (in particular chlorhexidine), which remain on the skin following evaporation of the alcohol, or excipients (including emollients), which decrease the evaporation time of the alcohol and can significantly increase product efficacy. In general, isopropyl alcohol is considered slightly more efficacious against bacteria and ethyl alcohol is more potent against viruses; however, this is dependent on the concentrations of both the active agent and the test microorganism. For example, isopropyl alcohol has greater lipophilic properties than ethyl alcohol and is less active against hydrophilic viruses (*e.g.*, poliovirus). Generally, the antimicrobial activity of alcohols is significantly lower at concentrations below 50 per cent and is optimal in the 60 to 90 per cent range. Little is known about the specific mode of action of alcohols, but based on the increased efficacy in the presence of water, it is generally believed that they cause membrane damage and rapid denaturation of proteins, with subsequent interference with metabolism and cell lysis. This is supported by specific reports of denaturation of *Escherichia coli* dehydrogenases and an increased lag phase in *Enterobacter aerogenes*, speculated to be due to inhibition of metabolism required for rapid cell division.

Clearly, the mechanism of action of glutaraldehyde involves a strong association with the outer layers of bacterial cells, specifically with unprotonated amines on the cell surface, possibly representing the reactive sites. Such an effect could explain its inhibitory action on transport and on enzyme systems, where access of substrate to enzyme is prohibited. Partial or entire removal of the cell wall in hypertonic medium, leading to the production of spheroplasts or protoplasts and the subsequent prevention of lysis by glutaraldehyde when these forms are diluted in a hypotonic environment, suggests an additional effect on the inner membrane, a finding substantiated by the fact that the dialdehyde prevents the selective release of some membrane-bound enzymes of *Micrococcus lysodeikticus*. Glutaraldehyde is more active at alkaline than at acidic pHs. As the external pH is altered from acidic to alkaline, more reactive sites will be formed at the cell surface, leading to a more rapid bactericidal effect. The cross-links thus obtained mean that the cell is then unable to undertake most, if not all, of its essential functions. Glutaraldehyde is also mycobactericidal. Unfortunately, no critical studies have as yet been undertaken to evaluate the nature of this action.

The bacterial spore presents several sites at which interaction with glutaraldehyde is possible, although interaction with a particular site does not necessarily mean that this is associated with spore inactivation. *E. coli, S. aureus,* and vegetative cells of *Bacillus subtilis* bind more glutaraldehyde than do resting spores of *B. subtilis;* uptake of glutaraldehyde is greater during germination and outgrowth than with mature spores but still lower than with vegetative cells. Low concentrations of the dialdehyde (0.1 per cent) inhibit germination, whereas much higher concentrations (2 per cent) are sporicidal.

The aldehyde, at both acidic and alkaline pHs, interacts strongly with the outer spore layers; this interaction reduces the release of dipicolinic acid (DPA) from heated spores and the lysis induced by mercaptoethanol (or thioglycolate)-peroxide combinations. Low concentrations of both acidic and alkaline glutaraldehyde increase the surface hydrophobicity of spores, again indicating an effect at the outermost regions of the cell. It has been observed by various authors that the greater sporicidal activity of glutaraldehyde at alkaline pH is not reflected by differences in uptake; however, uptake per se reflects binding and not necessarily penetration into the spore. It is conceivable that acidic glutaraldehyde interacts with and remains at the cell surface whereas alkaline glutaraldehyde penetrates more deeply into the spore.

This contention is at odds with the hypothesis of Bruch, who envisaged the acidic form penetrating the coat and reacting with the cortex while the alkaline form attacked the coat, thereby destroying the ability of the spore to function solely as a result of this surface phenomenon. There is, as yet, no evidence to support this theory. Novel glutaraldehyde formulations based on acidic rather than alkaline glutaraldehyde, which benefit from the greater inherent stability of the aldehyde at lower pH, have been produced. The improved sporicidal activity claimed for these products may be obtained by agents that potentiate the activity of the dialdehyde.

During sporulation, the cell eventually becomes less susceptible to glutaraldehyde (see "Intrinsic resistance of bacterial spores"). By contrast, germinating and outgrowing cells reacquire sensitivity. Germination may be defined as an irreversible process in which there is a change of an activated spore from a dormant to a metabolically active state within a short period. Glutaraldehyde exerts an early effect on the germination process. L-Alanine is considered to act by binding to a specific receptor on the spore coat, and once spores are triggered to germinate, they are committed irreversibly to losing their dormant properties.

Glutaraldehyde at high concentrations inhibits the uptake of L-[^{14}C]alanine by *B. subtilis* spores, albeit by an unknown mechanism. Glutaraldehyde-treated spores retain their refractivity, having the same appearance under the phase-contrast microscope as normal, untreated spores even when the spores are subsequently incubated in germination medium. Glutaraldehyde is normally used as a 2 per cent solution to achieve a sporicidal effect; low concentrations (<0.1 per cent) prevent phase darkening of spores and also prevent the decrease in optical density associated with a late event in germination. By contrast, higher concentrations (0.1 to 1 per cent) significantly reduce the uptake of L-alanine, possibly as a result of a sealing effect of the aldehyde on the cell surface. Mechanisms involved in the revival of glutaraldehyde-treated spores are.

There are no recent studies of the mechanisms of fungicidal action of glutaraldehyde. Earlier work had suggested that the fungal cell wall was a major target site, especially the major wall component, chitin, which is analogous to the peptidoglycan found in bacterial cell walls.

Glutaraldehyde is a potent virucidal agent. It reduces the activity of hepatitis B surface antigen (HBsAg) and especially hepatitis B core antigen ([HBcAg] in hepatitis B virus [HBV]) and interacts with lysine residues on the surface of hepatitis A virus (HAV). Low concentrations (<0.1 per cent) of alkaline glutaraldehyde are effective against purified poliovirus, whereas poliovirus RNA is highly resistant to aldehyde concentrations up to 1 per cent at pH 7.2 and is only slowly inactivated at pH 8.3. In other words, the complete poliovirus particle is much more sensitive than poliovirus RNA. In light of this, it has been inferred that glutaraldehyde-induced loss of infectivity is associated with capsid changes.

Glutaraldehyde at the low concentrations of 0.05 and 0.005 per cent interacts with the capsid proteins of poliovirus and echovirus, respectively; the differences in sensitivity probably reflect major structural variations in the two viruses. Bacteriophages were recently studied to obtain information about mechanisms of virucidal action. Many glutaraldehyde-treated *P. aeruginosa* F116 phage particles had empty heads, implying that the phage genome had been ejected. The aldehyde was possibly bound to F116 double-stranded DNA but without affecting the molecule; glutaraldehyde also interacted with phage F116 proteins, which were postulated to be involved in the ejection of the nucleic acid. Concentrations of glutaraldehyde greater than 0.1 to 0.25 per cent significantly affected the transduction of this phage; the transduction process was more sensitive to the aldehyde than was the phage itself. Glutaraldehyde and other aldehydes were tested for their ability to form protein-DNA cross-links in simian virus 40 (SV40); aldehydes (*i.e.*, glyoxal, furfural, prionaldehyde, acetaldehyde, and benzylaldehyde) without detectable cross-linking ability had no effect on SV40 DNA synthesis, whereas acrolein, glutaraldehyde, and formaldehyde, which formed such cross-links, inhibited DNA synthesis.

ANTHROPOLOGICAL WORK ON ALCOHOL

Anthropological work on alcohol and other drugs often challenges conventional assumptions about substance use and abuse. In this chapter, we delineate this theme by highlighting areas where the uniqueness of anthropological contributions to understanding drug use and abuse is clearly seen. We have not reviewed the entire field of anthropological research in the area; given the enormous body of literature, especially in alcohol studies, such a review would constitute a project beyond the scope of this book. Furthermore, several excellent reviews have been published on specific topics within this research domain. Anthropologists tend to take a different tack in approaching studies of substance use and abuse; consequently, their work is often controversial to policymakers and treatment providers. Controversy—and sometimes skepticism-frequently surround the approach and methods, interpretation of data and recommendations for policy.

This controversy has been recognized within this discipline itself. In certain instances, their orientation is congruent with that of colleagues in other disciplines. In recent years, in fact, some anthropologists have created or joined interdisciplinary research groups and are working as team researchers.

Anthropological interest in studies of drug use and abuse has been apparent for several decades. Until the 1970s, however, most of this research was a by-product of broader ethnographic studies of small societies in Latin America, Africa, Oceania and the North American Indian and Eskimo tribes. In this tradition, data collected on the use of mind-altering substances formed one component of an overall study, such as noting the use of plants for medicinal purposes or describing curing ceremonies where sacred plants were employed.

Cross-cultural research on substances has been conducted through the use of data housed in the Human Relations Area Files, by in-depth analysis of available ethnographic data, through special conferences in which drug use data were presented for a variety of societies, communities, or ethnic groups, or through projects in which the anthropologists conducted a controlled comparison or contrasted different drug use traditions cross-culturally. Thus, it is fair to conclude that as of the early 1970s, anthropology had not yet developed an explicit drug research tradition, especially with respect to abuse of drugs. With increased funding available for such studies and with the expansion of applied anthropology, the situation has changed dramatically over the past two decades.

In response to the AIDS epidemic, anthropologists have become central participants in a major research agenda on the sociocultural context of risk behaviours in an effort to develop effective prevention programmes. In fact, anthropologists have been quite successful in obtaining funding for AIDS- and HIV-related projects in large part because of the need to conduct qualitative studies in this area. Increased risk for HIV infection as related to alcohol and drug use has been an important aspect of this research. The field of drug studies has developed into several subspecializations among anthropologists concerned with family, AIDS, treatment and special populations (such as North American Indians, Hispanic groups, black Americans, the working class and women).

Up to the present, anthropological discussions of substance use and abuse are almost always treated separately with respect to alcohol and other drugs. This tendency reflects wider patterns in national and international policy and research, as well as treatment of substance abuse. For example, the National Institute on Alcohol Abuse and Alcoholism was established in 1970 separate from the National Institute on Drug Abuse and the World Health Organization continues to separate alcohol from the "illicit" drugs in most of its working conferences and publications. In anthropology, it is relatively rare for more than one drug to be encompassed in anthropological studies and publications.

Anthropologists in the substance use-abuse field have focused primarily on studies of alcohol, reflecting, in part, the relative order of usage of particular drugs throughout the world: first, ethanol; second, nicotine; third, caffeine; fourth, betel; and fifth, marijuana. Over the past twenty years, however, a substantial tradition has developed in cannabis research, mainly conducted in the Caribbean and Latin America. Similarly, there has been an increasing interest in tobacco studies and trends in publications and symposia at anthropology meetings seem to indicate that this line of research is gathering momentum. Some research has focused on the role tobacco plays in social relationships. Another theme has been the interaction of transnational tobacco companies seeking new markets in Third World countries. Several researchers with roots in early studies of religious experience and hallucinogens have specialized in the cultural context of the ingestion of hallucinogens such as peyote and mescaline in the past two decades.

Due in part to American society's definition of heroin, opium, marijuana and cocaine use as clearly deviant behaviour and in part to demand for legal and/or clinical intervention, studies on such drugs by anthropologists have attracted considerable attention. The street culture around the use of heroin and the use of methadone in the treatment of heroin addiction has, for example, drawn a kind of interest that is perhaps more curious than serious.

While such anthropological research tended to focus on heroin in the 1970s, it shifted to an interest in crack-cocaine in the 1980s and 1990s, as use patterns and the wider society's concern about the use and abuse of certain substances changed. The typical anthropologist's emic focus on the drug user's cognitive and social worlds is often at odds with the clinician's and policymaker's perception of the' problem. The clinical and policy domains of our society take a more etic stance with respect to use of these "illicit" drugs and thus express impatience, at the very least, with arguments that suggest the use of these substances is not necessarily deviant from the user's perspective. This perspective has characterized much of the anthropological research on alcohol and drugs during the 1990s.

Regardless of the position taken regarding the deviance of such drug use, it is becoming apparent that prevention and treatment programmes must do more than remove the drug in order to be lastingly effective. Drug use fits into a cluster of behaviours and beliefs and treatment agendas must deal with that reality in proposing alternative ways of life to a recovering addict. Thus, a lucid understanding of the cognitive and social worlds of drug users is highly pertinent to preventive and intervention efforts.

In addition to twenty years of research on heroin addicts anthropologists have more recently undertaken studies on cocaine, as cocaine has gained popularity as the drug of choice for many Americans. One such study examines the ways in which cocaine users interpret their environment and the resultant influences of the user subculture on drug use patterns.

Another combines interest in issues of multiple drug use by investigating cocaine use among methadone clients. Thus far, a small body of research on cocaine exists in anthropology. Because of the relatively widespread indigenous use of kava and betel in Oceania, anthropologists working there have in the course of ethnographic research reported on the use of these substances.

Anthropologists approach research in drug use and abuse by illuminating the cultural context in which they take place. For example, while American society has dramatically moved in an antitobacco direction, some anthropologists have recently called attention to the functional role of tobacco use. Consequently, anthropologists at times have been seen as rabble-rousers and troublemakers who rock the boat of cherished assumptions about the pathology or deviancy of drug use. For example, anthropologists have been taken to task for overstressing the functional role of alcohol in culture and ignoring its dysfunctional use.

Three decades ago, David Mandelbaum clearly articulated the anthropological slant on functional and dysfunctional use of alcohol: "Drunkenness cannot be understood apart from drinking in general and drinking cannot be understood apart from the characteristic features of social relations of which it is part and which are reflected and expressed in the act of drinking". More recently, Mac Marshall has similarly stressed the essential value in examining deviant drug use within the context of normal patterns: "All the ethnographic accounts show the necessity of understanding the variety of normal drinking styles in any social setting before attempting to deal with abnormal (or addictive) drinking".

As colleagues in other disciplines have taken a greater interest in the role of culture in drug use and abuse, they have understood and applied the concept of culture in ways that are sometimes discrepant with the concept in anthropology. Specifically, in the minds of many scholars outside anthropology, "culture" has become synonymous with presumed membership in a particular ethnic group, nationality, racial group, religious affiliation, class and so forth as related to particular drug use and abuse patterns.

This perspective stands in contrast to the more traditional anthropological view of culture as a dynamic process through which individuals and societies learn the sum total of their society's behaviours and associated belief systems, including those encompassing drug use practices and beliefs.

THE TERATOGENIC EFFECTS OF ALCOHOL

Alcohol is a teratogen. There has been no teratogenic agent yet studied in man which has shown a clear threshold effect, *i.e.* where the substance could be considered safe at a particular level, beyond which its teratogenic effect begins to take hold, and alcohol is no exception. That being the case, the teratogenic effects of alcohol can also induce fetal malformations both at the earliest, as well as at the lowest level of intake, its effectiveness spreading differentially over the whole spectrum of reproduction, affecting the developing fetus in varying degrees, in both extent and severity, depending on the dosage and timing.

This explains also why maternal alcohol consumption can affect the offspring through all gradations of teratogenesis, ranging from transient to very mildly affected, and from moderately affected right up to the full blown Fetal Alcohol Syndrome. During transient teratogenesis the impact of the agent on the fetal tissue may not inflict permanent damage, as the substance can be degraded by the mother and fetal tissue in time, depending on genetic influences, susceptibility and maternal nutritional status.

Alcohol Teratogenesis on Structural Development

Alcohol is a low molecular substance and is therefore quite capable of crossing the placental barrier and entering the fetus, causing the level of alcohol in the fetus to be approximate to that of the mother. In the first 21 days of the fetal development the preliminary cell organisation of the embryo begins to take place. If an excessive amount of alcohol is consumed before the blastocyst is embedded in the uterus, the impact can be so severe that the fetus is miscarried. By the end of the 36th day, often long before the woman even realises that she is pregnant, the neural tube is clearly present and open, and most of the rudimentary organs have already been formed, such as limbs, heart, brain, eyes, mouth, digestive tract etc. It is therefore obvious that if a teratogenic substance such as alcohol is consumed during this most critical period of rapid growth of cell development and organ formation, this can result in various forms of malformation in the newborn, such as defective heart, musculoskeletal abnormalities, mental handicap etc., without any specific outward signs of FAS. Even though it is considered that the first three months of gestation is the most critical period for alcohol-induced malformations to occur, both human and animal experiments have been able to demonstrate that the teratogenic effects of alcohol continues throughout the whole gestational period, affecting at the later stage particularly the brain development and function.

Alcohol Teratogenesis in Brain Development

Alcohol has the most detrimental effect on both brain development and function. The infant is not only born with a brain smaller in size, but the teratogenic effect of alcohol both reduces the number of brain neurons, as well as alters their distributution, resulting in mental deficiency in varying degrees, from milder behavioural problems to obvious mental handicap. Animal studies have shown that while many areas of the brain are affected by maternal alcohol exposure, its effects seem to be particularly detrimental on the hippocampus, where it produces marked changes in its mossy fibres and 20 per cent reduction in the pyramidal cells in the CA1 region, as well as a sparsity in the number of dendritic spines in the CA1 pyramidal neurons.

It has been therefore speculated that both the intellectual decrements and the behavioural deficits seen in infants born to mothers using alcohol during pregnancy may result directly from these specific hippocampal structural alterations. Hundreds of experiments have been able to confirm that maternal alcohol consumption indeed causes irreversible brain alterations in both human and non-human infants, which also include abnormalities in EEG patterns, abnormal visual evoked responses, and both defective auditory brain stem and spatial learning developments. Autopsy reports on deceased patients with FAS have shown widespread anomalies, many of which have been associated with disruption in the migration and integration of neural and glial cells during embryogenesis.

In addition, other autopsy studies have shown that the nature and degree of brain malformations in children of alcoholic and heavy drinking mothers are extremely variable, suggesting that a wide variety of broad spectrum neurologic, behavioural and intellectual deficits would therefore also be found in survivors. As only about half of the autopsied cases studied had enough physical characteristics to warrant a diagnosis of FAS, it seems now quite obvious that alcohol-related brain damage, as well as learning and behavioural deficits, do occur frequently in the absence of any external signs of FAS.

Fetal Alcohol Effects

The teratogenic effects of alcohol spreads differentially over the whole spectrum of reproduction, varying only in the extent and degree. At one end of the spectrum are the children warranting a firm diagnosis of FAS, and at the other end of the spectrum are the children who lack the common physical characteristics of FAS but who, nevertheless, have some subtle or marked physical and/or mental deficiencies by being exposed to varying amounts of alcohol in utero. It is now widely accepted that the classical diagnosis of FAS is totally inadequate as for every child born with FAS there are thousands of others whose lives are partially handicapped, or limited, by being exposed to alcohol during gestational development. These children, without sufficient physical stigmata for firm diagnosis of FAS, are now identified as suffering from Fetal Alcohol Effects (FAE).

Over the years a number of epidemiological studies have investigated both the physical and neurobehavioural effects of varying levels of prenatal alcohol exposure on both human and non-human infants. One of the foremost experts and contemporary researchers in this field is Dr Ann P. Streissguth, of the Department of Psychiatry and Behavioural Sciences at the University of Washington, who has researched, published and lectured widely on the teratogenic effects of maternal alcohol consumption From the year 1974 Dr Streissguth and her team began a seven and half year longitudinal, prospective, population-based study, examining the long term effects of moderate prenatal alcohol exposure on 486 infants born to mothers who had reported no major problems with alcohol but were, nevertheless, social drinkers *i.e.* reported consuming on average two or more drinks most days during pregnancy, or reported a "binge- pattern" of drinking *e.g.* consuming five or more drinks per any occasion in the month before pregnancy recognition. Only six out of the mothers interviewed felt that they might have used alcohol excessively during pregnancy.

The mothers were primarily white, married, middle class, and at low risk for adverse pregnancy outcome, and all were receiving prenatal care by the fifth month of pregnancy. This cohort sample of children, which included 261 boys and 225 girls, were first examined and evaluated on the first and second day after birth, and then approximately at eight and eighteen months, at four years, and at seven and at seven and a half years after birth.

FAE in Infants and Preschool Children

Most of these infants were born basically within normal limits as a group. However, the more the mothers had reported consuming alcohol, the poorer the overall performance of the newborns. Already, on the first day of life, the infants of the mothers who had been drinking more during pregnancy functioned significantly worse. They were usually born with lighter weight and were more jittery and tremulous. They had difficulties with habituation, which is the ability to turn off redundant stimuli, considered as a basic nervous system function of the newborn. On the second day of life they had a longer latency to begin sucking and had a weaker suck as measured on a pressure transducer with non-nutritive nipple. They also suffered from disrupted sleep patterns, low level of arousal, unusual body orientation, abnormal reflexes, hypotonia and excessive mouthing.

By eight months, and then onwards, these infants seemed to continue suffering from disrupted sleep-wake patterns, poorer balance and motor control, longer latency to respond, poorer attention, visual recognition and memory, decrements in mental development, spoken language and verbal comprehension, including lower IQ scores. As references indicate, these findings have also been confirmed independently by other workers.

FAE in Young School-age Children

After seven years Dr Streissguth and her team re-examined the cohort of 486 children. The results showed that learning problems and classroom behaviour which were most negatively related to moderate alcohol exposure in utero were: co-operation, sustained attention, retention of information, comprehension of words, impulsiveness, tactfulness, word recall and organisational skills, all indicating increased risk of learning disabilities. In fact, even though these children were within an average range of intelligence, their overall performance on arithmetic and reading tests were negative. It also became apparent that maternal drinking of two or more drinks per day on average, after statistically adjusting for appropriate covariates, was related to a 7-point decrement in IQ in these seven year olds.

Furthermore, that children of women who reported never drinking five or more drinks on any occasion in the month before pregnancy recognition, were on average one to three months behind in reading and arithmetic skills. In addition 24 per cent of these children were participating in special remedial programmes at school, compared to 15 per cent of children of abstainers. This represents 9 per cent excess of learning disabilities for children of mothers who had never been drinking five or more drinks on one occasion prior to pregnancy recognition. These children were also found to have higher learning problem score of 17 per cent compared to children of abstainers which was 7 per cent. The conclusion of this study was that two maternal alcohol use patterns have now been identified as being particularly detrimental to the offspring *i.e.* two or more drinks an average per day during pregnancy, and "binge-pattern" of alcohol consumption, *e.g.* five or more drinks on any occasion, particularly when consumed in the month so before pregnancy recognition, as both can lead to marked behaviour and learning disabilities in school-age children. It was also concluded that these alcohol-related behaviour and attention decrements seem to have been already clearly observable from an early infancy, long before academic learning had even occurred. Half an year later, Professor Streissguth and her team selected from the cohort study of 482 children, 384 subjects. The reason for the selection was because some tests were either added or modified after the initial testing had begun, therefore not all tests had been administered to all the 482 children. One of the tests added was Children's Memory Test blocks, which had been completed only by these 384 children. The results showed that low-level prenatal alcohol consumption is most strongly related to attention and memory deficits across both verbal and visual modalities, poor integration and quality responses, as well as to negative behaviour patterns involving distractibility, inflexibility, and poor organisational skills. Also inadequate perceptual motor functioning was apparent.

This wide pattern of performance deficits uniformly occurred despite the presence of average IQ, suggesting that maternal alcohol induced behaviour decrements in the offspring seem to be a more sensitive indicator of central nervous system damage than the IQ scale itself. The study concluded that maternal social drinking seemed to result in the offspring having similar, but less severe consequences than those seen in children born with FAS, indicating in both cases, the clear occurrence of alcohol-induced permanent and irreversible central nervous system damage during critical stages of fetal development.

FAE in adolescents and adults: As with children born with FAS, population based studies carried out on the offspring born to socially drinking mothers have shown that, on maturation, these children can still show subtle and permanent alcohol-related neurobehavioural deficits, IQ and achievement decrements, combined with various attention, memory and learning problems.

Alcohol And Male Reproduction

Alcohol is a direct testicular toxin. It causes atrophy of semeniferous tubules, loss of sperm cells, and an increase in abnormal sperms. Alcohol is also known to be a strong Leydig cell toxin, and it can have an adverse effect on the synthesis and secretion of testosterone. Alcohol can cause significant deterioration in sperm concentration, sperm output and motility. Semen samples of men consuming excessive amounts of alcohol have shown distinct morphological abnormalities. It has been also established that approximately 80 per cent of chronic alcoholic men are sterile and, furthermore, that alcohol is one of the most common causes of male impotence.

Alcohol and Nutritional Status

Numerous animal studies of experimental alcoholism, where nutritional status has been well controlled, have shown that the damage to the developing fetus, such as low birth weight, central nervous system impairment and congenital abnormalities, are caused as a direct consequence of the teratogenic effects of alcohol. In addition, some of these studies have also been able to show a clear continuum effect; the higher the blood alcohol of the mother, the greater the damage to the developing fetus. Even though the direct connection between alcohol intake and birth defects is now indisputable, other etiological factors associated with maternal drinking must also be considered as contributing to adverse pregnancy outcome.

The most important of these secondary factors is alcohol-induced malnutrition, as nutritional deficiencies occur frequently with alcohol intake, due to reduced appetite. However, in cases where nutritional food intake is adequate, alcohol still considerably reduces nutritional status by directly interfering with nutrition utilization, digestion and absorption, as well as greatly increasing urinary excretion of both vitamins and minerals. Alcohol-induced zinc depletion is particularly well documented. This could be of a particular importance, as some studies on human pregnancies have shown a positive correlation with reduced zinc status and low birth weight and fetal malformations, suggesting that inadequate zinc nutriture could also act independently as a teratogenic agent. In addition, folic acid deficiency, which results from alcohol-induced urinary excretion, has been linked directly with the occurrence of spina bifida(. Besides direct malnutrition, other secondary metabolic disturbances due to alcohol consumption may also contribute to an adverse pregnancy outcome, such as alcohol-induced hypoglycaemia, ketoacidosis, as well as various alterations in both lipid, and amino acid metabolism.

FUNCTIONAL DRUG: CANNABIS

Cannabis is a widespread drug of ancient vintage, usually classified as a hallucinogen based on its effects on mood and perception. Anthropologists have studied its use and its role in social structure in a number of cultures. Jamaica, for example, has a high rate of regular users of cannabis, making it an excellent place to study the issues that unfold around the use and abuse of this drug, known to Jamaicans as ganja.

In Jamaica, ganja use is integrally linked to all aspects of working-class social structure: cultivation, cash crops, marketing, economics, consumer-cultivator-dealer networks; interclass relationships and processes of avoidance or cooperation; parent-child, peer and mate relationships; folk medicine; folk religious doctrines; gossip sanctions; personality and culture; interclass stereotypes; legal and church sanctions; perceived requisites of behavioural changes for social mobility; and adaptive strategies.

Ganja figures strongly in the economic realm. On the lowest rung of the Jamaican socioeconomic ladder, poor families with few marketable skills make their living however they can, often engaging in several diverse economic activities. One of these activities may be the cultivation of cannabis. Vera Rubin and Lambros Comitas point out that vendors of cannabis typically lead a stable family life and are otherwise law abiding and conservative.

Working-class Jamaicans often believe that use of ganja makes work go more pleasantly and allows them to work harder; however, the middle and upper classes, who employ the lower classes and supervise their work, believe the drug is detrimental to work performance. Melanie Dreher has investigated the interplay of these differing views in the setting of a Jamaican sugar estate. Three farms were contrasted, all having different proportions of cannabis smokers to nonsmokers. Dreher examined productivity figures by categories of smokers and nonsmokers and found no significant differences in the work performance of smokers and nonsmokers. The results support neither the views of the workers nor the managers.

Dreher's study illustrates the increasing use of quantitative data in anthropological work on drug use, but it also points to the levels of subtlety and sophistication that can be added to research by the inclusion of qualitative material. This dimension of research is one of anthropology's strongest assets; it allows for deeper and more accurate understanding of complex questions and helps restrain impulses to hasty generalization.

Dreher has also studied the use of cannabis among Jamaican women. She examined the different patterns of ganja use among women in two similar Jamaican villages, where women in one village seemed more inclined to smoke cannabis than in the other. At the time of this study, smoking was contrary to norms for Jamaican women, although they routinely made cannabis teas for medicinal purposes. In the village with the higher rate of smokers, women were found to have more economic opportunity and thus more independence.

Women in the village with fewer smokers found it adaptive to conform to the norms, so that they would not alienate men who were potential husbands and sources of support. By taking a cross-cultural look at what is considered to be a growing problem in the United States, Dreher has undertaken applied research to shed light on an important policy question.

A common conceptual framework for the behaviour of drug users whose conduct is outside society's usual limits is the socialpsychological notion of deviance. Dreher argues that the behaviour of the women she studied is better understood anthropologically in terms of intracultural variation. Again, we see a contextual focus, a perspective that, instead of looking solely at individual action, allows for the connecting of individual action to sociocultural structure. Currently more tolerance for ganja smoking has developed among lower-class Jamaican women.

Use of the substance fits into a constellation of personal characteristics to which the term "roots daughter" has been given. Connected to the ideas generated by the Rastafarian religion about what is "natural" and African, a "roots daughter" is dignified, independent and intelligent. Findings of this later research indicate changing attitudes and thereby remind us of culture's dynamic nature. Anthropological research often probes the role of ritual in human action, as in the case of the Colombian work of William Partridge. He found that smoking of cannabis was associated with work life through the action of ritual. An important part of worker comradeship is sharing cannabis when one can afford it; those who have it share it with whomever may not have it at that time.

Workers who use but do not provide cannabis for sharing at work breaks are considered undependable and isolate themselves from the social networks from which work gangs are developed in the agricultural economy. In this fashion, meaning is shifted from one social context to another. In a domain similar to ritual, Rubin has looked at the first experience of smokers as a rite of passage, an experience that strongly influences whether boys become regular smokers. If their first experience is a good one, they will probably become regular smokers; if not, they tend not to become users.

The practice of learning about the cultural intricacies of studied groups makes the anthropological enterprise often more time-consuming than other research approaches. In the Costa Rican work of J. Bryan Page, for example, a command of "proper" Spanish was not adequate to the task of following conversations in the argot of the drug culture on the street. The arcane vernacular was found to be useful to the speakers; one use was the concealment of illegal economic activity, economic activity again motivated by need.

Research in Costa Rica involved life histories, participant observation and ethnographic interviews and subdivided smokers into different categories whose members found their smoking experiences to be shaped considerably by conditioned expectations and socioeconomic factors. Users who enjoyed economic stability and good social support found the smoking of cannabis to enhance activity; those users with less secure lives did not have universally pleasing experiences.

In a keynote address to the Alcohol and Drug Study Group, Dreher documented several theoretical and methodological contributions from anthropology, especially as distinct from the research forthcoming from sociology and social psychology. Not surprisingly, the use of cannabis in modern society is viewed by anthropologists as a "sociocultural phenomenon rather than an individual characteristic". The holistic and comparative approach, the use of society or community as the unit of analysis, the reliance upon ethnohistorical and ethnographic studies and the combination of qualitative and quantitative methods make anthropological research on cannabis unique.

The cannabis question is obviously a loaded issue in American society at the present and therefore it is not surprising that there is resistance to accepting the results of these "controversial" anthropological findings.

However, the same skepticism regarding these conclusions about the potential functional role of ganja use in certain cultural contexts is remarkably familiar to some of the resistance alcohol researchers in anthropology have encountered when they call into question certain widespread assumptions about the etiology and diagnosis of alcoholism.

The question of where to draw the line between acceptable usage—if usage of a particular drug is thought ever to be acceptable—and dysfunctional-pathological usage is extremely complicated. This message is probably one of the most important ones that anthropology offers, especially on controversial drugs such as cannabis.

SOCIOCULTURAL CONTEXT OF ALCOHOL USE

In 1940 Ruth Bunzel pioneered the way for anthropological research on alcohol through the publication of the results of a controlled comparison of the role of drinking in two different Central American societies where she had conducted in-depth ethnographic fieldwork. She had not set out to study drinking or alcoholism specifically. Applying psychoanalytic concepts, Bunzel connected drinking behaviour to its wider sociocultural context.

By identifying positive functions within these drinking patterns as well as explicating drunken behaviour, Bunzel found that drinking seemed to help lubricate social relations in the village of Chamula in Mexico. In comparison, in the Guatemalan village of Chichicas- tenango, alcohol provided a release from anxieties related to a stressful environment.

In the late 1950s Dwight B. Heath took up the gauntlet of this genre of research in his study of culture change following the Bolivian revolution of 1953. One area of investigation was change in drinking behaviour as an indication of shifting interethnic social relations between the peasants and the mestizos: "drinking is a useful index, being both highly visible and an integral part of the etiquette of relations between men". With this work and an earlier study among the Navajo Indians, Heath began the first sustained effort in anthropology to advance alcohol studies.

The number of ethnographies concerned with drinking practices and beliefs has proliferated since the 1950s. In Peru alone, three publications during the late 1960s and early 1970s focus on alcohol. In one example, Ozzie Simmons offers a description of the sociocultural integration of alcohol use within Lunahuana, a Spanish-speaking coastal village in Peru: "the meshing of drinking with a configuration of culture and social structure [gives] alcohol positive symbolic and functional roles" within the society. Similarly, Paul Doughty found that "the use of alcoholic drinks was highly patterned and integral to normal social interaction" within the mestizo community of Huaylas in highland Peru. Allan Holmberg summarizes that for the agricultural peasant village of Viru on the north coast of Peru, "traditional patterns of drinking are such an integral part of the value structure of Viru that they are not likely to change in the near future". A more recent example of basic ethnographic research out of which alcohol data developed but was not the central theme or intent of the research is Ndolamb Ngokwey's fieldwork among the Lele of Kasai in the Republic of Zaire in which he analyzes the drinking of palm wine: when it is drunk and within what contexts, types of wine drunk, drinking manners and its connection with health and illness. By concluding that "Lele rules and practices concerning palm wine reproduce cultural values, notions and categories", Ngokwey clearly articulates the sociocultural integration theme. He also raises another issue, which runs through much of the literature on alcohol (as well as other drugs): gender differences in consumption.

Gerald Mars, an anthropologist focusing on occupational issues, is another recent researcher whose studies uncovered distinctive patterns in drinking styles. Among longshoremen in Newfoundland, Canada, two distinct groups emerged: the "regular men" and the "outside men." The regular men were regularly hired and rehired to work on the docks at the port of St. John's, where one of the requirements to be a "regular man" was to learn the proper drinking behaviour for members of the group.

These patterns were so fundamental to the continuities between nonwork and work roles that in Mars's research the men rarely mentioned the importance of drinking roles without explicit questions because "I didn't think you counted drinking! Everyone always drinks with their buddies!".

The consumption of alcohol is always subject to rules and regulations and breaching those rules arouses strong emotional response. Currently we see such intense emotionality expressed through the Mothers Against Drunk Driving (MADD) campaign. MADD has garnered phenomenal support in American society, to the point of strongly influencing such formal legal controls as blood alcohol levels permissible for drivers and the penalties for exceeding those levels. Gender issues and alcohol and drug use and abuse have garnered considerable attention recently. Two edited books, for example, address differences between men and women with respect to the proper and improper use of alcohol and alcohol and drugs in different cultural settings.

This genre of research offers much promise regarding its relevance to gender studies generally and, more particularly, regarding an explicit concern that women's drinking and drug use patterns be better documented and understood. Typically the province of sociologists in the past, workplace-based research on alcohol consumption and abuse has been studied by anthropologists at the Prevention Research Centre.

Acculturation and Culture Change Studies

By the 1960s, anthropologists had developed an increased interest in applying their ethnographic and often emic approach in order better to understand the problematic use of alcohol. This was particularly the case with respect to American Indians. As part of the interdisciplinary team at the Tri-Ethnic Research Project at the University of Colorado, Theodore Graves contrasted SpanishAmerican, Anglo-American and Ute Indian groups living in the same area as a means for addressing the question: "Under what conditions is acculturation accompanied by symptoms of social and psychological disorganization and under what conditions it is not?". Stark differences in drinking practices and problems with alcohol as well as other types of social problems existed among these three cultural groups.

Graves and his colleagues combined ethnographic observations with structured and unstructured interviews conducted with a randomly selected sample from the three groups. He found that the relatively unacculturated SpanishAmericans and Indians evidence patterns of alcohol use and abuse distinct from each other. While the Spanish-Americans retained strong controls socially and psychologically, the Indians were weak on this dimension. Furthermore, while the unacculturated Indian groups displayed excessive drinking patterns and problems associated with heavy drinking, the Spanish as a group did not.

With this study, a long tradition in acculturation studies of American Indians and other ethnic groups in the United States was well underway. Paradoxically, acculturation is sometimes hypothesized to be a protective factor against dysfunctional drinking practices and at other times a risk factor. As Graves pointed out in 1967,

Acculturation is obviously not the unqualified evil that some observers regard it. When traditional cultural strategies for personal satisfaction have become inapplicable, a reorientation towards a set of new and potentially attainable goals appears to be a promising path to mental health. Furthermore, where traditional social and personal control systems are weak, acculturation may also serve to promote the development of new controls and thereby make the group better able to prevent disruptive individual behaviour.

Change in drinking patterns under conditions of culture change and/or acculturation has been an ongoing theme of alcohol research in the past three decades. The Institute of Applied Social and Economic Research (ISAER) Alcohol Project in Papua New Guinea is an example of an ambitious two-year research project directed by anthropologist Mac Marshall and undertaken to address a growing problem with alcohol there. A notable feature of the resulting conference and monograph is the fact that most presentations were made by scholars—including many anthropologists—who had conducted indepth ethnographic research at a particular field site on Papua New Guinea. While none of these researchers had intentionally undertaken a study of alcohol and culture, each discovered that alcohol consumption was important enough to collect data on the topic.

In 1982 Marshall edited a monograph on changes in drinking patterns in highland and coastal villages and urban areas of Papua New Guinea and problems associated with drinking. The significance of this project is seen in the fact that it constitutes "the first time anywhere that such a large body of ethnographic information has been assembled specifically in the service of public policy decisions on alcohol".

Drunken Comportment

Two issues—the extent to which alcohol consumption poses serious social problems and the extent to which available alcohol control policies help to stem the tide of those perceived problems—have historical reference in anthropology in the "drunken comportment" concept. In their 1969 book, psychologist Craig MacAndrew and anthropologist Robert Edgerton examine the phenomenon of drunkenness cross-culturally using ethnographic and ethnohistorical data to evaluate several assumptions held about behaviour under the influence of alcohol.

First, drawing upon ethnographic data reported from five societies, they evaluate evidence for the "disinhibiting" effects of alcohol. They found that "even during periods of extreme intoxication, the inhibitions that are normally in effect remain in effect. Drunken persons in these societies may stagger, speak thickly and become stuporous, without any corresponding display of changes or-the-worse. In a word, if alcohol were a 'superego solvent' for one group of people due to its toxic action, then the same disinhibiting effect ought to be evident in all people. In point of fact, however, it is not".

In short, MacAndrew and Edgerton call into question the common assumption hold by many people in American society that alcohol serves as an disinhibitor, a point of view often perpetuated by medical science. Thus, even in the midst of increasing cross-disciplinary cooperation, anthropologists continue to find themselves in a position to question seemingly entrenched viewpoints about alcohol and alcohol-related behaviours. Unfortunately, noncritical thinking about alcohol and alcoholism also characterizes some of the research on alcohol. As Heath has succinctly put it, "polemic masquerades as science in much of what is written on alcoholism".

The Disease Concept of Alcoholism

In understanding alcohol abuse—how it develops, how it is diagnosed and how to prevent it or intervene in it—alcohologists tend to rely upon simplistic oppositions such as the nature-nurture or genetic-environment contrasts. Similarly the disease concept is often contrasted with the moral model concept in which alcoholism is seen to stem from weaknesses of individuals, which keep them from controlling their drinking. Historically, American society experienced a long period in which the moral model reigned supreme—during the 150 years of strong temperance and prohibition movements. With the founding of the Yale Centre for Alcohol Studies after World War II under the leadership of E. M. Jellinek, the disease or biomedical model was developed as a counter to the moral model. The disease model, according to Jellinek, explains alcoholism as a progressive disease with clear symptoms and certain recognizable, inevitable phases. The insistence on fidelity to the disease concept in the treatment of alcoholism, however, is a more convenient, established theory than scientific understanding based upon scholarly work. In reality, American society has superimposed the disease model upon the moral weaknesses model in popular understanding of the etiology and nature of alcoholism. Thus, alcoholics can be held responsible for their addiction based on personality features while at the same time be excused based on a presumed physiological predisposition.

In addition to the confusion around the moral-medical distinction, many wellmeaning professionals, as well as the general public, operate with a vague idea of what is really meant by the disease concept and assume that whatever it is is clearly supported by scientific evidence. Contrary to some popular beliefs, the same symptoms are not always present in all alcoholics or in those seeking treatment for alcoholism. The course of alcoholism varies widely.

Also, from a disease point of view, we might expect that biomedical criteria would be used to diagnose alcoholism. In fact, however, mainly behavioural criteria are applied. Drinking patterns and behaviour under the influence of alcohol figure more prominently in diagnostic criteria established to distinguish moderate, heavy and alcoholic drinking than do strict biomedical indicators such as organ damage and withdrawal signs.

We recently surveyed a group of anthropologists as to their points of view about the disease concept of alcoholism. This is far from a neutral topic—some respondents commented on its sensitivity—but responses were frank. David Strug noted that he had never "felt comfortable with the disease concept of alcoholism based on a biomedical model. The cultural component will always remain a contextual constant that must be considered". In connecting the idea of alcoholism with disease, Merrill Singer concludes that it reflects "broader patterns of medicalization, privatization of suffering and politically-endorsed individualized problem-solving patterns.

While virtually all respondents expressed skepticism about the disease concept, Dwight Heath was the most clearly dubious among these reports: "If alcoholism is a disease, it is a most unusual one inasmuch as an individual can often bring an end to it by modifying his/her behaviour even in the absence of any other intervention. Most of the reasons commonly given for calling it a disease are fallacious".

Interestingly, these opinions are generally consistent with the World Health Organization's position that alcohol problems do not necessarily follow a coherent pattern throughout the general population and do not necessarily have to do with a physiological dependence on alcohol.

Anthropologists insist, as do some other alcohologists, that alcoholism is a complex phenomenon, perhaps having predisposing factors (genetic and physiological) in combination with a series of precipitating factors (including psychol-ogical, social and cultural) contributing to etiology. The need for a biocultural synthesis of studies of individual and cultural variation is compelling.

PHARMACEUTICAL DRUG

A pharmaceutical drug, also referred to as medicine, medication or medicament, can be loosely defined as any chemical substance intended for use in the medical diagnosis, cure, treatment, or prevention of disease.

PHARMACOTHERAPY

Pharmacotherapy is a science which focuses on the use of drugs to treat disease. This branch of the sciences involves almost every branch of medicine, and integrates a wide variety of sciences such as chemistry as well. Many people around the world benefit from pharmacotherapy each year, but pharmaceutical companies create new drugs just to make a huge profit such as Watson Pharmaceuticals. Furthermore, this is the primary reason why medications are expensive and prices are driven up yearly.

TYPES OF MEDICATIONS

For the Gastrointestinal Tract (Digestive System)

Upper digestive tract: Antacids, reflux suppressants, antiflatulents, antidopaminergics, proton pump inhibitors (PPIs), H2-receptor antagonistss, cytoprotectants, prostaglandin analogues Lower digestive tract: laxatives, antispasmodics, antidiarrhoeals, bile acid sequestrants, opioid.

For the Cardiovascular System

General: β-receptor blockers ("beta blockers"), calcium channel blockers, diuretics, cardiac glycosides, antiarrhythmics, nitrate, antianginals, vasoconstrictors, vasodilators, peripheral activators Affecting blood pressure (antihypertensive drugs): ACE inhibitors, angiotensin receptor blockers, α blockers Coagulation: anticoagulants, heparin, antiplatelet drugs, fibrinolytics, anti-hemophilic factors, haemostatic drugs Atherosclerosis/cholesterol inhibitors: hypolipidaemic agents, statins.

For the Central Nervous System

Drugs affecting the central nervous system include: hypnotics, anaesthetics, antipsychotics, antidepressants (including tricyclic antidepressants, monoamine oxidase inhibitors, lithium salts, and selective serotonin reuptake inhibitors (SSRIs)), antiemetics, anticonvulsants/ antiepileptics, anxiolytics, barbiturates, movement disorder (*e.g.*, Parkinson's disease) drugs, stimulants (including amphetamines), benzodiazepines, cyclopyrrolones, dopamine antagonists, antihistamines, cholinergics, anticholinergics, emetics, cannabinoids, and 5-HT (serotonin) antagonists.

For Pain and Consciousness (Analgesic Drugs)

The main classes of painkillers are NSAIDs, opioids and various orphans such as paracetamol, tricyclic antidepressants and anticonvulsants.

For Musculo-skeletal Disorders

The main categories of drugs for musculoskeletal disorders are: NSAIDs (including COX-2 selective inhibitors), muscle relaxants, neuromuscular drugs, and anticho-linesterases.

For the eye

- *General*: Adrenergic neurone blocker, astringent, ocular lubricant
- *Diagnostic*: Topical anesthetics, sympathomimetics, parasympatholytics, mydriatics, cycloplegics
- *Anti-bacterial*: Antibiotics, topical antibiotics, sulfa drugs, aminoglycosides, fluoroquinolones
- *Anti-fungal*: Imidazoles, polyenes
- *Anti-inflammatory*: NSAIDs, corticosteroids
- *Anti-allergy*: Mast cell inhibitors
- *Anti-glaucoma*: Adrenergic agonists, beta-blockers, carbonic anhydrase inhibitors/hyperosmotics, cholinergics, miotics, parasympathomimetics, prostaglandin agonists/prostaglandin inhibitors. nitroglycerin

For the Ear, Nose and Oropharynx

Sympathomimetics, antihistamines, anticholinergics, NSAIDs, steroids, antiseptics, local anesthetics, antifungals, cerumenolyti

For the Respiratory System

Bronchodilators, NSAIDs, anti-allergics, antitussives, mucolytics, decongestants corticosteroids, beta-receptor antagonists, anticholinergics, steroids

For Endocrine Problems

Androgens, antiandrogens, gonadotropin, corticosteroids, human growth hormone, insulin, antidiabetics (sulfonylureas, biguanides/metformin, thiazolidinediones, insulin), thyroid hormones, antithyroid drugs, calcitonin, diphosponate, vasopressin analogues

For the Reproductive System or Urinary System

Antifungal, alkalising agents, quinolones, antibiotics, cholinergics, anticholinergics, anticholinesterases, antispasmodics, 5-alpha reductase inhibitor, selective alpha-1 blockers, sildenafils, fertility medications

For Obstetrics and Gynecology

NSAIDs, anticholinergics, haemostatic drugs, antifibrinolytics, Hormone

Replacement Therapy (HRT), bone regulators, beta-receptor agonists, follicle stimulating hormone, luteinising hormone, LHRH gamolenic acid, gonadotropin release inhibitor, progestogen, dopamine agonists, oestrogen, prostaglandins, gonadorelin, clomiphene, tamoxifen, Diethylstilbestrol

For the Skin

emollients, anti-pruritics, antifungals, disinfectants, scabicides, pediculicides, tar products, vitamin A derivatives, vitamin D analogues, keratolytics, abrasives, systemic antibiotics, topical antibiotics, hormones, desloughing agents, exudate absorbents, fibrinolytics, proteolytics, sunscreens, antiperspirants, corticosteroids

For Infections and Infestations

antibiotics, antifungals, antileprotics, antituberculous drugs, antimalarials, anthelmintics, amoebicides, antivirals, antiprotozoals

For Nutrition

Tonics, iron preparations, electrolytes, parenteral nutritional supplements, vitamins, anti-obesity drugs, anabolic drugs, haematopoietic drugs, [[food product drug]s

For Neoplastic Disorders

Cytotoxic drugs, therapeutic antibodies, sex hormones, aromatase inhibitors, somatostatin inhibitors, recombinant interleukins, G-CSF, erythropoietin

For Euthanasia

An euthanaticum is used for euthanasia and physician-assisted suicide. Euthanasia is not permitted by law in many countries, and consequently medicines will not be licensed for this use in those countries.

LEGAL CONSIDERATIONS

Medications may be divided into over-the-counter drugs (OTC) which may be available without special restrictions, and prescription only medicine (POM), which must be prescribed by a licensed medical practitioner. The precise distinction between OTC and prescription depends on the legal jurisdiction. A third category, behind-the-counter medications (BTMs), is implemented in some jurisdictions.

BTMs do not require a prescription, but must be kept in the dispensary, not visible to the public, and only be sold by a pharmacist or pharmacy technician. The International Narcotics Control Board of the United Nations imposes a world law of prohibition of certain medications. They publish a lengthy list of chemicals and plants whose trade and consumption (where applicable) is forbidden. OTC medications are sold without restriction as they are considered safe enough that most people will not hurt themselves accidentally by taking it as instructed. Many countries, such as the United Kingdom have a third category of pharmacy medicines which can only be sold in registered pharmacies, by or under the supervision of a pharmacist.

For patented medications, countries may have certain mandatory licensing programmes which compel, in certain situations, a medication's owner to contract with other agents to manufacture the drug. Such programmes may deal with the contingency of a lack of medication in the event of a serious epidemic of disease, or may be part of efforts to ensure that disease treating drugs, such as AIDS drugs, are available to countries which cannot afford the drug owner's price. In some countries, government-regulated cannabis is available by prescription.

MODERN PHARMACOLOGY

For most of the nineteenth century, drugs were not highly effective, "if all medicines in the world were thrown into the sea, it would be all the better for mankind and all the worse for the fishes". During the First World War, developed the treating wounds with an irrigation, a germicide which helped prevent gangrene. In the inter-war period, the first anti-bacterial agents such as the sulpha antibiotics were developed. The Second World War saw the introduction of widespread and effective antimicrobial therapy with the development and mass production of penicillin antibiotics, made possible by the pressures of the war and the collaboration of British scientists with the American pharmaceutical industry.

Medicines commonly used by the late 1920s included aspirin, codeine, and morphine for pain; digitalis, nitroglycerin, and quinine for heart disorders, and insulin for diabetes. Other drugs included antitoxins, a few biological vaccines, and a few synthetic drugs. In the 1930s antibiotics emerged: first sulfa drugs, then penicillin and other antibiotics. Drugs increasingly became "the center of medical practice". In the 1950s other drugs emerged including corticosteroids for inflammation, rauwolfia alkloids as tranqulizers and antihypertensives, antihistamines for nasal allergies, xanthines for asthma, and typical antipsychotics for psychosis. As of 2008, thousands of approved drugs have been developed. Increasingly, biotechnology is used to discover biopharmaceuticals. In the 1950s new psychiatric drugs, notably the antipsychotic chlorpromazine, were designed in laboratories and slowly came into preferred use. Although often accepted as an advance in some ways, there was some opposition, due to serious adverse effects such as tardive dyskinesia. Patients often opposed psychiatry and refused or stopped taking the drugs when not subject to psychiatric control.

Governments have been heavily involved in the development and sale of drugs. In the Elixir Sulfanilamide disaster led to the establishment of the Food and Drug Administration, and the 1938 Federal Food, Drug, and

Cosmetic Act required manufacturers to file new drugs with the FDA. The 1951 Humphrey-Durham Amendment required certain drugs to be sold by prescription. In 1962 a subsequent amendment required new drugs to be tested for efficacy and safety in clinical trials. Until the 1970s, drug prices were not a major concern for doctors and patients. As more drugs became prescribed for chronic illnesses, however, costs became burdensome, and by the 1970s nearly every country required or encouraged the substitution of generic drugs for higher-priced brand names.

SYNTHETIC DISORDERS IN HUMANS

Within their coiled strands, chromosomes hide a world of information about how a person should look, how tall he should be, what should be his skin colour and also the diseases that he may be prone to (which are referred to as genetic disorders in human). They have this information coded in the form of nucleotide sequences that form DNA (dioxyribonucleic acid) molecules. A single molecule of DNA is coiled to form a single chromosome. These chromosomes are present in pairs. Human beings have 23 pairs of chromosomes or 46 chromosomes in all. Of them, 1 pair forms the sex chromosomes. The rest 22 pairs are the autosomes. Sex chromosomes are of two types - X and Y. A man has one X and one Y chromosome (XY) while a woman has two X chromosomes (XX). It is just the difference of one sex chromosome that decides an individual's gender.

The chemical information of every organism is specifically coded in the nucleotide sequences of genes - the unit of heredity. Each gene codes for an enzyme that plays an important role in various biochemical reactions. In other words, a gene contains a specific sequence of nucleotide bases that are responsible for the organization of amino acids in the correct order to form an enzyme. Any disruption in this sequence, causes genetic disorders in humans.

CAUSES OF DISORDERS IN HUMANS

Genetic diseases in humans are caused due to abnormalities in genes or chromosomes.

Such defects can be caused by the following mechanisms:

- *Mutations*: These are sudden inheritable changes in the nucleotide sequence of a gene.
- *Aneuploidy*: Aneuploidy is caused when there are abnormal number of chromosomes in an organism. This could be due to loss of a chromosome (monosomy) or presence of extra copy of a chromosome (trisomy, tetrasomy, etc.)
- *Deletions*: Loss of a part of chromosome as in the case of Jacobsen syndrome.
- *Duplications*: Duplication of a portion of chromosome that results in extra amount of genetic material.

- *Inversions*: Inversion of the nucleotide sequence because a portion of chromosome has broken off, got inverted and reattached at the original location of the chromosome.
- *Translocations*: When a portion of chromosome has got transferred on to some other chromosome. Sometimes translocation can take place between two chromosomes, in which case they interchange chromosome segments. However, in some cases a portion of a chromosome may simply get attached to another chromosome.

TYPES OF DISORDERS IN HUMANS

- *Autosomal Dominant Genetic Disorders*: These disorders are caused when an individual has inherited the defective gene from a single parent. This defective gene belongs to an autosome. Such an inheritance is also known as autosomal dominant pattern of inheritance.
- *Autosomal Recesive Genetic Disorders*: Such disorders manifest only when an individual has got two defective alleles of the same gene, one from each parent. These genetic disorders are inherited via the autosomal recessive pattern of inheritance.
- *Sex-Linked Disorders*: These are disorders related to sex chromosomes or genes in them.
- *Multi-factorial Genetic Disorders*: Such disorders are the result of genetic as well as environmental factors.

COMMON DISORDERS IN HUMANS

Achondroplasia

It is an autosomal dominant genetic disorder which is the most common genetic cause of dwarfism. Individuals suffering from achondroplasia vary from 4 feet to 4 feet 4 inches in height. They have disproportionately short limbs. However, there is no intellectual disability. In this disorder, the cartilage, specially in the long bones, fail to convert into bones. It is caused due to mutation in the FGFR3 gene (located on chromosome 4. which codes for a protein that regulates transformation of cartilage to bone). Although, an individual may inherit the disorder from an affected parent, the disorder is usually the result of a mutation in the sperm or egg of a healthy parent. Achodroplasia can be detected before birth with the help of prenatal ultrasound. There is no treatment for this disorder. However, limb extending surgeries can be done, although this is a controversial issue.

Achromatopsia

It is an autosomal congenital recessive disorder which is characterized by visual acuity loss, colourblindness, light sensitivity and nystagmus. It is also known as rod monochromatism. The symptoms are first noticed in children at the age of six months when they exhibit nystagmus and photophobic activities. Achromatopsia is of two forms. The more severe form is known as compete achromatopsia. Those who exhibit milder symptoms are known to suffer from partial achromatopsia. Using optical and visual aids are useful in improving vision of those suffering from achromatopspia.

Acid Maltase Deficiency

It is an autosomal recessive disorder, in which the defect is in the gene for the acid maltase enzyme, which leads to accumulation of glycogen stored in muscles. Glycogen build up, weakens the muscles of a patient suffering from this disorder. This may affect respiratory muscles resulting in respiratory failure. It is also known as the *Pompe Disease.* Although, in childhood and adolescence the symptoms show slow progress and are less severe, infantile forms cause death within first year, if not treated on time.

Albinism: Albinism is a congenital disorder in which there is little or completely no production of melanin in hair, skin and iris of the eyes. Hence albinos (people suffering from albinism) have light coloured skin, hair and eyes. It is caused due to inheritance of recessive alleles from parents. This disorder can't be cured. However, the symptoms can be alleviated with the help of surgical treatment, vision aids and using device that provide protection from sun.

Alzheimer's Disease: Alzheimer's disease is the most common form of dementia which is characterized by gradual memory loss, irritability, mood swings, confusion and language breakdown. Although, scientists are not unequivocal about the cause of this disease, the most widely accepted reason is the amyloid cascade hypothesis, that suggests excess production of a small protein fragment called ABeta (Aβ). Also known as Senile Dementia of the Alzheimer Type (SDAT) or simply Alzheimer's, this is a degenerative disease and scientists are yet to find its cure. However, balanced diet, mental exercises and stimulation are often suggested for prevention and manging of the disease.

Angelman Syndrome

It is a neurological disorder that was first described by a British pediatrician, Dr. Harry Angelman, in 1965. This disorder is marked by intellectual and developmental delays, severe speech impairment and problems in movement and balance, recurrent seizures and small heads. Children with Angelman syndrome typically have a happy demeanor. They are hyperactive with short attention span and show jerky hand movements. These children appear normal at birth.

This genetic disorder in human is a classical case of genetic imprinting, in which the disorder is caused due to deletion or activation of the maternally inherited chromosome 15. Its sister syndrome is the Prader-Willi syndrome in which there is a similar loss or inactivation of the paternally inherited chromosome 15.

Bardet-Biedl Syndrome

It is a pleiotropic recessive genetic disorder that is characterized by obesity, polydactyly, deterioration of rod and cone cells, mental retardation and defect in the gonads and kidney disease. It is difficult to diagnose Bardet-Biedl Syndrome, specially in the young. As no cure is yet known for the disorder, treatment is concentrated on specific organs and systems.

Barth Syndrome

A rare but serious sex linked genetic disorder, the Barth syndrome is caused due to mutations or alterations in the BTHS gene. The gene is located on the long arm of X chromosome. This disorder primarily affects the heart. Besides heart defects, Barth syndrome results in poor skeletal musculature, short stature, mitochondrial abnormalities and deficiency of white blood cells. There is no cure for this disorder. Treatment focuses on managing the symptoms and preventing infections.

Bipolar Disorder

Also known as manic depressive disorder or bipolar affective disorder, individuals suffering from bipolar disorder suffer from highly elevated moods, referred to as mania or episodes of severe depression. Research shows that both genetic as well as environmental factors are responsible for this disorder. Medicines as well as psychotherapy is found to be useful in dealing with the severe mood swings associated with the disorder.

Bloom Syndrome

Bloom syndrome is an autosomal recessive genetic disorder, which is characterized by a high frequency of breaks and rearrangements in the chromosomes of an effected person. Symptoms include short stature, butterfly shaped facial rash, high pitched voice, increased susceptibility to cancer, leukemia, respiratory illnesses and infections. Some may even show mental retardation. Like other genetic disorders in human, there is no treatment for Bloom's syndrome. All treatment is preventive in nature. This disease is more common in Ashkenazi Jews with a frequency of 1/100 individuals suffering from this disorder.

Colour Blindness

Colour blindness refers to the inability of differentiating among certain colours. This can be genetically inherited and can also be caused due to a damage to the eye, nerve or brain. As far as genetics is concerned, colour blindness is most commonly the result of mutations in the X chromosome. However, research has shown that mutation in 19 different chromosomes can cause colour blindness. There is no treatment to cure colour blindness. However, certain types of tinted filters and contact lenses may enable an individual to differentiate colours.

Cri-du-Chat Syndrome

Cri du Chat syndrome is caused due to deletion of short (p) arm of chromosome 5. Most cases of this disorder are not inherited. In such cases, they are caused due to spontaneous deletion of a segment of chromosome 5 during formation of egg or sperm or during early stages of fetal development. The syndrome gets the name from the characteristic high pitched cry of an infant that resembles the cry of a cat. Other symptoms are intellectual disabilities, delayed development, microcephaly (small head), low birth weight, typical facial features and weak muscle tones during infancy. No specific treatment is available for this disorder.

Cystic Fibrosis

Cystic fibrosis is an inherited disease of the glands that secrete mucus and sweat. Cystic fibrosis causes the mucus to become thick and sticky that clogs various organs of the body, that results in other complications. It mostly affects lungs, liver, pancreas, sinuses, intestines and sex organs. Cystic fibrosis also causes excess loss of salts through sweat that results in dehydration, tiredness, weakness and elevated heart rate. This is an autosomal recessive disorder in which the mutation is caused in the CFTR gene. At present, there is no cure for cystic fibrosis. However, doctors treat the symptoms using antibiotic therapy along with other treatments that would clear the mucus that accumulates in different organs.

Down Syndrome

Also called *Trisomy 21* it is a genetic disorder in human that is the result of extra copy of chromosome 21, that a child inherits from his/her parent. This extra genetic material causes delays in mental as well as physical development of a child. Physical peculiarities caused due to this syndrome include a narrow chin, a prominently round face, protruding tongue, short limbs, the Simian crease and poor muscle tone. Almost 1 in every 800 to 1000 births may have genetic abnormality. The incidence of this disorder increases with maternal age. Amniocentesis during pregnancy or birth can detect this abnormality. Karyotyping test of a child confirms this syndrome, if done after birth. No specific cure is available. However, treatment of the health problems and training and special education of such individuals is of great help.

Duchenne Muscular Dystrophy

It is an X linked recessive trait which is characterized by progressive degeneration of muscles that results in loss of ambulation and finally leading to death. It is one of the most prevalent muscular dystrophies that affect only males. Females are just the carriers and hence don't show the symptoms. The disorder is caused due to mutation in the gene DMD (located on X chromosome) that codes for protein dystrophin which is an important component of muscle tissue. Physical therapy is effective in lessening physical disabilities of children suffering from this disorder. Besides this, recent advancements in medicines are helping in extending lives of those affected. Stem cell research also shows promising developments in dealing with this form of muscular dystrophy.

Fragile X Syndrome

Also known as the *Martin-Bell syndrome* or *marker X syndrome,* the Fragile X syndrome is the most common cause of inherited form of mental retardation. It is the result of trinucleotide repeat disorder, in which, the the trinucleotide gene sequence CGG in the X chromosome is repeated several times. The result is intellectual disabilities, high levels of anxiety and hyperactivity like fidgeting, autistic behaviour like hand flapping, avoiding eye contact, shyness, mental retardation and attention deficit disorder and other symptoms. It is an X-linked dominant disorder that has no cure. Medicines, educational, behavioural and physical therapy are the only help available to individuals suffering from this disorder.

Galactosemia

A rare metabolic genetic disorder in human, galactosemia impairs body's ability to break down galactose. Alternately, it is also known as *Galactose-1-phosphate uridyl transferase deficiency (Galactosemia type I),Galactokinase deficiency (Galactosemia type II),Galactose-6-phosphate epimerase deficiency(Galactosemia type III).* Infants suffering from galactosemia show symptoms within a few days after birth or soon after they start nursing. The symptoms include yellowing of skin, eyes, diarrhea, vomiting, refusal to drink milk, malnourishment and also mental retardation. If milk or milk products are given to infants suffering from galactosemia, then accumulation of galactose in their system damages brain, eyes, liver and kidneys. Eliminating lactose and galctose from diet of an individual is the only way to treat classic galactosemia (galactosemia type I) which is the most common and most severe form of galactosemia. Although very different, galctosemia is often confused with lactose intolerance.

Hemophilia

This is a recessive X-linked genetic disorder in which the bodies of individuals lose the ability to coagulate blood or blood clotting. As the mutation is caused in X chromosome and the condition is recessive, the females are carriers and males suffer from the symptoms of hemophilia. However, under rare occasions, females may also suffer from hemophilia. There are two variations of the disorder. Hemophilia A which is more common than the other variation, Hemophilia B. Regular infusion of the coagulating factor that lacks in an individual, helps one control blood loss caused due to excessive bleeding.

Huntington's Disease

It is an autosomal dominant genetic disorder in the huntingtin gene, that produces a faulty protein instead of the normal "huntingtin" protein. The faulty protein causes damage to specific areas of the brain, that initially manifests as abnormal involuntary movements that become increasingly uncoordinated jerky movements.

As the disease progresses, decline on mental abilities (marked with dementia), behavioural and psychiatric problems become prominent. Although physical symptoms of Huntington's disease may manifest at any age, it most commonly occurs in individuals between 35 to 45 years old. In rare cases, when onset of the disease takes place as early as 20 years, the condition is referred to as *juvenile, akinetic-rigid* or *Westphal variant HD.* Although there is no cure, medicines help individuals to cope with emotional disabilities. Speech therapy, occupational and behavioural therapy also help individuals deal with the disabilities due to Huntington's disease.

Jackson-Weiss Syndrome

It is an autosomal dominant genetic disorder in which there are foot abnormalities, and premature fusion of bones in the skull lead to deformations of the facial features (widely spaced eyes, bulging forehead) and the skull. In this syndrome, the great toes are short and wide and turn away from the rest of the toes. Some toes may be fused or have some other abnormalities. The mutation is caused in the FGFR2 gene which is located in chromosome 10. Treatment involves corrective surgery for deformed bones in face and foot.

Klinefelter Syndrome

It is the most common sex linked genetic disorder. In which males have an extra X chromosome. Hence,this disorder is also known as *47, XXY* or *XXY syndrome.* The most common symptom is infertility. Besides this, males with the XXY syndrome have impaired physical, language and social developments. As these individuals produce less testosterone than other males, such teenagers may be less muscular and have less facial hair than their peers. The presence of the extra X chromosome can't be undone. However, testosterone replacement therapy, a variety of therapeutic options like behavioural, speech and occupational therapy and educational treatments are the options available for those suffering from Klinefelter's syndrome.

Krabbe Disease

Krabbe disease is a rare degenerative disorder of the nervous system. It occurs due to mutation in the GALC gene which results in deficiency of enzyme galactosylceramidase. Deficiency of this enzyme affects the development of the myelin sheath of nerve cells. This disorder is inherited in autosomal recessive pattern and manifests itself in babies of 6 months of age.

However, it can occur during adolescence or adulthood as well. Bone marrow transplant has help some who suffer from mild form of this disorder. Treatment is usually symptomatic and supportive.

Langer-Giedion Syndrome

Langer-Giedion syndrome is a genetic disorder in human that is caused due to deletion or mutation of at least two genes on chromosome 8. This is not an inherited disorder. It is caused due to random events during formation of reproductive cells (sperms and eggs) in individuals. This is a rare disorder that causes bone abnormalities and typical facial features. Individuals suffering form Langer-Giedion syndrome have multiple non-cancerous tumors in their bones that cause pain, restrict joint movement and exerts pressure on nerves, blood vessels, spinal cord and the tissues surrounding the tumors. Some intellectual disability may be associated with this disorder. External fixators can be used for facial and limbic reconstructions.

Lesch–Nyhan Syndrome

It is an X-linked recessive disorder which causes deficiency of the enzyme hypoxanthine-guanine phosphoribosyltransferase (HPRT). Lack of HPRT leads to accumulation of uric acid in body, which leads to gout and kidney problems, poor muscle control and mental retardation of moderate degree. A striking feature of this disorder is a child biting his lips and fingers. This self mutilating behaviour appears in second year of a child's life. Other than these features, an individual suffering from this disorder shows facial grimacing, involuntary writhing and repetitive movement of limbs that is characteristic of Huntington's disease. This disorder is alternatively also known as *Nyhan's syndrome, Juvenile gout* and *Kelley-Seegmiller syndrome*. Treatment for the Lesch-Nyhan syndrome is symptomatic.

Marfan Syndrome

Marfan syndrome is an inherited genetic disorder of the connective tissue, in which mutation is caused in the FBN1 gene that codes for the protein fibrillin-1. Marfan syndrome can be mild or severe. There is great variability in the features of this disorder that are associated with skeleton, skin and joints. However, confirmatory symptoms of the disorder are long limbs, dislocated lenses and dilation of the aortic root. Individuals suffering from Marfan syndrome or *Marfan's syndrome* usually have heart problems. They typically are tall and thin with slender, tapering fingers. Once diagnosed with Marfan's syndrome, regular visit to the cardiologist is required. Treatment depends upon the organ system that is affected. Regular check ups, medicines and surgery may be required to treat the symptoms of this disorder that is alternately also known as *leodosis*.

Muscular Dystrophy

Muscular dystrophy (MD) refers to a group of genetically inherited disorders of progressive degeneration of skeletal muscles. It also causes defects in muscle proteins and death of muscle cells and tissues. These disorders vary in severity and the extent and distribution of muscle weakness. Although the skeletal muscles are primarily affected, muscular dystrophy may impair functions of other systems of the body as well. While in some cases the symptoms appear in infancy of childhood, in certain instances muscular degeneration sets in during adulthood. *Duchenne MD* is the most common form of MD. Other common disorders are *Becker MD, Facioscapulohumeral MD*and *Myotonic MD.*

There is no specific treatment to cure or reverse MD. However, therapeutic options like speech therapy, respiratory therapy, speech therapy and corrective orthopedic surgery and orthopedic appliances are used to treat disabilities due to MD.

Myotonic Dystrophy

Myotonic dystrophy, also known as *dystrophia myotonica (DM)* is an autosomal recessive genetic disorder that is caused due to repetition of a trinucleotide sequence. It affects the muscles of the body and is a multi-system disorder. Other than progressive muscle wasting, there is formation of cataracts in the eye, cardiac conduction defects and hormonal imbalances. There are two variations of this disorder – DM 1 and DM 2. DM 1 is more severe than DM 2. In DM 1 the trinucleotide sequence repeat is located on chromosome 9 whereas in case of DM 2 the trinucleotide sequence repeat occurs in chromosome 3. Although this disorder can manifest at any stage of one's life, variability with respect to age on onset of the symptoms reduces within successive generation. Hence it is a good example of *anticipation.* There is no cure for this disorder. Nevertheless the affected organs can be treated to mange the symptoms.

Nail-Patella Syndrome

Nail-patella syndrome (NPS) is inherited via autosomal dominant pattern. It is a disorder that affects the joints, bones, fingernails and kidneys. It is most commonly characterized by lack of nail and knee caps. Bone deformations manifest in elbow and abnormally shaped hip bone. Research shows that individuals suffering from the NPS are susceptible to developing glaucoma and scoliosis. Other names for NPS are *hereditary onychoostedysplasia,iliac horn syndrome, Fong disease* or *Turner-Kiser syndrome.*

Neurofibromatosis

An autosomal dominant condition, neurofibromatosis (abbreviated NF) is a genetically inherited condition in which nerve tissue grow tumors, that may be benign or may cause medical complications by compressing nerves and tissues around them. Hence, in this disorder the bones, nervous system, the spine and the skin are affected. Tumors under the skin may appear as bumps and are associated with skin discolouration. Learning disabilities are also associated with this disorder. There are two types of this disorder. Neurofibromatosis type 1 is more common than neurofibromatosis type 2. While type 1 is caused due to mutation in chromosome 17, neurofibromatosis type 2 is the result of a mutation in chromosome 22. Due to lack of any cure, treatment is aimed at managing symptoms and complications. In case, the tumor becomes cancerous (as happens with 10 per cent of the cases), chemotherapy may be required. Surgery is resorted to, when the tumor compresses any organ or specific tissue of the body.

Noonan Syndrome

It is an autosomal dominant genetic disorder that may be inherited or arise due to spontaneous mutation in genes KRAS, PTPN11, RAF1, and SOS1. Individuals suffer from developmental disabilities that result in heart malformations, short stature, characteristic facial features, impaired blood clotting and indentation of the chest. Speech, language and learning disabilities are also common.

Triple X Syndrome

As the name suggests, trisomy refers to the genetic disorder which results in an extra copy of the X chromosome in females. This disorder is variably known as the *XXX syndrome, triplo-X, trisomy X,* and *47,XXX aneuploidy*.

This genetic disorder is not inherited. It is caused due to non-disjunction during cell division, that results in an extra copy of chromosome in reproductive cells. In some cases, the extra copy of X chromosome may be caused during cell division in early embryonic development. Women with this disorder.

Osteogenesis Imperfecta:

This is an autosomal dominant disorder of the connective tissue in which bones break easily and sometimes due to no apparent reason. Hence, it is also known as *brittle bone syndrome* or *Lobstein disorder*. Genetic mutation impairs synthesis of collagen - a protein that makes bones strong. Osteogenesis Imperfecta may also weaken muscles, cause brittle bones, curved spine and a impaired hearing.

Exercise, physical therapy, medicines and orthopedic devices are the only treatment available for people suffering from this disorder, as cure hasn't yet been found.

Patau Syndrome

Also known as *trisomy D* or *trisomy 13*, Patau syndrome is caused due to non-disjunction of chromosome 13 during meiosis, due to which, an affected individual inherits an extra copy of the chromosome. Robertsonian translocation can be another cause of this disorder. Like other genetic disorders in human that originate due to non-disjunction of chromosomes, the incidence of Patu syndrome increases with maternal age. The extra copy of chromosome results in kidney and heart defects, neurological problems, facial defects, polydactyly (having extra fingers) and deformed feet. Features of this disorder are present from birth and may be confused with Edward's syndrome. Hence, genetic testing is important to confirm diagnosis. While certain infants may be able to survive only for a couple of days, depending upon the severity of the conditions, those who survive with milder symptoms undergo treatment, focusing the particular disability that each individual suffers from.

Phenylketonuria

Phenylketonuria (abbreviated as PKU) is an autosomal recessive disorder that causes deficiency in the enzyme phenylalanine hydroxylase. The result of this deficiency is that instead of being metabolized to tyrosine, the amino acid phenylalanine gets converted into phenylketone (also known as phenylpyruvate). This compound is detected in urine. If left untreated, excess phenylketone can impair development of the brain. This will manifest as mental retardation, seizures or brain damage. A diet low in phenylalanine or other means that lowers amount of the compound helps in dealing with the disorder.

Porphyria

Porphyria is an inherited genetic disorder in which, synthesis of any one of the 8 enzymes involved in the process of synthesis of heme, is disrupted. Heme is linked to a chemical called protoporphyrin. Disruption in the heme biosynthetic pathway results in accumulation of porphyrin or its precursors in the body, that cause neurological or dermal problems. Acute porphyria affects the nervous system, whereas cutaneous porphyria or erythropoeitic porphyria primarily affects the skin.

This disorder is inherited in autosomal dominant pattern. Taking heme externally through a vein and medicines can alleviate the symptoms of this disorder.

Retinoblastoma

Retinoblastoma is a cancer of the retina, that affects children younger than 5 years. It can be genetic as well as non genetic. The genetic form, which is the cause in almost half of the cases of retinoblastoma, is the result of mutation in chromosome 13. Retinoblastoma usually affects one eye. Characteristic physical feature is whiteness of the retina, which is referred to as "cat's eye reflex" or leukocoria. Other symptoms include, deterioration in vision, eye pain, redness and irritation in the eye. Retiblastoma is curable if treated at an early stage. However, if not treated on time, cancer may spread out from the eye to other parts of the body.

Rett Syndrome

Rett syndrome is a neurological and developmental disorder that is inherited through X-linked dominant pattern. It occurs almost exclusively in females. For about a year of normal growth, girls with Rett's syndrome show clinical features that include decreased rate of head growth, small hands and feet, disabilities related to learning communication, coordination and speech. Affected girls lose control over purposeful use of hands and show repetitive movements like wringing of the hands and clapping.

10

Food Biotechnology

HEALTH AND NUTRITIONAL BENEFITS

A variety of healthier cooking oils derived from biotechnology are already on the market. Using biotechnology, plant scientists have decreased the total amount of saturated fatty acids in certain vegetable oils. They have also increased the conversion of linoleic acid to the fatty acid found mainly in fish that is associated with lowering cholesterol levels. Another nutritional concern related to edible oils is the negative health effects produced when vegetable oils are hydrogenated to increase their heat stability for cooking or to solidify oils used in making margarine. The hydrogenation process results in the formation of trans-fatty acids.

Biotechnology companies have given soybean oil these same properties, not through hydrogenation, but by using biotechnology to increase the amount of the naturally occurring fatty acid, stearic acid. Animal scientists are also using biotechnology to create healthier meat products, such as beef with lower fat content and pigs with a higher meat-to-fat ratio. Other health and nutritional benefits of crops improved through biotechnology include increased nutritional value of crops, especially those that are food staples in developing countries.

Scientists at Nehru University in New Delhi used a gene found in the South American plant amaranth to increase the protein content of potatoes by 30 per cent. These transgenic potatoes also contain large amounts of essential amino acids not found in unmodified potatoes. Other examples include golden rice and canola oil, both of which are high in vitamin A. The golden rice developers further improved rice with two other genes that increase the amount and digestibility of iron.

Biotechnology also promises to improve the health benefits of *functional foods*. Functional foods are foods containing significant levels of biologically active components that impart health benefits beyond our basic needs for sufficient calories, essential amino acids, vitamins and minerals. Familiar examples of functional foods include compounds in garlic and onions that lower cholesterol and improve the immune response; antioxidants found in green tea; and the glucosinolates in broccoli and cabbage that stimulate anticancer enzymes.

We are using biotechnology to increase the production of these compounds in functional foods. For example, researchers at Purdue University and the U.S. Department of Agriculture, created a tomato variety that contains three times as much of the antioxidant lycopene as the unmodified variety. Lycopene consumption is associated with a lower risk of prostate and breast cancer and decreased blood levels of "bad cholesterol." Other USDA researchers are using biotechnology to increase the amount of ellagic acid, a cancer protective agent, in strawberries.

PRODUCT QUALITY

We are also using biotechnology to change the characteristics of the raw material inputs so that they are more attractive to consumers and more amenable to processing. Biotechnology researchers are increasing the shelf life of fresh fruits and vegetables; improving the crispness of carrots, peppers and celery; creating seedless varieties of grapes and melons; extending the seasonal geographic availability of tomatoes, strawberries and raspberries; improving the flavour of tomatoes, lettuce, peppers, peas and potatoes; and creating caffeine-free coffee and tea.

Japanese scientists have now identified the enzyme that produces the chemical that makes us cry when we slice an onion. Knowing the identity of the enzyme is the first step in finding a way to block the gene to create "tearless" onions. Much of the work on improving how well crops endure food processing involves changing the ratio of water to starch. Potatoes with higher starch content are healthier because they absorb less oil when they are fried, for example. Another important benefit is that starchier potatoes require less energy to process and therefore cost less to handle.

Many tomato processors now use tomatoes derived from a biotechnology technique, somaclonal variant selection. The new tomatoes, used in soup, ketchup and tomato paste, contain 30 per cent less water and are processed with greater efficiency. A 1?2 per cent increase in the solid content is worth $35 million to the U.S. processed-tomato industry.

Another food processing sector that will benefit economically from better quality raw materials is the dairy products industry. Scientists in New Zealand have now used biotechnology to increase the amount of the protein casein, which is essential to cheese making, in milk by 13 per cent.

Biotechnology also allows the economically viable production of valuable, naturally occurring compounds that cannot be manufactured by other means. For example, commercial-scale production of the natural and highly marketable sweetener known as fructans has long eluded food-processing engineers. Fructans, which are short chains of the sugar molecule fructose, taste like sugar but have no calories. Scientists found a gene that converts 90 per cent of the sugar found in beets to fructans. Because 40 per cent of the transgenic beet dry weight is fructans, this crop can serve as a manufacturing facility for fructans.

SAFETY OF THE RAW MATERIALS

The most significant food-safety issue food producers face is microbial contamination, which can occur at any point from farm to table. Any biotechnology product that decreases microbes found on animal products and crop plants will significantly improve the safety of raw materials entering the food supply. Improved food safety through decreased microbial contamination begins on the farm. Transgenic disease-resistant and insect-resistant crops have less microbial contamination. Biotechnology is improving the safety of raw materials by helping food scientists discover the exact identity of the allergenic protein in foods such as peanuts, soybeans and milk, so they can then remove them.

Although 95 per cent of food allergies can be traced to a group of eight foods, in most cases we do not know which of the thousands of proteins in a food triggered the reaction. With biotechnology techniques, we are making great progress in identifying these allergens. More importantly, scientists have succeeded in using biotechnology to block or remove allergenicity genes in peanuts, soybeans and shrimp.Finally, biotechnology is helping us improve the safety of raw agricultural products by decreasing the amount of natural plant toxins found in foods such as potato and cassava.

Food Processing

Microorganisms have been essential to the food-processing industry for decades. They play a role in the production of the fermented foods. They also serve as a rich source of food additives, enzymes and other substances used in food processing.

MODERN FOOD TECHNOLOGY

At the beginning of this century certain aspects of modern food technology were already well developed but others, which had become commonplace by 1950, had scarcely developed at all. We may, therefore, conveniently begin our study by outlining the situation in 1900.

By that time the canning industry had been firmly established for meat, fish, fruit, and vegetables. Although great improvements were to be made both in the technique itself and in the quality of the product, canned food was widely used for both its convenience and its cheapness. The fact that it would keep almost indefinitely was a most important point in its favour — as much for the housewife anxious to hold a reserve against unexpected demand as for the military victualler meeting the requirements of an army in the field.

Refrigeration, too, was firmly established as an important means of preservation, and large cargoes of frozen meat were coming into Europe from the new sources of supply in Australia, New Zealand, and North and South America. The quality left something to be desired, however, partly because of imperfections in the freezing process, which affected the taste, and partly because of the inferior quality of much of the product, reared under conditions very different from those on European farms. Nevertheless it was acceptable and, because of its cheapness, it found a ready market. As opposed to canned products, however, which would keep more or less indefinitely, refrigerated products had a relatively short life because the cost of fuel made it uneconomic to freeze for long periods.

The magnitude of the trade is indicated by the fact that in 1900 New Zealand exported four million carcasses of mutton and lamb and the Argentine more than two million. There was also a rapidly growing trade in frozen beef: 20 000 tons from the Argentine in 1900. The freezing of fish was less successful, however, and in the fish trade refrigeration was generally on a short-term basis sufficient to cover transit from the increasingly distant fishing grounds to the ports. Fruit, vegetables, and eggs did not at that time lend themselves to preservation in this way.

The processing of dairy products, too, was being gradually moved from the farm to the factory. By 1850 cheese factories were established in both the USA and Australia, but in Britain and elsewhere in Europe they met fierce opposition from farmers and dairymaids alike: the one feared for his position as the traditional cheese producer and the latter for her employment.

Nevertheless, the first English cheese factory — at Longford, Derby — was in operation in 1870, under the guidance of American advisers, and within five years there were ten such factories, processing the milk of 8000 cows. Production of factory-made cheese increased steadily, but the larger farmers, able to buy modern equipment and assimilate new techniques, contrived to hold their own.

In butter-making the move to the factory was fostered by the invention of the centrifugal cream-separator by the Swedish engineer de Laval in 1877, which made possible great savings in labour and space. While the traditional dairy-farming countries continued to be major suppliers, the new producers, notably Australia and New Zealand, made rapidly growing contributions.

All producers found a rival, however, in margarine. Invented in the 1860s by the French chemist H. Mège-Mouriés, its large-scale manufacture had been firmly established, especially in Holland, by 1900. An important discovery here, made in 1899, was that certain oils, too soft for margarine manufacture, could be hardened by hydrogenation in the Sabatier-Senderens reaction. This process began to be worked on an industrial scale about 1910.

Milk is another important perishable food whose handling changed rapidly. With the extension of the railways and the growth of the urban areas, the towns increasingly drew their supplies from the countryside in bulk. The milk travelled in tinned-steel churns and was delivered from smaller churns at the doorstep. Some surplus milk was condensed or dried. Pasteurization was introduced about the turn of the century, originally with the commercial objective of prolonging the life of the milk and later it was used more widely for the sake of health in general and for the prevention of tuberculosis in particular.

In the first half of this century the trends indicated above continued, with emphasis on preservation and factory processing. Development was facilitated by a better understanding of the scientific principles involved. In addition, some quite new processes came into general use, especially after the Second World War. The most important of these were freeze-drying, which allowed easy reconstitution of many products, and deep-freezing, which went hand in hand with the availability of domestic cold-storage lockers. Before studying these innovations, however, we may conveniently consider progress in established food technology.

DAIRY PRODUCTS

Throughout the first half of this century milk and products derived from it remained of major importance. At mid-century world milk production was around 180 million tons annually, and from this were made some three million tons of butter and two million tons of cheese. Most countries remained more or less self-supporting in respect of these commodities: few imported more than 10 per cent of their over-all needs. The position of the United Kingdom was quite different, however: in 1939 she imported 80 per cent of her butter and 50 per cent of her cheese. The main exporters were New Zealand, by far the largest, followed by Australia, Denmark, and Holland. While technological developments, notably mechanization of existing methods and centralization of manufacture, ensured that quantities were sufficient, this was not always the case so far as quality was concerned. Factory-made butter was the least affected, in that the process of its manufacture is relatively simple. With cheese, however, the decline was very evident, and mainly the result of commercial factors. The full flavour of a cheese is realized only as the result of a long ripening process governed by a complex system of enzymes. If this process is curtailed, the product, though unimpaired from the nutritional point of view, may be relatively tasteless and of poor texture; such curtailment did occur in the cheese factories and stores, though it would be idle to suggest that all farmhouse cheese was perfect.

Much of the decline was promoted by changing demand. Bread and cheese was the traditional meal of the industrial worker in the factory and the demand was for quantity rather than quality. To satisfy this, quickripening moist cheeses were produced on the basis of rapid turnover and larger profits. Less cheese was stored in lofts, to ripen, but more in cold-stores where microbiological action was suspended. Worse, means were discovered of making cheese from pasteurized milk. At first the process denatured milk proteins and prevented coagulation with rennet, but this was overcome by using a somewhat lower pasteurization temperature. The result was 'processed' cheese, first manufactured in Switzerland at the beginning of the century and later made elsewhere in Europe and in North America. From the purely practical point of view it has much to commend it: although its moisture content is higher than that of real cheese, thus making it poorer value for money, it keeps indefinitely and there is no wasteful rind. Its disadvantage is its taste and texture.

We have already noted the advent of margarine as a rival to butter. The original product was in effect little more than an emulsion of purified beef tallow with milk, from which the fatty fraction was then precipitated by addition of ice and worked to a buttery consistency. It was not very palatable but it was cheap and nutritious. In the twentieth century vegetable fats and oils were introduced and the skim milk used was subjected to lactic fermentation.

Great improvements were made in the emulsification process, and incorporation of vitamins A and D (compulsory in Britain during the Second World War) was common. By 1950 the palatability of margarine and its nutritional value had been greatly enhanced. By careful control of the manufacturing process, products could even be made suitable for different climatic conditions. World production in 1938 amounted to more than one million tons, roughly one-third that of butter.

Milk contains approximately 87 per cent water; it is not surprising therefore that there has long been an interest in drying (or condensing) it, not only as a means of indefinite preservation but also to reduce its bulk. Most early processes depended on direct heating: the milk was allowed to run over a hot roller and the dry residue was then scraped off. Unfortunately this causes considerable change in flavour and the lost water is not easily reabsorbed when desired. Nevertheless, it provides a product useful for various purposes in the food industry.

During the Second World War spraydrying processes were developed in which a spray of skimmed milk was allowed to fall through a rising current of hot air; by the time the droplets reached the bottom of the plant they have turned to powder. Again, however, there was difficulty in reconstituting the product. In 1946 this was overcome by an agglomeration process developed in the USA by the Instant Milk Company. Spray-drying was also used to dry eggs.

There were some important changes in the process of pasteurization. From the 1920s this was done in bulk rather than in bottles and consisted in holding the milk for half an hour at about 65°C; it was then rapidly cooled. During the Second World War, however, this method began to be replaced by the so-called HTST (High Temperature Short Time) process in which the working temperature was about 72°C but the operating time only a quarter of a minute. It was followed by the UHT (Ultra High Temperature) process in which the milk is superheated to about 150°C for only a few seconds. The short heating periods of these last two processes demanded the development of elaborate heat exchangers.

MEAT AND FISH

For meat and fish, and for fruit, vegetables, and condensed milk, canning continued to be the most important means of preservation. The basic principle remained unchanged but there were significant developments in practice, organization, and, of course, in scale; in 1950 the USA, which had become the world's greatest producer of canned foods, used some 10 000 million cans.

At the same time developments in bacteriology — a science which had not even been born when Nicolas Appert introduced his preservation process in 1810 on a purely empirical basis — made it possible to conduct the process more efficiently and, above all, more safely. An essential requirement was proper understanding of the heat-transfer process: it was essential that the whole of the food mass was held for long enough at the required temperature.

The essence of canning is to destroy by heat all the micro-organisms that cause food to go bad and then to seal the container completely so that there can be no subsequent contamination until it is opened. Contamination, should it occur, is usually harmless but certain organisms can produce severe, and sometimes fatal, food poisoning. This was not uncommon in the early days of canning but is now very rare.

In Appert's original process the food to be preserved was put into loosely stoppered glass bottles which were then immersed in boiling water for what experience indicated was 'long enough'. The stoppers were then driven home and luted round the edge. This procedure presented two hazards: some organisms can survive the temperature of boiling water and there might be leakage round the stopper. The first hazard was overcome by replacing the water by a solution of calcium chloride, which boils at a much higher temperature (246°F), and later, in the 1870s, by using autoclaves in which the boiling-point of water is raised by application of pressure; this was a technique widely adopted for sterilizing surgical instruments. By the end of the century large reliable autoclaves were in general use. They remained in normal practice until the 1960s, when the process of flame sterilization began to be used. In this, the cans are spun rapidly (two revolutions per minute) while a gas flame is applied to their surface. Under these conditions heat transfer is so rapid that there is no local overheating.

The risk of contamination after sterilization was largely overcome by using cans made of tinplate instead of glass containers. The end discs were soldered on and a small hole was left in the upper disc to be closed by soldering when heat treatment was complete. It was a slow and laborious process: even a good artisan could make no more than about 600 cans a day.

The development of can-making machinery was, therefore, essential to the expansion of the industry: by mid-century such machines were capable of making 600 cans in two minutes. The major development here was the advent of the sanitary or open-top can about 1905. This was made from seamed tubing double-seamed at top and bottom; the need for soldering was eliminated.

For most of the period in question, tinplate was the standard material for making cans, the thickness of the tin, which is expensive, being progressively reduced: in 1950 a thickness of 0.0001 inch was normal. Such thin films were easily damaged and any imperfection could be a site of corrosion; to avoid this it became customary to lacquer the tinplate. After the Second World War the cheapness of aluminium led to its use in making cans, especially for vegetables and drinks; for the latter, the pull-ring opener was introduced. At the same time, steel coated with an exceedingly thin film of chromium/chromium oxide began to replace tinplate; this, too, was lacquered.

The storage and transport of refrigerated food was well established by the end of the nineteenth century, but it was generally believed that if food was cooled below -2°C it would suffer irreversible and deleterious changes. In 1929, however, Clarence Birdseye showed that quick freezing to lower temperatures is satisfactory for many products, mainly because the ice crystals formed are much smaller. This opened the way to the enormous frozen-food industry which, as a complement, demanded the general adoption of domestic deep-freeze lockers in addition to refrigerators.

A very important innovation in this area was accelerated freeze-drying, originally developed in Sweden in the 1930s for drying biological materials of pharmaceutical importance. It is based on the fact that at low pressures ice will sublime — that is to say, it passes directly into the vapour phase without melting. The dehydrated residue is left in a powder form which readily reabsorbs water to reconstitute the material in something very like its natural form.

Freeze-drying has been used for many vegetables; for meat and fish; and for coffee. During the Second World War it was used for the purification of penicillin, but the first commercial plant for food preservation seems to have been opened in Russia in 1954. It is an expensive process in terms of energy consumption but against this it can be argued that the product can be stored indefinitely without further demand for energy, such as is necessary for refrigerated goods.

MILLING AND BAKING

We have already mentioned the new pattern of wheat production and the problems arising from the introduction of new types of wheat having physical properties different from those to which millers were accustomed. A compensation, however, was that increasing use of the combine harvester -by 1938 50 per cent of American wheat was harvested in this way — gave cleaner grain.

At the same time, the increasing mechanization of doughmaking for bread — and of the making of cakes and biscuits too — demanded production of a uniform product. New milling processes had to take account, in particular, of two characteristics of the grain: its hardness and its shape, especially that of its longitudinal crease. The steel rollers that replaced the traditional mill stones served a double purpose. First, a set of fluted rollers crushed the grain and effected some separation of bran from flour. Then the grain was put through a series of pairs of reduction rolls, one rotating at a slightly different rate from its companion in order to introduce a shearing movement, before being sieved. The most important changes in this process as the century advanced were a great reduction in the size of the rollers, relative to throughput, and the introduction of air-flotation techniques, rather than sieving, for separating flour grains of different degrees of fineness.

The proportion of the wheat that could be converted into flour varied, and so, other things being equal, there was a tendency to favour new varieties offering a high extraction rate. Other things were not always equal, however; thus many of the more productive strains proved to contain excessive amounts of amylase, an enzyme that breaks starch down into sugar.

Much bread is still made by the traditional method of setting the dough aside for two or three hours to let it ferment or rise. Shortly after the Second World War, however, American bakeries adopted a dough development process based on mechanical manipulation, which cut this preliminary stage to a few minutes, and saved much space and labour. This development made it possible to bake continuously, instead of in batches, in tunnel ovens.

Production was also speeded by the use of improvers. Flour, like wine, improves with keeping, as a result of spontaneous chemical changes. In the nineteenth century, when flour was imported from abroad, the maturing process took place during the voyage but when Britain and other European importers bought grain from North America and did their own milling this no longer happened. To overcome this difficulty, certain chemical improvers were developed which accelerated the ageing process; they also had the advantage of producing a more uniform type of flour and one that led to a more easily workable dough.

WATER PASTEURIZATION TECHNIQUES

Water quality and human health have been closely linked throughout history. However, it was not until the last quarter of the 19th century that pioneering work by Robert Koch and Louis Pasteur established the germ theory of infectious disease. With the understanding that fecal-borne bacteria, viruses, and protozoans were responsible for most water-borne diseases, it was possible to develop sanitation and water treatment practices which provided people with a safe water supply. In industrial countries safe water is now taken for granted.

In developing countries however, the burden of disease caused by contaminated water and a lack of sanitation continues to be staggering, particularly among young children.

Diarrhea is caused by microbes entering the mouth, most often from contaminated water. According to the United Nations Children's Fund (UNICEF) diarrhea is the most common childhood disease in developing countries. Dehydration resulting from diarrhea is the leading cause of death in children under the age of five, annually killing an estimated five million children. Diarrhea is also the most common cause of child malnutrition, which can lead to death or permanently impaired mental and physical development.

UNICEF estimates that 60 per cent of rural families and 23 per cent of urban families in developing countries are without safe water. In some areas all water supplies may be contaminated. If a water source is suspected of being unsafe, the most common recommendation is to boil the water. This recommendation is seldom followed for several understandable reasons, the most important being the time and the amount of scarce fuel it would require.

Contrary to what many people believe, it is not necessary to boil water to make it safe to drink. Also contrary to what many people believe, it is usually not necessary to distill water to make it safe to drink. Heating water to 65° C (149° F) will kill all germs, viruses, and parasites. This process is called pasteurization and its use for milk is well known though milk requires slightly different time temperature combinations. One obvious problem that arises with pasteurization is the question of how to tell when and if the water has reached the right temperature. Solutions to this problem will be covered in the next section. Pasteurization will not help if water is brackish or chemically contaminated. In this document we describe several pasteurization techniques applicable to developing countries. Pasteurization is not the only technique that can be used to make water safe to drink. Chlorination, ultra-violet disinfection, and the use of a properly constructed, properly maintained well are other ways of providing clean water that may be more appropriate, particularly if a large amount of water is needed. Conversely, if a relatively small amount of water is needed, pasteurization systems have the advantage of being able to be scaled down with a corresponding decrease in cost. As always, the selection of the right system should be based on local conditions.

This document describes techniques used to pasteurize water, but it is also necessary to educate people about the need for clean water and how to keep their water clean. Among many people in the developing world clean water is not perceived as being important. Also, since many people do not understand how germs are transmitted, many cased have been reported where people unthinkingly recontaminate their clean water by putting it into a contaminated container.

BASIC METHODS OF SOLAR WATER PASTEURIZATION-SOLAR COOKERS

A simple method of pasteurizing water is to simply put blackened containers of water in a solar box cooker, an insulated box made of wood, cardboard, plastic, or woven straw. One popular type of solar box cooker is made of aluminized cardboard and has a solar collection area of about 58 cm by 48 cm (23 inches by 19 inches). It has a reflective lid that increases the sunlight collected.

With this device a yield of 4 to 12 liters (1 to 3 gallons) per day is achieved in the field. Each person requires about 4 liters (1 gallon) of water per day, about half of which is for drinking and the other half is for dish washing and brushing one's teeth. The cost for this device is on the order of $20, US, depending on how easily available the basic materials are.

Other types of solar cookers can be used. A recent development in solar cookers is the solar panel cooker, which consists of reflective panels that concentrate sunlight on the food. The food is in an oven roasting bag to reduce heat loss. Replacing the food with a darkened container of water makes a solar water pasteurizer. While the cost of these panel cookers is low, not more than 2 liters of water can be pasteurized at a time, though in the right climate several batches per day can be pasteurized. Regardless of the type of solar cooker used, a way of knowing that the water reached the pasteurization temperature is needed. An inexpensive device that does this was developed. It is a plastic tube with both ends heated, pinched, and sealed, and with a particular type of soybean fat in one end that melts at 154° F.

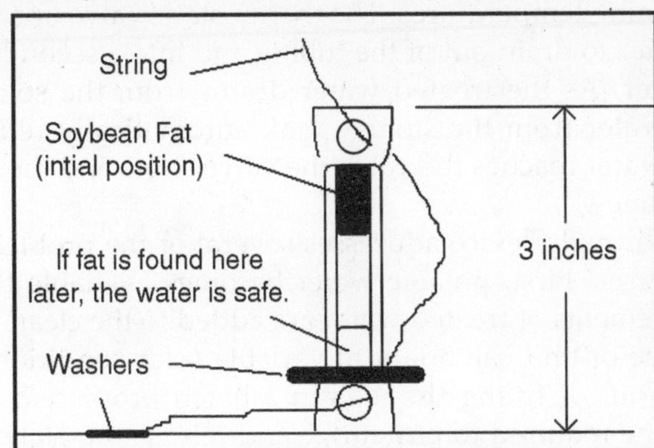

Fig. A Water Pasteurization Indicator. The Indicator would Sit at an Angle in the Bottom of a Water Container.

The tube itself is buoyant, but is weighted with a washer so it sinks to the bottom (coolest) part of the water, with the fat in the high end of the tube.

If the fat is found in the low end of the tube at any time after, the water reached the proper temperature, even though the water may have since cooled down. A nylon string makes it easy to take the tube out without recontaminating the water. The tube is reused by flipping it over and sliding the string through the other way. This device also works with fuel-heated water. Since heating the water to the pasteurization temperature rather than the boiling point reduces the energy required by at least 50 per cent, the fuel savings offered by this simple device alone is considerable.

This device works anytime when water is pasteurized in batches regardless of the source of the heat. If one were burning fuel to pasteurize pots of water the pasteurization indicator would still be usable, as long as one didn't get the nylon string too close to the fire. Since heating the water to the pasteurization temperature rather than the boiling point reduces the amount of energy required by at least 50 per cent, the fuel savings offered by this simple device is considerable.

FLOW-THROUGH PASTEURIZATION DEVICES

In order to produce more water PAX World Service produced a flow-through unit which consists of 15 meters (50 feet) of black-painted tubing coiled within a standard solar box cooker. One end of this tubing is connected to a thermostatic valve and the other to a storage tank for the untreated water supply. This storage tank also contains a sand/gravel/charcoal filter that does the preliminary filtering.

The small amount of water (about 1.5 liters) within the tubing allows rapid heating of the water to the valve's opening temperature of 83.5° C (182° F). This is well above the required temperature, but the valve is derived from a mass-produced automotive radiator thermostat valve, so there is a limited selection of opening temperatures. The thermostatic valve opens allowing the pasteurized water to drain out of the tubing and into a second storage vessel for treated water. As the treated water drains from the solar box cooker, contaminated water from the storage tank automatically refills the tubing. Once this cool water reaches the valve the valve shuts and the pasteurization process begins anew.

This flow-through device addresses several of the problems inherent in the batch processes. First, potable water becomes available throughout the day as new increments of treated water are added to the clean storage vessel. Second, this type of unit can adapt to variable solar conditions which takes the guesswork out of filling the jugs in a batch process. If the insolation increases the time required to pasteurize and release the water in the tubing decreases, thus supplying increments of treated water at a faster rate. If insolation decreases the residence time in the solar box cooker will increase, but it will still be pasteurized which may not be the case in a batch unit where the user overestimated the amount of water which could be treated for that day. Although this is a respectable increase, much more dramatic improvements can still be achieved by recycling the heat in the outgoing pasteurized water. Once the water has been pasteurized and released from the solar box cooker the energy in this water can be used to preheat the incoming water. Since the temperature of the water entering the solar box cooker is higher, it takes less time to finish the pasteurization process, allowing more water to be treated. Also, the flow resistance of the heat exchanger smoothes the flowrate of the water. A simple device which accomplishes this preheating is a counter-current heat exchanger. The hot water flows on one side of a metal plate, while on the other side of the plate cooler fluid flows in the opposite direction. The energy from the hot water is transferred to the cold water, thus preheating the incoming contaminated water by lowering the temperature of the outgoing pasteurized water.

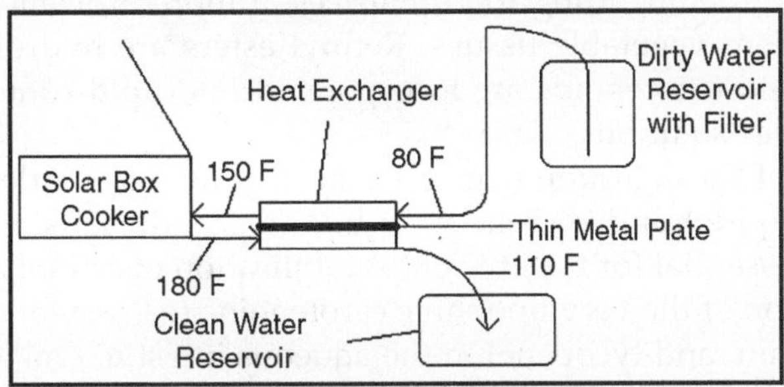

Fig. PAX -Style Water Pasteurizer with Heat Exchanger. Typical Temperatures are Shown in Degrees Fahrenheit..

There are many ways of building a counter-current heat exchanger. Both a tubular version and a flat version have been tested using various configurations and materials, with experimental results favoring the less expensive flat version, though the tubular version is easier to construct from purchased parts. The flat plate unit allows between 75 per cent and 80 per cent of the energy to be reused in preheating the incoming water, and roughly four to five times more water will be pasteurized over a flow-through unit without a heat exchanger. This corresponds to about 80 to 96 liters (20 to 24 gallons) of treated water per day, which is a ten to twelve-fold improvement over the original solar box cooker batch method. An additional benefit is that the chance of burns is greatly reduced because the outflowing water is much cooler reducing the burn hazard. The cost of the heat exchanger itself is on the order of $15 US, making the cost of the complete PAX system about $65. Thus for an increase in cost of about 15 per cent the heat exchanger provides about a 400 per cent increase in water output.

ROLE OF VITAMIN A IN HUMAN METABOLIC PROCESSES

Vitamin A (retinol) is an essential nutrient needed in small amounts by humans for the normal functioning of the visual system; growth and development; and maintenance of epithelial cellular integrity, immune function, and reproduction. These dietary needs for vitamin A are normally provided for as preformed retinol (mainly as retinyl ester) and pro-vitamin A carotenoids.

Overview of Vitamin A Metabolism

Preformed vitamin A in animal foods occurs as retinyl esters of fatty acids in association with membrane-bound cellular lipid and fat-containing storage cells. Pro-vitamin A carotenoids in foods of vegetable origin are also associated with cellular lipids but are embedded in complex cellular structures such as the cellulose-containing matrix of chloroplasts or the pigment-containing portion of chromoplasts. Normal digestive processes free vitamin A and carotenoids from embedding food matrices, a more efficient process from animal than from vegetable tissues. Retinyl esters are hydrolysed and the retinol and freed carotenoids are incorporated into lipid-containing, water-miscible micellar solutions.

Products of fat digestion (e.g., fatty acids, monoglycerides, cholesterol, and phospholipids) and secretions in bile (e.g., bile salts and hydrolytic enzymes) are essential for the efficient solubilisation of retinol and especially for solubilisation of the very lipophilic carotenoids (e.g., aa- and bb-carotene, bb-cryptoxanthin, and lycopene) in the aqueous intestinal milieu.

Micellar solubilisation is a prerequisite to their efficient passage into the lipid-rich membrane of intestinal mucosal cells (i.e., enterocytes). Diets critically low in dietary fat (under about 5-10 g daily) or disease conditions that interfere with normal digestion and absorption leading to steatorrhea (e.g., pancreatic and liver diseases and frequent gastroenteritis) can therefore impede the efficient absorption of retinol and carotenoids. Retinol and some carotenoids enter the intestinal mucosal brush border by diffusion in accord with the concentration gradient between the micelle and plasma membrane of enterocytes.

Some carotenoids pass into the enterocyte and are solubilized into chylomicrons without further change whereas some of the pro-vitamin A carotenoids are converted to retinol by a cleavage enzyme in the brush border. Retinol is trapped intracellularly by re-esterification or binding to specific intracellular binding proteins. Retinyl esters and unconverted carotenoids together with other lipids are incorporated into chylomicrons, excreted into intestinal lymphatic channels, and delivered to the blood through the thoracic duct. Tissues extract most lipids and some carotenoids from circulating chylomicrons, but most retinyl esters are stripped from the chylomicron remnant, hydrolysed, and taken up primarily by parenchymal liver cells.

If not immediately needed, retinol is re-esterified and retained in the fat-storing cells of the liver (variously called adipocytes, stellate cells, or Ito cells).

The liver parenchymal cells also take in substantial amounts of carotenoids. Whereas most of the body's vitamin A reserve remains in the liver, carotenoids are also deposited elsewhere in fatty tissues throughout the body.

Usually, turnover of carotenoids in tissues is relatively slow, but in times of low dietary carotenoid intake, stored carotenoids are mobilised. A recent study in one subject using stable isotopes suggests that retinol can be derived not only from conversion of dietary pro-vitamin carotenoids in enterocytes - the major site of bioconversion - but also from hepatic conversion of circulating pro-vitamin carotenoids. The quantitative contribution to vitamin A requirements of carotenoid converted to retinoids beyond the enterocyte is unknown.

Following hydrolysis of stored retinyl esters, retinol combines with a plasma-specific transport protein, retinol-binding protein (RBP). This process, including synthesis of the unoccupied RBP (apo-RBP), occurs to the greatest extent within liver cells but it may also occur in some peripheral tissues. The RBP-retinol complex (holo-RBP) is secreted into the blood where it associates with another hepatically synthesised and excreted larger protein, transthyretin.

The transthyretin-RBP-retinol complex circulates in the blood, delivering the lipophilic retinol to tissues; its large size prevents its loss through kidney filtration. Dietary restriction in energy, proteins, and some micronutrients can limit hepatic synthesis of proteins specific to mobilisation and transport of vitamin A. Altered kidney functions or fever associated with infections can increase urinary vitamin A loss.

Holo-RBP transiently associates with target-tissue membranes, and specific intracellular binding proteins then extract the retinol. Some of the transiently sequestered retinol is released into the blood unchanged and is recycled. A limited reserve of intracellular retinyl esters is formed, that subsequently can provide functionally active retinol and its oxidation products (i.e., isomers of retinoic acid) as needed intracellularly. These biologically active forms of vitamin A are associated with specific cellular proteins which bind with retinoids within cells during metabolism and with nuclear receptors that mediate retinoid action on the genome. Retinoids modulate the transcription of several hundreds of genes.

In addition to the latter role of retinoic acid, retinol is the form required for functions in the visual and reproductive systems and during embryonic development. Holo-RBP is filtered into the glomerulus but recovered from the kidney tubule and recycled.

Normally vitamin A leaves the body in urine only as inactive metabolites which result from tissue utilisation and as potentially recyclable active glucuronide conjugates of retinol in bile secretions. No single urinary metabolite has been identified which accurately reflects tissue levels of vitamin A or its rate of utilisation. Hence, at this time urine is not a useful biologic fluid for assessment of vitamin A nutriture.

Biochemical Mechanisms for Vitamin A Functions

Vitamin A functions at two levels in the body. The first is in the visual cycle in the retina of the eye; the second is in all body tissues systemically to maintain growth and the soundness of cells. In the visual system, carrier-bound retinol is transported to ocular tissue and to the retina by intracellular binding and transport proteins. Rhodopsin, the visual pigment critical to dim-light vision, is formed in rod cells after conversion of all-*trans* retinol to retinaldehyde, isomerization to the 11-*cis*-form, and binding to opsin.

Alteration of rhodopsin through a cascade of photochemical reactions results in ability to see objects in dim light. The speed at which rhodopsin is regenerated relates to the availability of retinol. Night blindness is usually an indicator of inadequate available retinol, but it can also be due to a deficit of other nutrients, which are critical to the regeneration of rhodopsin, such as protein and zinc, and to some inherited diseases, such as retinitis pigmentosa.

The growth and differentiation of epithelial cells throughout the body are especially affected by vitamin A deficiency (VAD). Goblet cell numbers are reduced in epithelial tissues. The consequence is that mucous secretions with their antimicrobial components diminish. Cells lining protective tissue surfaces fail to regenerate and differentiate, hence flatten and accumulate keratin.

Both factors - the decline in mucous secretions and loss of cellular integrity - diminish resistance to invasion by potentially pathogenic organisms. The immune system is also compromised by direct interference with production of some types of protective secretions and cells. Classical symptoms of xerosis (drying or nonwetability) and desquamation of dead surface cells as seen in ocular tissue (i.e., xerophthalmia) are the external evidence of the changes also occurring to various degrees in internal epithelial tissues.

Current understanding of the mechanism of vitamin A action within cells outside the visual cycle is that cellular functions are mediated through specific nuclear receptors. These receptors are activated by binding with specific isomers of retinoic acid (i.e., all-*trans* and 9-*cis* retinoic acid). Activated receptors bind to DNA response elements located upstream of specific genes to regulate the level of expression of those genes. The synthesis of a large number of proteins vital to maintaining normal physiologic functions is regulated by these retinoid-activated genes. There also may be other mechanisms of action that are as yet undiscovered .

THE ROLE OF IRON IN HUMAN METABOLIC PROCESSES

Iron has several vital functions in the body. It serves as a carrier of oxygen to the tissues from the lungs by red blood cell haemoglobin, as a transport medium for electrons within cells, and as an integrated part of important enzyme systems in various tissues. The physiology of iron has been extensively reviewed.

Most of the iron in the body is present in the erythrocytes as haemoglobin, a molecule composed of four units, each containing one heme group and one protein chain. The structure of haemoglobin allows it to be fully loaded with oxygen in the lungs and partially unloaded in the tissues (e.g., in the muscles). The iron-containing oxygen storage protein in the muscles, myoglobin, is similar in structure to haemoglobin but has only one heme unit and one globin chain. Several iron-containing enzymes, the cytochromes, also have one heme group and one globin protein chain.

These enzymes act as electron carriers within the cell and their structures do not permit reversible loading and unloading of oxygen. Their role in the oxidative metabolism is to transfer energy within the cell and specifically in the mitochondria. Other key functions for the iron-containing enzymes (e.g., cytochrome P450) include the synthesis of steroid hormones and bile acids; detoxification of foreign substances in the liver; and signal controlling in some neurotransmitters, such as the dopamine and serotonin systems in the brain. Iron is reversibly stored within the liver as ferritin and hemosiderin whereas it is transported between different compartments in the body by the protein transferrin.

IRON REQUIREMENTS

Basal Iron Losses

Iron is not actively excreted from the body in urine or in the intestines. Iron is only lost with cells from the skin and the interior surfaces of the body - intestines, urinary tract, and airways. The total amount lost is estimated at 14 μg/kg body weight/day. In children, it is probably more correct to relate these losses to body surface. A non-menstruating 55-kg women loses about 0.8 mg Fe/day and a 70-kg man loses about 1 mg.

The range of individual variation has been estimated to be ±15 per cent.Earlier studies suggested that sweat iron losses could be considerable, especially in a hot, Humid climate. However, new studies which took extensive precautions to avoid the interference of contamination of iron from the skin during the collection of total body sweat have shown that these sweat iron losses are negligible.

Growth

The newborn term infant has an iron content of about 250-300 mg (75 mg/kg body weight). During the first 2 months of life, haemoglobin concentration falls because of the improved oxygen situation in the newborn infant compared with the intrauterine foetus. This leads to a considerable redistribution of iron from catabolised erythrocytes to iron stores. This iron will cover the needs of the term infant during the first 4-6 months of life and is why iron requirements during this period can be provided by human milk, that contains very little iron. Because of the marked supply of iron to the foetus during the last trimester of pregnancy, the iron situation is much less favourable in the premature and low-birth-weight infant than in the term infant.

An extra supply of iron is therefore needed in these infants even during the first 6 months of life. In the full-term infant, iron requirements will rise markedly after age 4-6 months and amount to about 0.7-0.9 mg/day during the remaining part of the first year. These requirements are therefore very high, especially in relation to body size and energy intake. In the first year of life, the full-term infant almost doubles its total iron stores and triple its body weight. The change in body iron during this period occurs mainly during the first 6-12 months of life. Between 1 and 6 years of age, the body iron content is again doubled.

The requirements for absorbed iron in infants and children are very high in relation to their energy requirements. For example, in infants 6-12 months of age, about 1.5 mg of iron need to be absorbed per 4.184 MJ and about half of this amount is required up to age 4 years. In the weaning period, the iron requirements in relation to energy intake are the highest of the lifespan except for the last trimester of pregnancy, when iron requirements to a large extent have to be covered from the iron stores of the mother. The rapidly growing weaning infant has no iron stores and has to rely on dietary iron.

It is possible to meet these high requirements if the diet has a consistently high content of meat and foods rich in ascorbic acid. In most developed countries today, infant cereal products are the staple foods for that period of life. Commercial products are regularly fortified with iron and ascorbic acid, and they are usually given together with fruit juices and solid foods containing meat, fish, and vegetables. The fortification of cereal products with iron and ascorbic acid is important in meeting the high dietary needs, especially considering the importance of an optimal iron nutrition during this phase of brain development.

Menstrual Iron Losses

Menstrual blood losses are very constant from month to month for an individual but vary markedly from one woman to another.The main part of this variation is genetically controlled by the content of fibrinolytic activators in the uterine mucosa even in populations which are geographically widely separated (Burma, Canada, China, Egypt, England, and Sweden). These findings strongly suggest that the main source of variation in iron status in different populations is not related to a variation in iron requirements but to a variation in the absorption of iron from the diets. (This statement disregards infestations with hookworms and other parasites.) The mean menstrual iron loss, averaged over the entire menstrual cycle of 28 days, is about 0.56 mg/day.

The frequency distribution of physiologic menstrual blood losses is highly skewed. Adding the average basal iron loss (0.8 mg) and its variation allows the distribution of the total iron requirements in adult women to be calculated as the convolution of the distributions of menstrual and basal iron losses. The mean daily total iron requirement is 1.36 mg. In 10 per cent of women it exceeds 2.27 mg and in 5 per cent it exceeds 2.84 mg. In 10 per cent of menstruating (still-growing) teenagers, the corresponding daily total iron requirement exceeds 2.65 mg, and in 5 per cent of the girls it exceeds 3.2 mg/day. The marked skewness of menstrual losses is a great nutritional problem because personal assessment of the losses is unreliable. This means that women with physiologic but heavy losses cannot be identified and reached by iron supplementation. The choice of contraceptive method greatly influences menstrual losses. The methods of calculating iron requirements in women and their variation were recently re-examined.

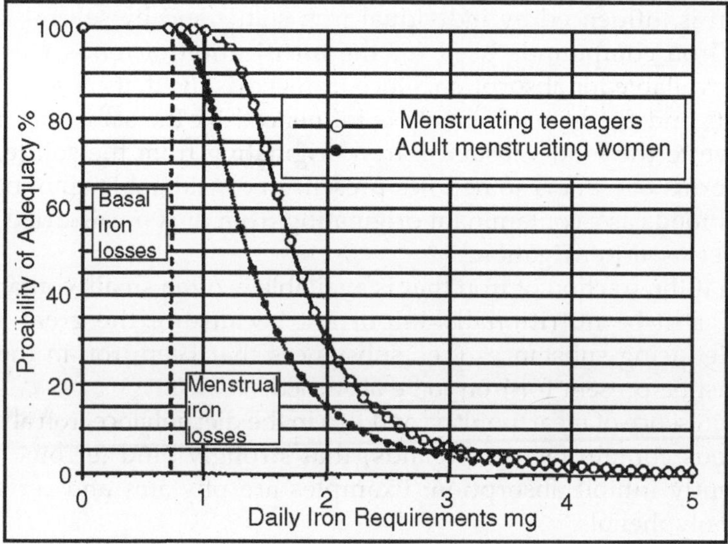

Fig. Distribution of Daily Iron Requirements in Menstruating Adult Women and Teenagers: the Probability of Adequacy at Different Amounts of Iron Absorbed.

Note: Left: basal obligatory losses that amount to 0.8 mg; right: varying menstrual iron losses. This graph illustrates that growth requirements in teenagers vary considerably at different age and between girls. In postmenopausal women and in physically active elderly people, the iron requirements per unit of body weight are the same as in men. When physical activity decreases as a result of ageing, blood volume and haemoglobin mass also diminish, leading to a shift of iron from haemoglobin and muscle to iron stores. This implies a reduction of the daily iron requirements. Iron deficiency in the elderly is therefore seldom of nutritional origin but is usually caused by pathologic iron losses.

IRON ABSORPTION

With respect to the mechanism of absorption, there are two kinds of dietary iron: heme iron and non-heme iron. In the human diet the primary sources of heme iron are the haemoglobin and myoglobin from consumption of meat, poultry, and fish whereas non-heme iron is obtained from cereals, pulses, legumes, fruits, and vegetables. The average absorption of heme iron from meat-containing meals is about 25 per cent.

The absorption of heme iron can vary from about 40 per cent during iron deficiency to about 10 per cent during iron repletion. Heme iron can be degraded and converted to non-heme iron if foods are cooked at a high temperature for too long. Calcium is the only dietary factor that negatively influences the absorption of heme iron and does so to the same extent that it influences non-heme iron.

Non-heme iron is the main form of dietary iron. The absorption of non-heme iron is influenced by individual iron status and by several factors in the diet. Iron compounds used for the fortification of foods will only be partially available for absorption. Once iron is dissolved, its absorption from fortificants and food contaminants is influenced by the same factors as the iron native to the food substance. Iron originating from the soil (e.g., from various forms of clay) is sometimes present in considerable amounts on the surface of foods as a contaminant originating from dust on air-dried foods or from water used in irrigation.

Even if the fraction of iron that is available is often small, contamination iron may still be nutritionally important because of the great amounts present.Reducing substances (i.e., substances that keep iron in the ferrous form) must be present for iron to be absorbed.

The presence of meat, poultry, and fish in the diet enhance iron absorption. Other foods contain factors (ligands) that strongly bind ferrous ions that subsequently inhibit absorption. Examples are phytates and certain iron-binding polyphenols.

Inhibition of Iron Absorption

Phytates are found in all kinds of grains, seeds, nuts, vegetables, roots (e.g., potatoes), and fruits. Chemically, phytates are inositol hexaphosphate salts and are a storage form of phosphates and minerals. Other phosphates have not been shown to inhibit non-heme iron absorption. In North American and European diets, about 90 per cent of phytates originate from cereals. Phytates strongly inhibit iron absorption in a dose-dependent fashion and even small amounts of phytates have a marked effect.

Bran has a high content of phytate and strongly inhibits iron absorption. Whole-wheat flour, therefore, has a much higher content of phytates than does white wheat flour. In bread some of the phytates in bran are degraded during the fermentation of the dough. Fermentation for a couple of days (sourdough fermentation) can therefore almost completely degrade the phytate and increase the bio-availability of iron in bread made from whole-wheat flour. Oats strongly inhibit iron absorption because of their high phytate content, that results from native phytase in oats being destroyed by the normal heat process used to avoid rancidity. Sufficient amounts of ascorbic acid can counteract this inhibition. By contrast, non-phytate-containing dietary fibre components have almost no influence on iron absorption.

Almost all plants contain phenolic compounds as part of their defence system against insects, animals, and humans. Only some of the phenolic compounds (mainly those containing galloyl groups) seem to be responsible for the inhibition of iron absorption. Tea, coffee, and cocoa are common plant products that contain iron-binding polyphenols. Many vegetables, especially green leafy vegetables (e.g., spinach), and herbs and spices (e.g., oregano) contain appreciable amounts of galloyl groups, that strongly inhibit iron absorption. Consumption of betel leaves, common in areas of Asia, also has a marked negative effect on iron absorption.

Calcium, consumed as a salt or in dairy products interferes significantly with the absorption of both heme and non-heme iron. Because calcium and iron are both essential nutrients, calcium cannot be considered to be an inhibitor in the same way as phytates or phenolic compounds. The practical solution for this competition is to increase iron intake, increase its bio-availability, or avoid the intake of foods rich in calcium and foods rich in iron at the same meal.

The mechanism of action for absorption inhibition is unknown, but the balance of evidence strongly suggest that the inhibition is located within the mucosal cell itself at the common final transfer step for heme and non-heme iron.

Recent analyses of the dose-effect relationship show that no inhibition is seen from the first 40 mg of calcium in a meal. A sigmoid relationship is then seen, reaching a 60 per cent maximal inhibition of iron absorption by 300-600 mg calcium. The form of this curve suggests a one-site competitive binding of iron and calcium. This relationship explains some of the seemingly conflicting results obtained in studies on the interaction between calcium and iron.

For unknown reasons, the addition of soy protein to a meal reduces the fraction of iron absorbed. This inhibition is not solely explained by the high phytate content of soy protein. However, because of the high iron content of soy proteins, the net effect on iron absorption of an addition of soy products to a meal is usually positive. In infant foods containing soy proteins, the inhibiting effect can be overcome by the addition of sufficient amounts of ascorbic acid. Some fermented soy sauces, however, have been found to enhance iron absorption.

Enhancement of Iron Absorption

Ascorbic acid is the most potent enhancer of non-heme iron absorption. Synthetic vitamin C increases the absorption of iron to the same extent as the native ascorbic acid in fruits, vegetables, and juices. The effect of ascorbic acid on iron absorption is so marked and essential that this effect could be considered as one of vitamin C's physiologic roles. Each meal should preferably contain at least 25 mg of ascorbic acid and possibly more if the meal contains many inhibitors of iron absorption.

Therefore, a requirement of ascorbic acid for iron absorption should be taken into account when establishing the requirements for vitamin C, that are set only to prevent vitamin C deficiency (especially scurvy). Meat, fish, and seafood all promote the absorption of non-heme iron. The mechanism for this effect has not been determined.

It should be pointed out that meat also enhances the absorption of heme iron to about the same extent. Meat promotes iron nutrition in two ways: it stimulates the absorption of both heme and non-heme iron and it provides the well-absorbed heme iron.

Epidemiologically, the intake of meat has been found to be associated with a lower prevalence of iron deficiency. Organic acids, such as citric acid, have in some studies been found to enhance the absorption of non-heme iron. This effect is not observed as consistently as is the effect of ascorbic acid. Sauerkraut and other fermented vegetables and even some fermented soy sauces enhance iron absorption. The nature of this enhancement has not yet been determined.

Iron Absorption from Meals

The pool concept in iron absorption implies that there are two main pools in the gastrointestinal lumen - one pool of heme iron and another pool of non-heme iron - and that iron absorption takes place independently from these two pools. The pool concept also implies that the absorption of iron from the non-heme iron pool results from all ligands present in the mixture of foods included in a meal.

The absorption of non-heme iron from a certain meal not only depends on its iron content but also, and to a marked degree, on the composition of the meal (i.e., the balance among all factors enhancing and inhibiting the absorption of iron).

The bio-availability can vary more than 10-fold among meals with a similar content of iron, energy, protein, fat, etc. Just the addition of certain spices (e.g., oregano) or a cup of tea may reduce the bio-availability by one-half or more.

However, the addition of certain vegetables or fruits containing ascorbic acid may double or even triple iron absorption, depending on the other properties of the meal and the amounts of ascorbic acid present.

HUMAN METABOLIC PROCESSES REQUIRING IODINE

At present, the only physiologic role known for iodine in the human body is in the synthesis of thyroid hormones by the thyroid gland. Therefore, the dietary requirement of iodine is determined by normal thyroxine (T_4) production by the thyroid gland without stressing the thyroid iodide trapping mechanism or raising thyroid stimulating hormone (TSH) levels. Iodine from the diet is absorbed throughout the gastrointestinal tract. Dietary iodine is converted into the iodide ion before it is absorbed. The iodide ion is bio-available and absorbed totally from food and water. This is not true for iodine within thyroid hormones ingested for therapeutic purposes.

Iodine enters the circulation as plasma inorganic iodide, which is cleared from circulation by the thyroid and kidney. The iodide is used by the thyroid gland for synthesis of thyroid hormones, and the kidney excretes iodine with urine. The excretion of iodine in the urine is a good measure of iodine intake. In a normal population with no evidence of clinical iodine deficiency either in the form of endemic goitre or endemic cretinism, urinary iodine excretion reflects the average daily iodine requirement. Therefore, for determining the iodine requirements, the important indexes are serum T_4 and TSH levels (indicating normal thyroid status) and urinary iodine excretion.

SIGNIFICANT SCIENTIFIC INFORMATION

All biologic actions of iodide are attributed to the thyroid hormones. The major thyroid hormone secreted by the thyroid gland is T_4 (tetra-iodo-thyronine). T_4 in circulation is taken up by the cells and is de-iodinated by the enzyme 5' prime-mono-de-iodinase in the cytoplasm to convert it into tri-iodo-thyronine (T_3), the active form of thyroid hormone. T_3 traverses to the nucleus and binds to the nuclear receptor. All the biologic actions of T_3 are mediated through the binding to the nuclear receptor, which controls the transcription of a particular gene to bring about the synthesis of a specific protein.

The physiologic actions of thyroid hormones can be categorised as:

- Growth and development
- Control of metabolic processes in the body.

Thyroid hormones play a major role in the growth and development of brain and central nervous systems in humans from the 15th week of gestation to age 3 years. If iodine deficiency exists during this period and results in thyroid hormone deficiency, the consequence is derangement in the development of brain and central nervous system. These derangements are irreversible, the most serious form being that of cretinism. The other physiologic role of thyroid hormone is to control several metabolic processes in the body. These include carbohydrate, fat, protein, vitamin, and mineral metabolism. For example, thyroid hormone increases energy production, increases lipolysis, and regulates neoglucogenesis, and glycolysis.

Table. The Spectrum of Iodine Deficiency Disorders.

Stage in life	Effects
Adult	Goitre with its complications Hypothyroidism Impaired mental function
Child and Adolescent	Goitre Juvenile hypothyroidism Impaired mental function Retarded physical development
Foetus	Abortions Stillbirths Congenital anomalies Increased perinatal mortality Increased infant mortality Neurological cretinism: mental deficiency, deaf mutism, spastic diplegia, and squint Myxedematous cretinism: mental deficiency and dwarfism Psychomotor defects
Neonate	Neonatal goitre Neonatal hypothyroidism

POPULATION AT RISK

Iodine deficiency affects all stages of human life, from the intra-uterine stage to old age. However, pregnant women, lactating women, women of reproductive age, and children younger than 3 years are considered to be at high risk. During foetal and neonatal growth and development, iodine deficiency leads to irreversible damage to the brain and central nervous system.

DIETARY SOURCES

The iodine content of food depends on the iodine content of the soil in which it is grown. The iodine present in the upper crust of earth is leached by glaciation and repeated flooding and is carried to the sea. Sea water is, therefore, a rich source of iodine. The seaweed located near coral reefs has an inherent biologic capacity to concentrate iodine from the sea.

The reef fish which thrive on seaweed are rich in iodine. Thus, a population consuming seaweed and reef fish has a high intake of iodine, as the case in Japan. The amount of iodine intake by the Japanese is in the range of 2-3 mg/day. In several areas of Asia, Africa, Latin America, and parts of Europe, iodine intake varies from 20 to 80 mg/day. In the United States and Canada and some parts of Europe, the intake is around 500 mg/day.

Table. Average Iodine Content of Foods (in ìg/g).

Food	Fresh basis		Dry basis	
	Mean	Range	Mean	Range
Fish (fresh water)	30	17-40	116	68-194
Fish (marine)	832	163-3180	3715	471-4591
Shellfish	798	308-1300	3866	1292-4987
Meat	50	27-97	-	-
Milk	47	35-56	-	-
Eggs	93	-	-	-
Cereal grains	47	22-72	65	34-92
Fruits	18	10-29	154	62-277
Legumes	30	23-36	234	223-245
Vegetables	29	12-201	385	204-1636

IODINE FORTIFICATION

Iodine deficiency is present in almost all parts of the developed and developing world, and environmental iodine deficiency is the main cause of iodine deficiency disorders. Iodine is irregularly distributed over the earth's crust, resulting in acute deficiencies in areas such as mountainous regions and flood plains. The problem is aggravated by accelerated deforestation and soil erosion. Thus, the food grown in iodine-deficient regions can never provide enough iodine for the people and livestock living there. The iodine deficiency results from geologic rather than social and economic conditions. It cannot be eliminated by changing dietary habits or by eating specific kinds of foods but must be corrected by supplying iodine from external sources.

It has, therefore, been a common practice to use common salt as a vehicle for iodine fortification for the past 75 years. Salt is consumed at approximately the same level throughout the year by the entire population of a region. Universal salt iodisation is now a widely accepted strategy for preventing and correcting iodine deficiency disorders. There are areas where consumption of goitrogens in the staple diet (e.g., cassava) affects the proper utilisation of iodine by the thyroid gland. For example, in Congo, Africa, as a result of cassava diets there is an overload of thiocyanate. To overcome this problem, appropriate increases in salt iodisation are required to ensure the recommended dietary intake.

The iodisation of salt is done either by spraying potassium iodate or potassium iodide in amounts that ensure a minimum of 150 μg iodine/day. Both of these forms of iodine are absorbed as iodide ions and are completely bio-available. Other methods of iodine prophylaxis are also used: iodised oil (capsule and injections), iodised water, iodised bread, iodised soya sauce, iodoform compounds used in dairy and poultry, and certain food additives. Iodine loss occurs as a result of improper packaging, Humidity and moisture, and transport in open trucks and railway wagons exposed to sunlight. To compensate for these losses, higher levels of iodine are used during the production of iodised salt. Losses during the cooking process vary from 20 per cent to 40 per cent depending on the type of cooking used. To ensure the consumption of recommended levels of iodine, the iodine content of salt at the production level should be monitored with proper quality assurance programmes. Regular evaluation of the urinary iodine excretion pattern in the population consuming iodised salt or exposed to other iodine prophylactic measures would help the adjusting of iodine intake.

PESTICIDES USED IN ORGANIC FARMING

Organic farmers use no synthetic pesticides. Pesticides are poisonous compounds that are designed to kill elements of nature that damage crops, including insects and fungal pests. As such, they have the potential to cause a variety of medical problems. Information about the effects of pesticides has primarily been obtained from studies of pesticide applicators and farmworkers.

Common reactions to pesticides include nausea, lung and eye irritation, and temporary nerve damage. Long-term and chronic medical problems have been harder to verify, but farmers using more pesticides seem to be at greater risk than nonfarmers for some forms of cancer and amyotropic lateral sclerosis (Lou Gehrig's disease), and tests on animals indicate that some pesticides may cause birth and immune-system defects. It is suspected that many pesticides may interfere with hormone systems that regulate functions like reproduction.

"For consumers in general, the unsettling truth is that no one really knows what a lifetime of consuming the tiny quantities of pesticides found on foods might do to a person. The effect, if any, is likely to be small for most individuals-but may be significant for the population at large". In The Organic Gourmet: Feast of Fields, Tracy Kett, a writer, caterer, and promoter of organic agriculture, noted, "Several studies have found that the possible short- and long-term human health effects of agricultural chemicals are numerous and range from respiratory problems in field workers (despite hazardous working conditions, farmers and their employees often work without any kind of safety clothing or respiratory protection) to severe allergic and asthmatic reactions, reproductive disorders, cancer and degenerative diseases in both consumers and farm workers."

Most of these chemicals have no purpose. In Eating with the Seasons, Paula Bartimeus, a professional nutritionist and regular writer for health magazines, said that "although the Government states that the levels of pesticides used are well within the safety margins, random testing has proved otherwise. Many farmers mix chemicals before spraying, and the combined and cumulative effects of such concoctions are still uncertain". The author of Designer Poisons, Marion Moses, lists some precautionary statements from pesticide labels: "Avoid inhalation or contact with eyes or skin. Harmful if swallowed. May cause eye irritation." "Avoid eye contact. Wash thoroughly after handling." "Corrosive to eyes. Causes eye damage. Do not get in eyes. Harmful if swallowed. Avoid contact with skin or clothing." "Do not apply to excessively sunburned or damaged skin. May cause skin reaction in rare cases." "Do not get in eyes, on skin, or clothing. Harmful if swallowed, inhaled or absorbed through skin". The Environmental Protection Agency (EPA) regulates pesticides, but there is concern that the regulation is inadequate.

A number of groups concerned about the use of pesticides believe that the EPA "has not tested all the ingredients in these chemicals and does not require companies to disclose or label their so-called inert ingredients." Moreover, the EPA fails to "take into consideration the effects of combinations of pesticides (only the effects of individual pesticides have been studied)."

Further, there is no acknowledgment that almost all pesticides "drift from their point of application". If a pesticide, applied according to the package label, doesn't result in residues higher than a safe level, that safe level becomes the legal limit. But the legal limit can sometimes exceed the level that safety alone would require if the government decides the benefit to farmers outweighs the risk to consumers". Another concern is that "because government standards for pesticide levels are based on the height and weight of an adult man, it's unclear whether pesticide consumption at those levels might be harmful to a child". A five-year study completed by the National Academy of Sciences found that children were more vulnerable to pesticides than adults. Nutritionist Susan M. Kleiner offered several reasons. "Children eat more food in proportion to their body weight than do adults, so their exposure to pesticides in greater.

Children eat more fruit...compared with adults. Children will often eat a lot of one or two specific foods. If that food has pesticide residues, it will increase their exposure to pesticides". The Environmental Working Group (EWG), a leading content provider for public interest groups with a special interest in the effect pesticides have on children, arranged for laboratory evaluations of baby foods produced by Gerber, Heinz, and Beech-Nut.

The testers "found 16 different pesticides in eight baby food products tested. Of those 16 pesticides, three are probably human carcinogens, five are possible human carcinogens, eight are neurotoxins, five are endocrine disruptors, and five are categorized as oral toxicity 1 chemical, the most toxic designation. More than half of all samples (53 per cent) contained detectable levels of pesticides, 18 per cent of samples had two or more pesticides in them, and one sample contained three different pesticides". The EWG obtained data from the U.S. Food and Drug Administration on the pesticide content of 42 fruits and vegetables.

"More than half of the total dietary risk from pesticides in these foods was concentrated in just 12 crops. The pesticides that were found in these foods are classified by the Environmental Protection Agency (EPA) as probable human carcinogens, nervous system poisons and endocrine system disrupters."

The 12 most contaminated foods, in order of levels of contamination, were strawberries, bell peppers (green and red) and spinach (tied), cherries (U.S.), peaches, cantaloupe (Mexican), celery, apples, apricots, green beans, grapes (Chilean), and cucumbers. It should not be surprising that there has been a boom in the sale of organic baby food. At least one school system-Berkeley, California-is offering pesticide-, herbicide-, and hormone-free foods to its students. "For now, schools are buying their goods from Berkeley's twice-a-week farmers' market, from where organic farmers and vendors are regularly invited to visit with students to explain their natural farming techniques.

Educators hope to integrate gardening know-how with classroom education, and, in the future, individual schools plan to start up organic gardens of their own". One cannot just wash off the pesticides before preparing the food, because some pesticides are not water soluble. Also, "Many chemicals routinely used in farming can penetrate the fruit or vegetable skin and therefore can't be washed off. Peeling helps remove them but eliminates valuable nutrients in the process. And a wax coating not only seals in pesticides, but the wax itself may contain fungicides, coloring agents and other potentially unhealthy substances".

LIMITATIONS OF ORGANIC PRODUCTS

Organic foods, however, are not necessarily healthful or safe. A high-fat food that is made from organic products remains high in fat. "You can buy organic chocolate bars, ice cream and cookies-all made with ingredients that are pesticide-, chemical-, antibiotic- and hormone-free-but they'll be laden with fat, sugar and calories". Moreover, the manure that organic farmers may use could contain bacteria. "The only real difference between organic and nonorganic food is in the growing-and that's not a big enough difference to protect your health from bacteria." Like other forms of food production, there may be compromises in the healthfulness of the organic food environment. In a January 1999 issue of the Humanist, Lisa Hamilton wrote that the major producers "may dump obscene amounts of quick-to-leach chicken manure on poor soil, carrying hazardous nitrates directly to groundwater, or blast crops with the nonexclusive, plant-based pesticide Pyrethrum, killing most insects in range, including beneficial ladybugs and lacewings".

Nevertheless, a May 2000 survey of 1,029 men and women commissioned by the National Centre for Public Policy Research, a Washington, DC, nonprofit education foundation, found that the majority of the public believe that products labeled "USDA Certified Organic" would be "safer," "better," and "healthier for consumers" than nonorganic foods. Responding to the results of this study, John Carlisle, director of the Environmental Policy Task

Force at the National Centre for Public Policy Research, said that misperceptions about the benefits and risks of organically certified products would be heightened by this USDA proposed label.

"Clearly, consumers want the USDA to amend this rule to include specific language on the USDA proposed seal to inform consumers that organic certification is based on production methods and conveys no assurance of food safety, nutrition or other quality". As a result of this study, the National Centre for Public Policy Research advised against the creation of a USDA organic seal that would only impart misinformation and create confusion.

A letter from John E. Frydenlund, the director of the taxpayer watchdog group Citizens against Government Waste, to the programme manager of the federal government's National Organic Programme stated that "there is no scientific evidence that organic food is safer, healthier, or, in any way, better than non-organic food." Moreover, Frydenlund continued, there are "inadequate safeguards to protect consumers from the health dangers caused by composted and raw manure used in organic production.

This inadequacy is exacerbated by a failure to provide adequate information informing consumers of the particular health dangers that exist in the consumption of organic foods". There are some additional limitations to organic farming. "Because organic farmers eschew pesticides and herbicides used by conventional farmers, they cannot use modern conservation tillage techniques that have been extraordinarily successful in reducing soil erosion. As a result, compared to conventional farming, organic farming is woefully inadequate in controlling soil erosion. But the most glaring environmental disadvantage of organic farming is its exorbitant need for land. Organic farming is only about half as productive as conventional farming, which means organic farming requires far more land to feed the world than modern methods". In an interview published in the April 2000 issue of Reason, Nobel Peace Prize winner Norman Borlaug, an agronomist, noted, "Even if you could use all the organic material that you have-the animal manures, the human waste, the plant residues-and get them back on the soil, you couldn't feed more than four billion people.

In addition, if all agriculture were organic, you would have to increase cropland area dramatically, spreading out into marginal areas and cutting down millions of acres of forests". These are essentially the beliefs of Dennis T. Avery, an expert in international agriculture, in Saving the Planet with Pesticides and Plastic. "With present knowledge levels, no responsible authority or organization should recommend either organic farming or traditional low-yield farming systems as a broad-gauge alternative to high-yield agriculture. In fact, slashing farm chemical usage is likely to produce more soil erosion, more human cancer, and less wildlife habitat. At present, organic farming could not even sustain the fertility of our existing cropland, or protect it effectively from erosion".

Index